Advances in Science, Technology & Innovation

IEREK Interdisciplinary Series for Sustainable Development

Advances in Science, Technology & Innovation (ASTI) is a series of peer-reviewed books based on important emerging research that redefines the current disciplinary boundaries in science, technology and innovation (STI) in order to develop integrated concepts for sustainable development. It not only discusses the progress made towards securing more resources, allocating smarter solutions, and rebalancing the relationship between nature and people, but also provides in-depth insights from comprehensive research that addresses the **17 sustainable development goals (SDGs)** as set out by the UN for 2030.

The series draws on the best research papers from various IEREK and other international conferences to promote the creation and development of viable solutions for a **sustainable future and a positive societal** transformation with the help of integrated and innovative science-based approaches. Including interdisciplinary contributions, it presents innovative approaches and highlights how they can best support both economic and sustainable development, through better use of data, more effective institutions, and global, local and individual action, for the welfare of all societies.

The series particularly features conceptual and empirical contributions from various interrelated fields of science, technology and innovation, with an emphasis on digital transformation, that focus on providing practical solutions to **ensure food, water and energy security to achieve the SDGs**. It also presents new case studies offering concrete examples of how to resolve sustainable urbanization and environmental issues in different regions of the world.

The series is intended for professionals in research and teaching, consultancies and industry, and government and international organizations. Published in collaboration with IEREK, the Springer ASTI series will acquaint readers with essential new studies in STI for sustainable development.

ASTI series has now been accepted for Scopus (September 2020). All content published in this series will start appearing on the Scopus site in early 2021.

Bao-Jie He · Joni Jupesta · Gloria Pignatta
Editors

Resilient Horizons: Building Sustainable Environments for Climate Adaptation and Health

Second Edition

Springer

Editors
Bao-Jie He
School of Architecture and Urban Planning
Chongqing University
Chongqing, China

Joni Jupesta
Institute for the Advanced Study of Sustainability
United Nations University
Tokyo, Japan

Gloria Pignatta
School of Built Environment
University of New South Wales
Sydney, NSW, Australia

ISSN 2522-8714 ISSN 2522-8722 (electronic)
Advances in Science, Technology & Innovation
ISBN 978-3-031-46108-8 ISBN 978-3-031-46109-5 (eBook)
https://doi.org/10.1007/978-3-031-46109-5

This Springer imprint is published by the registered company Springer Nature Switzerland AG
The registered company address is: Gewerbestrasse 11, 6330 Cham, Switzerland

Paper in this product is recyclable.

Scientific Committee

The Editors warmly thank all the Reviewers who have contributed their authority to the double-blind review process, to ensure the quality of this publication.

Preface

Climate on the Earth has been ever-changing, due to anthropogenic activities that enhance greenhouse gas (GHG) emissions. Climate change has caused a series of extreme environmental, economic and social consequences. At an average warming level of 1.0–1.1 °C above pre-industrial level, the Earth has already been undergoing a variety of natural disasters such as extreme heat, flooding, bushfire and drought. Scientists have pointed out that beyond an average warming threshold of 1.5 °C, climate-related disasters will be aggravated, exceeding the level the society can stand. Moreover, interrelated with many other mega challenges such as urbanization, population increase and economic growth, climate change has significantly deteriorated the sustainable development of world.

The society is now falling into an emergency situation. No time is left for any hesitation for addressing climate change challenges. The fact is that GHG concentration is keeping increasing and by 2030–2040 the global warming is likely to reach the critical threshold of 1.5 °C. The transformation of the traditional ways of environmental modifications, in various fields including transportation, industry, building, energy generation, agriculture and land use and forestry, towards sustainability is one of the most rapid, cost-effective and efficient pathways to actually reduce GHG emissions for climate change mitigation. Sustainable environmental planning and design is also the critical adaptive way to reduce climate-induced threats and risks in the current era. However, it calls for the complete support from all professions of the society.

This conference on the Climate Change and Environmental Sustainability covers four topics including Climate Change Mitigation and Adaptation, Sustainable Urban-Rural Planning and Design, Decarbonisation of Built Environment and Climate-Related Governance and Challenges. Such four topics can contribute to (i) measurement of climate change and assessment of environmental, economic and social impacts; (ii) understanding of the mechanisms for effective environmental planning and sustainable resource management; (iii) advancement of strategies and techniques for climate change mitigation and adaptation; (iv) the improvement of the involvement and integration of different stakeholders across international, national, state, city and community level for environmental sustainability and (v) demonstration of pilot cases and exemplars that can help address climate change issues.

This conference, in alignment with the most current international and national advocations and actions such as the Paris Agreement, United Nations Sustainable Development Goals and Carbon Neutrality, is a step towards empowering decision-makers and energy stakeholders to join forces and proactively address the challenges of climate change. By bringing together international researchers who are working on topics relevant to climate change and environmental sustainability to share their latest accomplishments and research findings, voices can be amplified and actions towards a more resilient, liveable, sustainable future taken. This conference is expected to serve as an important resource to inform people and provide them with a comprehensive understanding of the possible opportunities for environmental sustainability to address climate change.

Chongqing, China — Bao-Jie He

Acknowledgments

The Conference Chair, Co-chairs and Editors dedicate this book to all conference attendees from the industry, governmental organizations and academia who have been actively engaged with conference preparation, keynote speech moderation and presentation.

Many thanks go to Prof. Dr. E. Didem Evci Kiraz from the Department of Public Health, Adnan Menderes University Aydin (Türkiye); Dr. Joni Jupesta from the Institute for the Advanced Study of Sustainability, United Nations University (Japan) and Dr. Rongrong Cheacharoen from the Metallurgy and Materials Science Research Institute, Chulalongkorn University (Thailand) for their fantastic keynote speech.

Special appreciation is given to Dr. Beta Paramita from Universitas Pendidikan Indonesia (Indonesia), Dr. Huimin Liu from the Wuhan University (China), Dr. Yifu Ou from the University of Hong Kong (China) and Dr. Yuan Liang from the Hong Kong Baptist University (China) for their sessional moderation.

About The Conference

The Climate Change and Environmental Sustainability (abbreviated as "CCES") Conference is a conference series which actively responded to a series of mega challenges such as climate change, urbanization, population increase and economic growth. The CCES Conference was launched by Prof. Bao-Jie He at the School of Architecture and Urban Planning, Chongqing University, China, in collaboration with International Experts for Research Enrichment and Knowledge Exchange (IEREK), Egypt.

The CCES Conference, in alignment with the Paris Agreement, United Nations Sustainable Development Goals and Carbon Neutrality, advances a broad scope of research topics on climate change, decarbonization and urban sustainability. It covers four main topics including (i) Climate Change Mitigation and Adaptation, (ii) Sustainable Urban-Rural Planning and Design, (iii) Decarbonisation of Built Environment and (iv) Climate-Related Governance and Challenges (Fig. 1). This conference is expected to serve as an important resource to inform people and provide them with a comprehensive understanding of the possible opportunities for environmental sustainability to address climate change.

The 2nd International CCES Conference was held online on 5–7 October 2022, chaired by Prof. Bao-Jie He from the School of Architecture and Urban Planning, Chongqing University (China); Dr. Gloria Pignatta from the School of Built Environment, University of New South Wales (Australia); Prof. Linchuan Yang from the Department of Urban and Rural Planning, Southwest Jiaotong University (China) and Prof. Zhaowu Yu from the Department of Environmental Science and Engineering, Fudan University (China).

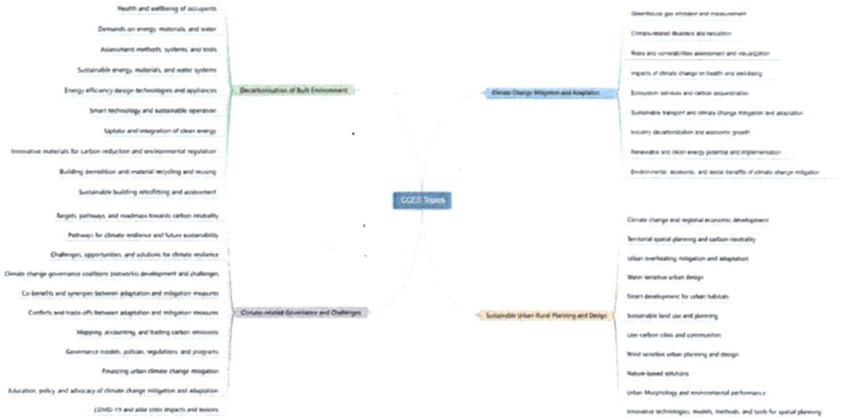

Fig. 1 Key topics in four sectors

This edition has received submissions from about 60 countries and regions around the world, 290+ registrations and approximately 90 abstract submissions from the following countries and regions: Albania, Australia, Bangladesh, Benin, Botswana, Cameroon, Canada, Cape Verde, China, Colombia, Dominican Republic, East Timor, Egypt, Ethiopia, Germany, Ghana, Haiti, India, Indonesia, Iran, Italy, Japan, Jordan, Kenya, Malawi, Malaysia, Mexico, Mali, Morocco, Mozambique, Nepal, Nigeria, Pakistan, Papua New Guinea, Peru, Philippines, Russia, Singapore, Slovakia, Slovenia, Somalia, South Africa, Sudan, Tanzania, Thailand, Tunisia, Türkiye, Uganda, United Arab Emirates, UK, USA, Vietnam and Zimbabwe.

Contents

Introduction

An Era to Address Climate Change for Urban Resilience

Bao-Jie He, Gloria Pignatta and Joni Jupesta

Abstract

The Earth is undergoing climate change, and cities are challenged by climate-related disasters. After progressive efforts of debating facts, reasons, and responders to climate change, the world enters an era of actually implementing climate change mitigation and adaptation. However, there are still many uncertainties. First, climate change is accelerating and climate-related disasters are more frequent, severe, and intense than ever before. Second, the complicated linkages between climate change and a series of environmental sustainability demands make decision makers difficult to decouple them. To overcome this megachallenge, the 2nd International Conference on Climate Change and Environmental Sustainability was held to widely collect experience and lessons tailoring to different contexts, in order to better promote the transformation towards climate change mitigation and adaptation implementation. A total of 14 conference papers were accepted for publication. The book contains four parts including "Climate Change Mitigation and Adaptation", "Climate-related Governance and Challenges", "Decarbonization of Built Environment", and "Sustainable Urban–Rural Planning and Design". The contents in this book are expected to provide pathways to climate change mitigation and adaptation, whilst highlighting environmental resilience.

Keywords

Climate actions · Mitigation and adaptation · Environmental resilience · Pathways · Built environment

1 Climate Change and Disasters

An era of industrialization and urbanization is accompanied with extensive modifications of atmospheric and physiomorphological environments. However, this story is not illegible, and scientists have made progressive and long-term efforts to convince all possible stakeholders worldwide that climate change is an indisputable fact (Stage-1). Among broad and deep debates and dialogues, but without any awareness to act, the global warming accelerated. The warming rate above the pre-industrial level before 1980 was 0.08 °C per decade, while the warming rate was more than doubled after 1981 with an alarming digit of 0.18 °C per decade (NOAA, 2020). Unfortunately, the past forty years witnessed another round of debates on why climate changes

B.-J. He (✉)
Centre for Climate–Resilient and Low–Carbon Cities, School of Architecture and Urban Planning, Chongqing University, Chongqing 400045, China
e-mail: baojie.unsw@gmail.com

Key Laboratory of New Technology for Construction of Cities in Mountain Area, Ministry of Education, Chongqing University, Chongqing 400045, China

Institute for Smart City of Chongqing University in Liyang, Chongqing University, Liyang 213300, Jiangsu, China

State Key Laboratory of Subtropical Building Science, South China University of Technology, Guangzhou 510641, China

Network for Education and Research On Peace and Sustainability (NERPS), Hiroshima University, Hiroshima 739-8530, Japan

G. Pignatta
School of Built Environment, Faculty of Arts, Design & Architecture, University of New South Wales, Kensington, NSW 2052, Australia

High Performance Architecture Research Cluster, Faculty of Arts, Design & Architecture, University of New South Wales, Kensington, NSW 2052, Australia

City Futures Research Centre, Faculty of Arts, Design & Architecture, University of New South Wales, Kensington, NSW 2052, Australia

J. Jupesta
Institute for the Advanced Study of Sustainability, United Nations University, Shibuy 150-8925, Tokyo, Japan

B.-J. He et al. (eds.), *Resilient Horizons: Building Sustainable Environments for Climate Adaptation and Health*, Advances in Science, Technology & Innovation, https://doi.org/10.1007/978-3-031-46109-5_1

(Stage-2) occur and who should be responsible for the climate change (Stage-3), and the fact that human beings lost the best opportunity to counterbalance climate change. By 2022, the warming in many regions reached 1.1–1.2 °C (IPCC, 2023).

A warming level of 1.1–1.2 °C leaves no time to hesitate in implementing climate change mitigation and adaptation. Climate change at this level causes significant disasters such as extreme heat, wildfires, flooding, droughts, and hurricanes. For instance, European citizens had a deadliest extreme heat experience in 2003 with about 70,000 losing their lives (Robine et al., 2008). There were more than 55,000 heat-related deaths during the extreme heat event in Russia, 2010 (Shaposhnikov et al., 2014). In 2022, more than 61,000 people lost their lives in Europe due to lasting heat wave events (Ballester et al., 2023). Beyond these, extreme heat is a driver to extreme drought and wildfires. For instance, the 2019–2020 Australian bushfire was a megafire lasting for more than one year, killed about 480 million animals, and damaged 24.3 million hectares (Binskin et al., 2020). The EU 2022 wildfire was the second worst across the European continental since 2000, with 45 countries attacked and 1.6 million hectares burnt (EU, 2023).

Accelerating warming, moreover, drives climate-related disasters to attack regions worldwide more frequently and intensively. Scientists report that climate-related disasters will severely surpass the current level when the warming level exceeds 1.5 °C. There is an only 0.3–0.4 °C gap between current warming level and critical threshold. However, the emission of greenhouse gases into the atmospheres remains at an intensive level without a hint to reduce. Without no doubt, more significant climate-related disasters and a much deadlier era are coming. Various types of climate-related disasters become the new normal. For instance, 2023 is a record-breaking year. Asian countries such as India, Bangladesh, China, Thailand and Vietnam, and European countries such as Spain and Portugal had experienced the hottest April ever recorded. Therefore, human beings have no choice to enter an era to implement climate change mitigation and adaptation (Stage-4).

2 An Era to Address Climate Change for Urban Resilience

Implementing climate change mitigation and adaptation can enhance the resilience of human settlements against climate change impact, especially in cities which are the main settlements of human beings under urbanization. Climate change mitigation involves responsible strategies aimed at addressing shifting climatic conditions by reducing the release of greenhouse gas emissions into the atmosphere. This necessitates a range of mitigation approaches centered around decarbonization, encompassing sectors such as construction and building, energy, and transportation. On the other hand, adaptation involves protective actions taken to ensure the well-being, comfort, health, safety, and lives of people by minimizing their exposure to climate-related disasters. Adaptation actions play a crucial role in complementing mitigation strategies, providing temporary safeguards against the adverse effects of disasters. While the ultimate goal of mitigation is achieving net-zero greenhouse gas emissions, this requires sustained long-term efforts and faces significant implementation challenges. Consequently, adaptation strategies become the primary focus for addressing immediate, medium-term, and even long-term climate impacts.

Cities are also the stands of various megatrends such as urbanization, population growth, population ageing and economic activities. However, a series of urban physio-morphological modifications and production activities during such processes drive climate change. For instance, economic growth is built upon intense greenhouse gas emissions, and cities have stronger carbon emission intensities than rural areas. Decoupling carbon emissions and economic growth is a subtle challenge if renewables are not fully adopted to replace fossil fuels. Urban construction and operation also lower the resilience to climate change. The extensive use of concrete and cement to harden roads to a large extent weakens water seepage and absorption capacity of soils, increasing the likelihoods of urban flooding. Cities built upon concrete and cement also increase the capacity of absorbing, trapping, and storing excess heats, thereby causing local warming and threatening human thermal comfort, especially in summer seasons.

Therefore, urban planners, designers, and managers should give priorities to not only economy, environment and wellbeing, but also existing and emerging disasters and threats generated by climate change. Both mitigation and adaptation actions should be tailored to specific city contexts (e.g., urbanization levels, economic situation, environmental challenges, urban development patterns, and population structure). The Sustainable Development Goals proposed by the United Nations have defined 17 Goals that all decision makers should implement for the development and prosperity for all people and the planet. Whilst climate action is one of them, such 17 goals are all threatened by climate change, and some of them are drivers to climate change mitigation and adaptation. For instance, climate change threatens a set of living resources like productive materials, food accessibility, health and wellbeing, gender equality, water security, among others. The promotion of affordable and clean energy, resilient infrastructure, and responsible consumption and production is actually the pathway to mitigation and adaptation actions.

3 Pathways to Climate Change Actions and Urban Resilience Improvement

Nexus of climate change actions and urban resilience requires a comprehensive and holistic system of pathways. Specific urban contexts make it more difficult to develop "one-size-for-all" plans and actions. There is a need to broadly collect experience and lessons tailoring to different contexts, in order to better promote the transformation towards climate change mitigation and adaptation implementation. Accordingly, the authors of this Introduction chapter actively supported the 2nd International Conference on Climate Change and Environmental Sustainability held in Chongqing University (China) in collaboration with the University of New South Wales (Australia), Southwest Jiaotong University (China), and Fudan University (China). This conference was aimed at bringing together researchers and practitioners who are engaged in climate change and environmental sustainability to share thoughts, findings, and actions towards a more resilient, livable, sustainable future. After critically screening abstract submissions and blindly reviewing full submissions, a total of 14 conference papers were accepted and edited into the book titled "Climate Change and Environmental Resilience". Such 14 papers are divided into four parts, including "Climate Change Mitigation and Adaptation" (Part 1), "Climate-related Governance and Challenges" (Part 2), "Decarbonization of Built Environment" (Part 3), and "Sustainable Urban–Rural Planning and Design" (Part 4).

The first part consists of four chapters, covering topics on recycling wastewater to address increasingly severe drought challenges, reusing abandoned industrial buildings, resilience of plant ecosystem to climate change, and climate determinants of health. The chapter "A Practical Urban Habitat for Living in the Extreme Drought Conditions of the Future, Using Advanced Wastewater Recycling Technologies" describes the design of a climate resilient habitat of "Tataouine", a prototype "living laboratory" featuring nutrient restoration by microalgae and aquatic angiosperms and the sequential wastewater treatment using microbial and physio-chemical processes. The "Tataouine" is expected to be an integrated, self-sufficient, and a climate-resilient habitat to manage wastewater for domestic needs and cultivation of both edible and livestock feed crops under highly risky climate conditions. The chapter "Risk Assessment of Green Reuse of Abandoned Industrial Buildings: A Case of Steel Mill in Xi'an City" explores the green reuse of an abandoned industrial buildings, during urban renewal, by analyzing possible risks in institutional, economic, management, technical, and environmental dimensions. An empirical assessment of a steel mill showcased that economic risk was the most prominent, while the environmental risks was the lowest. This study was important for relevant stakeholders to understand the risks and make proper decisions for successful green reuse implementation. The chapter "Adaptation of Plant Ecosystems to Rapid Climate Change in The Ural Region" focuses on highly productive broad-leaved species which can fix free carbon in the Ural region, given increasingly sensitive climatic conditions. The authors explored the impacts of regional climate change (e.g., in temperature and precipitation accumulation) on the occurrence of bushfires and the restoration of coniferous species, so as to identify the most adaptive species for planting. The chapter "Climate Determinants of Health" aims to find out the most accurate, most concise, most accessible, and clearest data-based determinants to assess the relationship between climate and health, given the urgency of dealing with climate-related health impacts. This chapter clarifies six aspects of determinants related to social determinants, individual characteristics, community features, climate hazards, vulnerability, and exposure. These determinants are advantageous in revealing the actual extent of the relationship between climate and health. They guide academic studies and practices aimed at protecting people, communities, cities, and global health from the effects of climate change. Furthermore, they demonstrate that climate-related health impacts are not solely the ultimate responsibility and administration of a single sector.

The second part includes three chapters covering topics of climate-related human security and sustainability, personal norm to climate actions, and urban energy use. The chapter "Effects of Climate Change on Human Security and Sustainable Development" analyzes climate-related threats to human sustainability, in aspects of self-worth of the individual, freedom from fear, politics and international relations on the one hand, and the relationship between the environment and the basic necessities of life such as sustainable economic livelihood, healthcare, food security and forced migrations on the other. Suggestions on sustainable environmental governance at the international level and policy reform are also proposed to ensure sustainable development. The chapter "Role of Value Orientation and Belief in shaping Indian Pre-Service Teachers' Personal Norms to address Climate Change" highlights the chain effects of people's values and beliefs on human conduct and thereby the consequences of climate change. Based on the survey among pre-service teachers in Kerala (India), this chapter reveals the strong positive associations between climate change and (i) value orientation and belief, (ii) value orientation and personal norms, and (iii) belief and personal norms. The results can provide environmental managers with basis to regulate people's environmental value orientations to enhance climate change actions in the direction of environmental stewardship and sustainability. The chapter "Urban Energy Consumption in The City of Naples (Italy): A Geographically Weighted Regression Approach" explores the relationship between urban characteristics (i.e. socio-economic, physical, functional, and environmental) and residential energy consumption based on a statistical-spatial methodology, in order to support proper local-scale decisions on how to realize low-carbon cities. Based on the model of normalized residential energy consumption, it is possible to identify the areas with strong energy consumption intensities, and reveal the associated drivers. Solutions of refurbishing low-quality buildings, raising awareness and increasing green areas are suggested to be implemented in Arenella and Scampia (Italy), given low education and significant presence of impervious surfaces in such two neighborhoods.

The third part includes five chapters covering performance of high-performance construction materials, built environment decarbonization monitoring method, low-carbon building retrofitting, roof decarbonization design, and green roof decarbonization design. The chapter "Mechanical Properties and Impact Resistance of Steel Fiber Reinforced Fly Ash and Nano-silica Concrete" explores the Steel Fiber Reinforced Concrete with Fly Ash, showing that Nano silica addition is very effective in the development of high early strength of HVFAC as well as the impact resistance. The adoption of high-performance construction materials is meaningful to reduce the amount of construction materials which are the main source of embodied carbon in the building sector. The chapter "Nonintrusive Load Monitoring System for Decarbonization of Built Environment Based on Deep Learning Algorithm" presents a nonintrusive load monitoring algorithm and platform based on a deep neural network model, in order to allow end users to check the electricity situation of their home or office building in real time through web or mobile clients. The model is conducive to reducing carbon emissions and costs through identifying energy-saving suggestions, predicting energy consumption, and monitoring power redundancy. The chapter "Sustainable Renovation on Aosta Residential Building for Carbon Neutrality" combines onsite investigation and energy simulation to analyze outdoor and indoor environments, and building performance of a social housing building in Aosta (Italy), under a range of conditions, in order to determine the needs and focus of building retrofitting for operational energy use reduction. Suggestions of steel louvre blades, high-performance window profiles, and triple glazing are proposed to integrate passive cooling strategies, optimize the envelope, and reduce energy needs for heating and cooling. The chapter "Sensitivity Analysis Using Standardized Regression Coefficients of Roof Design Variables for Energy Performance in Residential Buildings" explores the sensitivity of energy use intensity and thermal discomfort hours to roof design variables of a house prototype in Alexandria (Egypt). The results show that roof geometrical transformations result in the maximum energy use intensity variation of 17.3 kWh·m$-^{2}$ and 11% in indoor thermal comfort. Moreover, gable roofs, domes, and vaults can achieve better energy and thermal comfort performance. The chapter "Investigating the Impacts of Biochar Addition to Substrate on Mitigating Carbon Dioxide Emissions in Green Roofs" investigates the effect of applying biochar mixture into soil substrates on green roof decarbonization. Based on 36-h field experiments, the results verify the decarbonization potential of biochar application, and the inclusion of biochar can also help control rainwater runoff and alleviate urban eutrophication.

The fourth part includes two chapters covering ESG (Environmental, Social, and Corporate Governance) risk management, and sustainable housing design in responses to COVID-19. The chapter "Features of ESG Risk Management in The Implementation of Megaprojects in The Belt and Road Initiative" explores the ESG risks of implementing a China's water management megaproject: Neelum–Jhelum Dam in Pakistan, under the Belt and Road Initiative. The results showcase that the potential of ESG risk management is not fully realized, possibly due to the difficulty of controlling the use of funds to solve environmental and social problems, and the duration of the procedures for approving decisions by the relevant ministries in the host countries. The chapter "Sustainability, and Post-Covid 19 Era: Changing Housing Design and Models in the Cities and Istanbul" conducts a transdisciplinary investigation into the intersection of architectural design and urban planning in the post-pandemic era. This chapter examines newly designed housing models in the post-pandemic era, with a particular emphasis on Istanbul (Turkey) as a paradigmatic global city, finding that the Covid-19 pandemic has had a profound influence on the cities and housing markets, leading to a reconfiguration of housing models and typologies, as well as heightened importance on sustainability and environmental issues.

4 Concluding Remarks

A series of climate-related disasters become the new normal, while cities and communities have not well prepared for climate change mitigation and adaptation. The close linkages of environmental resilience to urban modification and urban operation make it more complicated and difficult to directly address climate change and secure environmental resilience.

This edition of the book presents topics on "Climate Change Mitigation and Adaptation", "Climate-related Governance and Challenges", "Decarbonization of Built Environment", and "Sustainable Urban–Rural Planning and Design". The contents are expected to provide lessons and experiences in achieving climate change mitigation and adaptation, while ensuring environmental sustainability.

Acknowledgements This work was supported by the National Natural Science Foundation of China (No. 42301339) and Fundamental Research Funds for the Central Universities (No. 2023CDJXY-008).

References

Ballester, J., Quijal-Zamorano, M., Méndez Turrubiates, R. F., Pegenaute, F., Herrmann, F. R., Robine, J. M., Basagaña, X., Tonne, C., Antó, J. M., & Achebak, H. (2023). Heat-related mortality in Europe during the summer of 2022. *Nature Medicine, 29*, 1857–1866. https://doi.org/10.1038/s41591-023-02419-z

Binskin, M., Bennett, A., & Macintosh, A. (2020). *Royal commission into natural disaster arrangements*. Commonwealth of Australia (p. 115). Australian Capital Territory.

EU. (2023). *The EU 2022 wildfire season was the second worst on record*. EU Science Hub. Available at: https://joint-research-centre.ec.europa.eu/jrc-news-and-updates/eu-2022-wildfire-season-was-second-worst-record-2023-05-02_en#:~:text=Last%20year%2C%20wildfires%20in%2045,average%20of%20fire%2Dravaged%20area

IPCC. (2023). Summary for policymakers. In *Climate change 2023: Synthesis report. Contribution of working groups I, II and III to the sixth assessment report of the intergovernmental panel on climate change* (pp. 1–34) [Core Writing Team, H. Lee & J. Romero (Eds.)]. IPCC. https://doi.org/10.59327/IPCC/AR6-9789291691647.001

NOAA. (2020). *Annual 2020 global climate report*. National Centers for Environmental Information. National Oxygen and Acetylene Association. Available at: https://www.ncei.noaa.gov/access/monitoring/monthly-report/global/202013

Robine, J. M., Cheung, S. L. K., Le Roy, S., Van Oyen, H., Griffiths, C., Michel, J. P., & Herrmann, F. R. (2008). Death toll exceeded 70,000 in Europe during the summer of 2003. *Comptes Rendus Biologies, 331*(2), 171–178. https://doi.org/10.1016/j.crvi.2007.12.001

Shaposhnikov, D., Revich, B., Bellander, T., Bedada, G. B., Bottai, M., Kharkova, T., Kvasha, E., Lezina, E., Lind, T., Semutnikova, E. & Pershagen, G. (2014). Mortality related to air pollution with the Moscow heat wave and wildfire of 2010. *Epidemiology, 25*(3), 359–364. https://doi.org/10.1097/EDE.0000000000000090

Monitoring of Sustainable Built Environment Solutions for Climate Adaptation

A Practical Urban Habitat for Living in the Extreme Drought Conditions of the Future, Using Advanced Wastewater Recycling Technologies

Simrat Kaur, Fatema Diwan and Brad Reddersen

Abstract

Injudicious uses of water and climatic changes are fostering the depletion of freshwater reserves, resulting in persistent megadroughts on multiple continents. Accompanied by wildfires, these megadroughts are emptying the lakes, reservoirs, and aquifers; resulting in dormant dams and hydropower failures. Unprecedented wildfires and dust storms destroy habitats, which have not yet evolved to be climate resilient. The parched land, not supporting the production of nutritious food, would lead to an increase in food prices in the global market, resulting in a high hunger index. Alternate options are mandatory for high water-demanding crops, namely, corn, maize, and soybeans, which have become primary sources of livestock feed and are causing mass destruction of forest biomes in tropical countries. Addressing these issues requires, designing an integrated, self-sufficient, and climate-resilient habitat that would incorporate practical means of managing water for domestic needs and cultivation of both, edible and livestock feed crops. This paper describes the design of a climate-resilient habitat named as "Tataouine" which is currently under construction in near-desert conditions in Arizona, United States. Tataouine is a prototype "living laboratory", that stand by the principle of infinite uses of scarce water resource, through integration of ancient and modern technologies. It features nutrient restoration by microalgae and aquatic angiosperms and sequential wastewater treatment using microbial and physio-chemical processes, followed by the revitalization of the treated water to make it fit for both non-potable and potable consumption. Tataouine represents an action plan for drought preparedness using a proprietary microbial-based "PolyBio" wastewater recirculation system that supports sustainable and resilient living for the future high-risk communities.

Keywords

Climate-resilient housing · Drought · Environmental footprints · Microalgae · Microbes · PolyBio · Wastewater

1 Introduction

As the most overused natural resource, the rate of water consumption is twice the human population growth rate (Human Development Report, 2006). As a consequence, half of the world's population will lack access to clean water by 2050. Drought, as one of the major natural disasters occurs due to variations in the climatic conditions along with overuse and abstractions of both surface and groundwater reserves. The hydrological and meteorological droughts cause substantial reductions in freshwater resources and soil moisture due to rainfall deficit which affects key economic activities such as agriculture, forestry, timber production, and hydropower generation.

The persistence of meteorological megadrought is assessed using a standardized precipitation index (SPI) of large timescale (Kim et al., 2022). The twenty-first century has witnessed an increasing trend of drought, both in terms of intensity and duration across many continents. The record-breaking drought prevailing in Southwestern North America for the last 22 years has matched the severity of

S. Kaur (✉) · B. Reddersen
Climate Survival Solutions India Pvt. Ltd., Siliguri, West Bengal, India
e-mail: simrat@climatesurvivalsolutions.in; kaursimrat.drl@gmail.com

F. Diwan
Climate Survival Solutions India Pvt. Ltd., Hyderabad, India
e-mail: fatema@climatesurvivalsolutions.in

B.-J. He et al. (eds.), *Resilient Horizons: Building Sustainable Environments for Climate Adaptation and Health*, Advances in Science, Technology & Innovation, https://doi.org/10.1007/978-3-031-46109-5_2

the late 1500s megadrought (Williams et al., 2022). Many drought events had been occurring in Central Chile; the one that occurred from 2010 to 2016 is among the global representatives of the megadrought case (Kim et al., 2022). The temperate zones of the world which are facing the extreme events due to global warming will be severely affected by droughts. For example, the excessive heat waves and warming due to rising land and ocean temperatures in different parts of the Europe show continent's vulnerability to persistent droughts (Ionita et al., 2021).

Earth's largest tropical wetland Pantanal in South America has been affected by the worst drought and fires causing huge biodiversity losses. Compared to the last 70 years, the 2020 drought in Pantanal was recorded as the most extreme and widespread (Libonati et al., 2022). The scarcity of summer rainfall during 2019–2020 caused prolonged and extreme drought which severely impacted the hydrology of the Pantanal and consequently led to the fall of hydrometric levels in the river Paraguay (Marengo et al., 2021). Anthropogenic activities such as global warming, deforestation, land cover change, hunting, beekeeping, and cattle grazing are the major causes of mega-fires associated with extreme dry spells and droughts in Pantanal (Voiland & Observatory, 2022). The tropical biomes of Pantanal are the biodiversity hotspots but have also become wildfire hotspots, over the recent decades due to increased anthropogenic activities, particularly, industrial agriculture and cattle ranching (Kumar et al., 2022).

High temperatures, low humidity, and strong winds are characteristics of fire weather and are important drivers of wildfires worldwide, particularly the forest biomes of western USA, the Mediterranean, and southeast Australia (Richardson et al., 2022). The compound effects of higher daily maximum temperatures and lower daily relative humidity increase the frequency of fire weather days in the Northern and Southern Hemisphere regions of the USA and Australia. However in geographical locations such as India and northeast Russia, the high relative humidity overrides the maximum daily temperatures, and thus lowering the frequency of fire weather in these regions (Richardson et al., 2022).

Drought and related climate hazards like heat waves and wildfires causing catastrophic burning of various ecosystems and biomes have become the costliest of natural disasters. These unprecedented climatic events have occurred at irreplaceable environmental costs due to considerable loss of biodiversity. In the African continent, the economies depend on rain-fed agriculture; therefore, drought causes suffering to millions of people due to crop failures, starvation, and heat waves. High seasonal variations in intensity, frequency, and duration of extreme climates including drought, pose a great challenge in developing largely adapted cultivars that are drought resistant, particularly the rain-fed crops such as beans, maize, wheat, potato, cassava, and rice, which are the staples in east African countries like Rwanda (Uwimbabazi et al., 2022). Agriculture including livestock farming dominates the usage of clean water because of the intensive irrigation required for water-demanding crops. Yet, the main drivers of water scarcity in the agriculture sector differ in various regions of the world. For example, the unjustified cultivation of water-intensive crops such as cotton and almonds, respectively, in the states of Arizona and California, USA has led to the overexploitation of their natural freshwater reservoirs. The shrinking of lakes due to over abstractions for cotton cultivation in Arizona is causing severe water shortages in the domestic sector.

The phenomenon of increase in the severity of contrasting hydrological events namely flooding and drought has been referred to as the acceleration of the terrestrial component of the hydrological cycle, which has a compound effect on global food production (Chagas et al., 2022). On other hand, the shifting land use pattern is negatively affecting, surface and groundwater aquifer recharge efficiencies. The surface irrigation diverts water from the natural aquifers such as rivers and lakes for distribution to the agriculture fields. Due to the permeable nature of the agricultural soils, much of the irrigated water gets infiltrated to recharge the underlying aquifers, which may have either negative or positive impacts on the quality and quantity of groundwater, depending upon the geographical location and soil properties (Rotiroti et al., 2019). Negatively, the contaminants derived from surface irrigation due to the intensive agriculture practices deteriorate the quality of groundwater, making it unsuitable for successive irrigation and the drinking usage. The shallow water table with highly permeable alluvial deposits in the floodplains along river banks, where the surface water and agriculture lands are interconnected, is highly vulnerable to point source nitrate and pesticide contaminations (Bartzas et al., 2015). Intensive use of nitrate fertilizers in conventional agriculture and due to its mobile and persistent characteristics, the irrigation return flow carries high loadings of nitrate which makes it unfit for drinking purposes (Jafari et al., 2012). In semi-arid regions, contamination and over-abstraction of groundwater aquifers pose direct threat to water security and thus there is need for alternative approaches to mitigate the dual impacts of agriculture on groundwater depletion and pollution.

Pollution and poor management of water resources are among the other major reasons for water scarcity during natural drought events. Although conservation (particularly of domestic usage) may act as the foremost social and regional strategy to endure the impacts of the water crisis;

it is inequitable. The lower income group populations suffer and bear the harsh consequences of drought such as heat waves, inadequate potable water, and food shortages due to poor agriculture yields. As an alternate strategy, the arid regions of the world are obtaining water from alternate sources such as the atmospheric water and treated wastewater. Harvesting the atmospheric water requires modern equipments, and its efficiency is dependent on the regional conditions. On the other hand, the recovery of treated wastewater is full-proof technology based on integrated biological and physio-chemical methodologies which are highly adoptable, even at the community levels.

The nutrient (inorganic and organic) rich sewage is being generated in huge volumes on a daily basis from domestic households. Many countries in the world lack adequate numbers of functional and efficient treatment plants that can tackle the daily sewage inputs. As a result, the inefficacious or zero treatment of municipal sewage becomes a burden on the freshwater habitats causing their pollution and inevitable environmental issues such as eutrophication. In the events of unprecedented torrential rainfalls and storms due to climate change, even developed countries such as the United Kingdom (BBC, 2022) dump billion liters of untreated polluted water due to system failure of sewage treatment plants into their rivers. This suggests that water management through the centralized or conventional wastewater treatment utilities around the globe is a failure; therefore, additional alternatives must be exercised, right at the community levels.

The decentralized wastewater treatment (DEWATS) offers a holistic approach toward three targeted global issues namely environmental pollution, climate change, and resource recovery. A resilient and adaptive approach rather than crisis-management approach is needed for the practical and feasible implementation of resource and water recovery from wastewater. Resource recovery from wastewater is a crucial and foremost driver for establishing a circular economy that aims to minimize the water footprints of all major economic sectors. Overexploitation of the non-lasting nutrient reserves, particularly the phosphate must be stopped in order to sustain our economies. The reductions in raw materials consumption are only possible, when alternate supplies are made available. This can be achieved when wastewater treatment processes are not just conditioned for disposal rather, conversion and transformation into unlimited nutrient supply, clean water, and green energy. Apart from the common digester gas combustion, wastewater treatment plants also have high potentials for heat generation to meet the growing energy demands (Kollmann et al., 2016). The challenges involved in the integration of resource (phosphate, nitrogen, bioplastics, exopolymers, aerobic granular sludge, manure, biochar), water, and energy recovery into existing wastewater treatment plants are attracting the interests of all stakeholders, business partners, policymakers, and end-users (Zarei, 2020).

It is difficult to upgrade the existing water treatment facilities which are fundamentally designed to meet the quality guidelines for effluent waters set by the regional pollution control boards in different parts of the world. Under existing water management policies and practices, human civilization is losing the most important life-supporting element on the planet, the water. Persistent drought in many geographical regions is a dangerous sign for the water crisis of magnitude that would turn into the next global war. As drought is directly related to the water-food-energy nexus, we need to build resilience around all three resources to survive under the harsh conditions of the future planet.

In this paper, we have discussed the resilient approach of implementing practical solutions for converting wastewater into a clean source of potable water, nutrients, and renewable energy to meet the ends of the water-energy-food nexus. The living laboratory and prototype of our self-sustaining climate-proof habitat called "Tataouine" is being built in the arid environments of Arizona, USA. This is a habitat with an on-site or decentralized sewage sanitation treatment system to recover nutrients and energy. The focus of self-sustaining and resilient habitat is to minimize the resource utilization and preservation of natural soil, water, and biological resources.

2 Impacts of Drought on Urban Habitats and Intersectoral Systems

The current path and global trends of urbanization are unsustainable, as they will offset the demand and supply of natural resources. Of the total 17 Sustainable Development Goals (SDG) established by the United Nations Assembly in 2015, the SDG 11 deals with the sustainable cities and communities. One of the major challenge in achieving SDG 11 is the rapid pace of urban expansion and change in land-use patterns across the globe (Chakraborty et al., 2022). Humans are unceasingly encroaching the natural lands. inclusive of all of its natural resources and biodiversity. Biodiversity including flora and fauna is crucial in maintaining the vitality and renewability of the natural resources. The natural green cover of an un-disturbed land harbors rich biodiversity which play vital roles in maintaining the regional hydrological cycle, bringing precipitation, and reducing spatial heating of the land under its cover. These functions are essential to mitigate the regional impacts of the warming and the increasing temperatures. Therefore, destroying the green land covers and forests for urban expansion of the cities will create ripple effects that would intensify the impacts of climate change and global warming. In this scenario, building a gray infrastructure would not alone suffice the need to overcome the growing demands for water, food, and energy resources.

Alternatively, the Urban Gray and Blue Infrastructure (GBI) could provide ecosystem services, namely, the restoration of water streams, carbon sequestration, water retention and regulation, thermal comfort, and biodiversity enhancement which could address the complex socio-ecological challenges associated with urbanization (Puppim de Oliveira et al., 2022). Establishing stable sources of urban water supply during the unprecedented waves of drought on a global level is crucial as drought affect large number of global population. The actions by the governments are not sufficient to tackle the rapidly increasing urban water scarcities. Hence, the role of private stakeholders is crucial for increasing the preparedness to mitigate the impacts of drought (Dilling et al., 2019).

In the light of climate change and unprecedented and recurrent events of extreme drought-associated water shortages, the water-energy-land nexus dynamics would need careful consideration to tackle the sufferings in the rising urban settlements. Due to increased urban land use, the demand for water is unsustainably growing. Furthermore, the climate change is threatening the water supply in such overcrowded urban areas which suffer variability due to increased evapotranspiration and lower precipitation (Cremades et al., 2021). The measures are conceived to mitigate the impacts of an on-going drought event rather than proactive planning of the methods to overcome the consequences of preceding drought events by incorporating the climate data and adopting holistic preventive methodology (Cremades et al., 2021).

In urban-agricultural intersectoral systems, conflicts could arise due to the shared water resources and increasing demands in events of drought. Thus, managing conflicts in such systems pose a great challenge. The unsustainable monoculture cropping systems are not only detrimental to soil fertility but also lead to the destruction of soil biomes and associated biodiversity. Most often, monoculture crops such as corn and soybean are water intensive and the impacts of drought on these crops are deficits in domestic water supplies. Globally, the policies around such water-intensive and feedstock oriented crops favor their production instead of mitigating the drought-led water shortages; thus the constraints of drought are faced locally (Zipper et al., 2017) by the urban populations. On the contrary, urban water usage represents flexible consumption which could effectively be curtailed at times of intense drought conditions by imposing restrictions and preventive measures to ensure the protection of agricultural interests (Zipper et al., 2017). However, policies should take into account, whether the crops grown in agriculture production systems are consumed locally or are merely for the export benefits. This is crucial as it accounts for huge virtual water footprint through the export of irrigated crops. This directly causes sufferings to the local masses. Therefore, water resource management requires an interdisciplinary approach to address the sustainability challenges posed in a socio-environmental system (Zipper et al., 2017).

3 Water Management and Treatment Systems to Build Climate-Resilient and Eco-Friendly Habitat of Tataouine

The amalgamation of the water treatment process and resource recovery represents a holistic and sustainable approach to tackle the issues of drought-associated water management and mitigation planning. Water is a vital resource which is the most consumed resource. The frequent shortages of water due to unsustainable use practices and climate change highlight the fact that there is an immediate need for its reuse and recycling from wastewater. By transitioning from pollutant removal toward resource recovery, the wastewater is no more considered as a waste stream but rather as a renewable resource. By establishing more circular resource flows, wastewater can contribute to sustainable development. The redesigning of wastewater treatment plants from mere treatment facilities into water resource factories is emerging as opportunity for establishing the circular economy. Most water management utilities still focus on wastewater collection and treatment rather than resource recovery. Thus, there is a need for water management utilities across the globe to strategically plan the transition from treatment toward resource recovery via decentralized treatment systems. The principal component of wastewater is water which is around 99% and the remaining 1% accounts for the suspended, colloidal, and dissolved solids. The conventional agriculture practices rely on heavy inputs of phosphorous (P), a nutrient that would be exhausted in less than 100 years. Both domestic and agriculture sectors generate P-rich wastewaters, which is mostly discharged without an efficient nutrient treatment, leading to the loss of precious and scarce nutrients. In order to sustain the conventional food production and to overcome the global nutrient supply crisis, the P must be recycled and recovered from the wastewaters.

4 Global Warming and Implications on Various Environmental Footprints

Apart from the direct impacts of climate change on water shortages, the indirect implications of a warmer world include increased water demands for domestic usages during extreme heat waves, particularly in the Northern Hemisphere. Among low-income households in developing countries, the household use of water as a substitute for electricity to cope with high temperature days is increasing (Qin et al., 2022). In countries such as Bangladesh where salinity intrusion and severe risks of elevating sea levels are clear markers of climate change impacts, the reduction in river flow directly hampers the electricity production at power plants, and salinity intrusion in rivers causes infrastructure damages (Shahid, 2012).

The rapid urbanization is changing the consumption patterns all over the world. Over the last two decades, the per capita water footprint associated with food consumption is increasing more among the urban resident as compared to their rural counterparts (Shi et al., 2022). An indicator of direct and indirect use of water by consumers or producers, the water footprints are categorized as blue water footprint (consumption of freshwater resources), green water footprint (rainwater utilization for irrigation), and greywater footprint (an indicator of pollution)(Hoekstra & Mekonnen, 2012). Similar to water footprint, the unexercised and illegal utilization of natural resources by humans for several purposes are generating various other forms of environmental footprints. Each footprint represents specific classes of pressures connected to process, product, or activity from the lifecycle assessments viewpoint. These humans generated footprints (Čuček et al., 2015) (Fig. 1) have interconnected impacts on the environment stability and the sustainable future development. For instance, the carbon, energy, and nitrogen footprints are mainly generated by the burnings of fossil fuels which are primarily responsible for global warming.

The ocean warming shows strong impacts on the biological thermal optima for aquatic species. The species inhabiting low-latitudes are more sensitive than high latitude counterparts; and thus unable to survive under the predicted 3 °C rise in ocean temperatures (Rummer et al., 2014). The global ocean warming in conjugation with high CO_2 concentrations impacts oceanic flora by stimulating the growth of toxin producing algal blooms, which are becoming more frequent with oceanic acidification (Beardall et al., 2009). Beyond the biogeographical shifts of the marine species, global warming is also causing thermal habitat loss that enhance the risk of mass extinctions within centuries (Reddin et al., 2022). Besides carbon, the nitrogen and phosphorous are the major elements required for the growth of phytoplankton. The increasing phosphorus and greywater footprints are not only threatening the non-renewable supplies of these resources in nature but are also responsible for the loss of biodiversity in aquatic habitats.

The studies done decades ago had shown that the terrestrial species of the tropics, which were already living close to their thermal optima, are at the greater risks of extinction due to the global warming (Deutsch et al.,

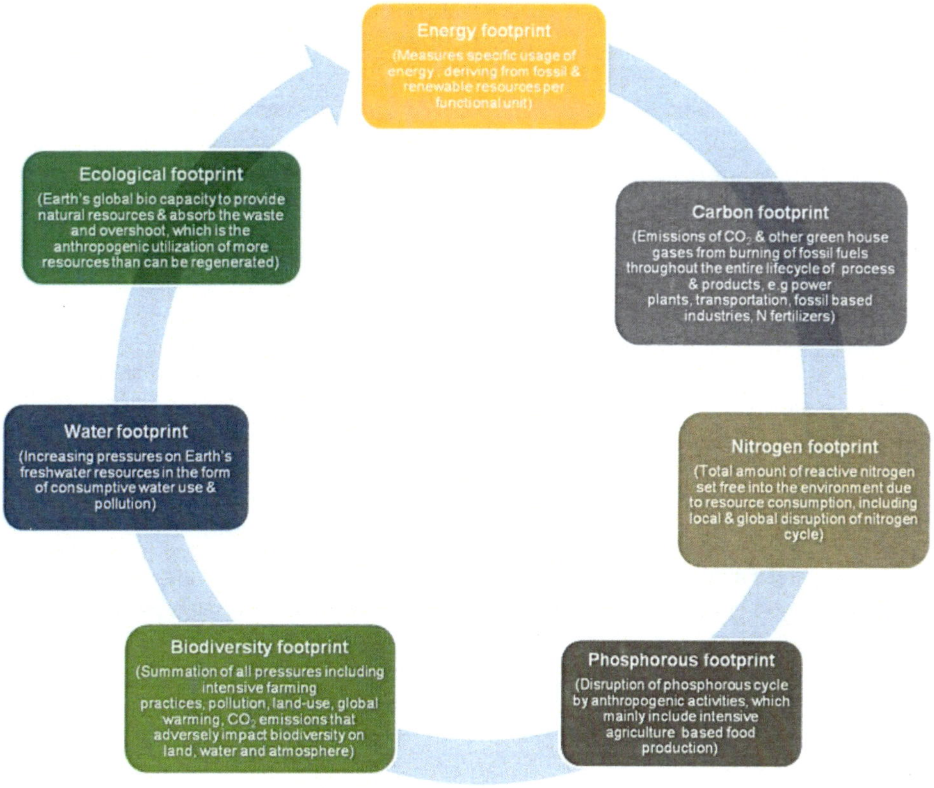

Fig. 1 Various forms of environmental footprints impact the sustainable development and ecosystem integrity

2008). The loss of biodiversity caused due to anthropogenic activities namely global warming, intensive agriculture, land-use change, nutrient pollution, and deforestation impact primary productivities and nutrient cycling in an ecosphere. The soil microbes including bacteria and fungi (which perform various complementary biogeochemical functions) modulate plant diversity-productivity relationships (Xi et al., 2022). Microbes have inhabited the Earth for over 3 billion years and are highly resilient and evolving than humans. Under a climate change scenario, the increase in extreme events such as rising temperatures, storms, and increasing humidity will impact the mobilization of microbial spores; and the production of toxins by harmful cyanobacteria blooms. The spread of airborne microorganisms can also occur from the wastewater treatment plants as in the case of aerosolized *Legionella* bacteria in the Netherlands that caused the Legionnaires' disease (Casadevall, 2021). The warming climate alters the microbial communities and their physiological activities in the wastewater treatment plants which are additionally susceptible to corrosion and damage from storms (Casadevall, 2022). Therefore, there is an immediate need to shift the conventional sewage treatment plants into the micro or neighborhood level decentralized wastewater treatment and management systems. We have discussed the "PolyBio" wastewater recirculation system which is designed for a typical 5-person home using most of the materials that should last at least 10 years before needing replacement. The major components of "PolyBio" should last 50 years or more using resilient materials. Unlike the conventional sewage treatment plants, which are mostly open, the wastewater treatment under the "PolyBio" system is much more concealed and in the closed loops.

The drought and biodiversity loss are interconnected and these relationships cannot be understood in isolation. Biodiversity plays an important role in community functioning including drought resistance and resilience to adapt to future events of water shortages (Ruijven & Berendse, 2010). A large body of literature studies has documented that, water shortages or chronic drought and subsequent wetting events shift soil microbial communities, particularly in herbaceous ecosystems (Xi et al., 2022). Microbial diversity plays a critical role in establishing resilience in plant communities during periods of drought in natural terrestrial ecosystems. Therefore, we must reduce our biodiversity footprints to allow the natural ecosystems to cope up with stressful environments and build resilience during chronic drought events. Similarly, the nitrate, phosphate, and water footprints are directly affecting the availability of freshwater by overshooting the loads of these nutrients and other pollutants in natural aquifers. As consequences of our ecological footprints, we are promptly crossing the planetary boundaries which would not enable sustainable living on this planet.

5 Minimizing Water Footprint to Combat Drought Impacts

Greywater footprint is the water polluted by individuals, society, farmers, industries, and companies. The polluted water is then being discharged into the freshwater sources such as rivers through sewage, agriculture runoff, and groundwater sources through seepage (Fig. 2).

The use of synthetic fertilizers and agrochemicals (pesticides, insecticides), during the entire crop yielding process, account for the water pollution creating the greywater footprint. The excess loads of nitrogen (N) and phosphate (P) fertilizers from crop fields ultimately get carried away into the freshwater bodies causing eutrophication, which threatens the aquatic ecosystem. Excessive N and P leads to the growth of harmful algal blooms which have negative and toxic impacts on human and animal lives. The eutrophication in aquifers and water streams also results in coastal water pollution. The greywater footprint encompasses all anthropogenic activities which alters the natural physiochemical properties of freshwater by the excessive dumping of the pollutants. However, in context of wastewater treatment management, the greywater does not contain water from the toilets and is free from the fecal matter and urine. Some definitions also exclude the kitchen wastewater as it contains the oily and greasy contaminants that require differential process of treatment. Although greywater is less contaminated and polluted, it is deemed unsuitable for restricted non-potable reuse in its crude form. Upto 80% of the total wastewater generated from the household is the greywater, with variable compositions that depend on the culture, living status, locations, and health conditions. The demand and consumption of the finite freshwater resource will soon outnumber the supply. Therefore, the greywater recycling and reuse have tremendous potential to contribute toward the urban sustainability by increasing the capacity of clean water supply. In addition, it allows the humanity to maintain future living within the range of planetary boundaries for non-renewable resources including water and nutrients such as phosphate.

While the world is struggling with the household, agricultural, and industrial water pollution in freshwater resources, humanity generates more than 60% of greywater in households, which if treated systematically, would mitigate the impacts of drought on domestic sectors. The decentralized wastewater treatment systems (DEWATS) are

Fig. 2 The blue water footprint of various socio-economic sectors and generation of greywater footprint

an on-site recycling of greywater generated in households, residential complexes, institutions (schools, colleges), and restaurants. This system of water management considerably reduces the burden on municipality wastewater treatment plants and additionally reduces the dependence on freshwater resources including groundwater.

6 On-Site or Decentralized "PolyBio" Water Circulation Systems at Tataouine Habitat in Arizona, USA

In wastewater resource management, the preliminary task is the proper segregation of the wastewater stream into blackwater and greywater. For the segregation, it is important to establish dual plumbing and sewerage system for gray and black water collection. The definition of greywater is based on geographical locations as it may or may not include the water from the kitchen source. The pipes carrying greywater under pressure should be properly labeled as per the state or the Department of Environmental Quality guidelines to eliminate confusion between greywater and drinking water pipes. In the case of Tataouine, the Arizona Department of Environmental Quality (ADEQ) guide is followed. According to ADEQ, the greywater is defined as "wastewater, collected separately from your sewage flow that originates from a clothes washer, bathtub, shower or sink, but not from a kitchen sink, dishwasher or toilet". Greywater is distinguished from black water, "which is the wastewater from toilets, kitchen sinks, and dishwashers". The treated Black water is not reused in the home because of the high risk of contamination by bacteria, viruses, and other pathogens.

The sewage system should be at least 10 m away from the habitable building. The treatment of both black and greywater would be carried out at the source of collection or in other words it would be the decentralized system of wastewater treatment. The two criteria that determine the treatment methods/processes in a decentralized treatment system are the following:

(a) Characteristics of sewage (black and gray) water—the composition of sewage water varies with the type of habitat and the cultural habits of the occupants (Tataouine habitat is aimed at building a new civilization that heals the planet and nurtures biodiversity and humans. The occupants of Tataouine would practice nature friendly habits and thus, the usage of environment unfriendly and toxic chemicals are replaced with eco-friendly natural-based products. Some of eco-friendly alternatives to personal care products are soap berries or Indian soap plant, activated charcoal, and shrub-like tree called *Acacia concinna*, all of which have cleaning properties). The various physio-chemical parameters of wastewater are turbidity (measure the level of suspended solids), chemical oxygen demand (level of pollutants that require additional biochemical degradation process), biological oxygen demand (level of pollutants that are degraded by various microbial processes), pH and concentration of fecal coliform and other pathogenic microbes.

(b) Intended reuse of the treated water—there are various classes of reclaimed water that are obtained when wastewater goes through different levels of treatment processes. As per the Arizona Administrative code, there are the following classes:

a. Class A+ reclaimed waterwater that has undergone secondary treatment, filtration, nitrogen removal, and disinfection. The criteria to follow are: *before disinfection*—24 h average turbidity after filtration is ≤2 NTU and should not exceed 5 NTU at any time; *after disinfection and before discharge*—no detectable in 4/7 of daily reclaimed water samples with a maximum fecal measure of <23/100 ml in a single sample. The total nitrogen should be <10 mg/l.

b. Class A reclaimed waterwater that has undergone secondary treatment, filtration, and disinfection. The criteria to follow are: *before disinfection*—24 h average turbidity after filtration is ≤2 NTU and should not exceed 5 NTU at any time; *after disinfection and before discharge*—no detectable in 4/7 of daily reclaimed water samples with maximum fecal measure of <23/100 ml in a single sample.

c. Class B+ reclaimed waterwater that has undergone secondary treatment, nitrogen removal, and disinfection. The criteria to follow *after disinfection and before discharge*—concentration of fecal coliform in 4/7 of daily reclaimed water should be <200/100 ml with a maximum concentration of fecal in a single sample <800/100 ml. The total nitrogen should be <10 mg/l.

d. Class B reclaimed waterwater that has undergone secondary treatment and disinfection. The criteria to follow *after disinfection and before discharge*—concentration of fecal coliform in 4/7 of daily reclaimed water should be <200/100 ml with a maximum concentration of fecal in a single sample <800/100 ml.

e. Class C reclaimed waterwater that has undergone secondary treatment in a series of waste stabilization ponds, including aeration with or without disinfection, and total retention time in the stabilization pond should be at least 20 days. The criteria to follow *after treatment and before discharge*—concentration of fecal coliform in 4/7 of daily reclaimed water should be <1000/100 l with a maximum concentration of fecal in a single sample <4000/100 ml.

All wastewater in the Tataouine habitat is recycled, thus the sewage undergo all necessary steps to reclaim water of the highest quality, possibly above the A+ reclaimed water category. Besides recycling water, the objective is to capture various components of the wastewater such as biochemical energy, nutrients, and drinking water through a series of biological and physio-chemical treatment processes as highlighted in Fig. 3.

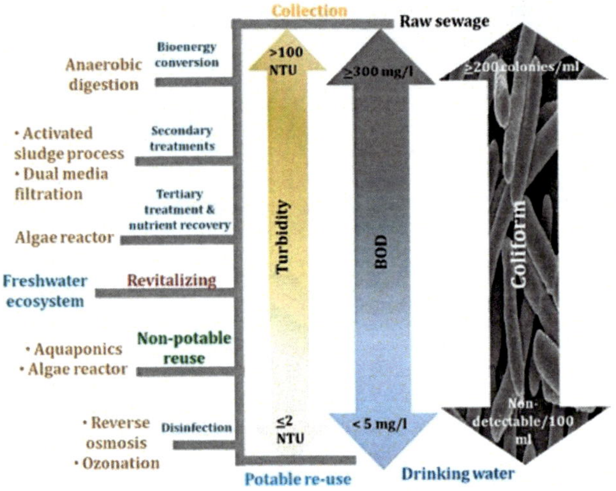

Fig. 3 The process flow for recycling of wastewater to obtain and recover energy, nutrients, and drinking water through sequential reduction of pollutant and pathogenic loads

7 Turning the Unidirectional Flow of Wastewater to a Circular Loop for Resource Recovery to Overcome Water Scarcity

Much of the freshwater utilized in daily life in several activities ultimately ends up in the oceans and is lost forever. However, if the wastewater (freshwater after its intended use) is treated appropriately and re-introduced in the system forming a circular loop, rather than following a unidirectional flow, is a promising approach to conserve much of the freshwater.

The wastewater must be considered as a "resource" rather than a waste. The circular economy-based water resource management system not only utilizes and treats the wastewater but also forms an impressive food production system. The "PolyBio" Systems will generate energy, food, and clean water, the cost of which can be deferred by the value-added output that is mostly absent with conventional large-scale wastewater treatment plants. Wastewater and water treatment plants that serve entire cities of significant size are very expensive to design, construct, and maintain and communities often struggle to cover the costs. A neighborhood or village "PolyBio" System that is fully integrated enables the circular utilization of nutrients and water and should cover its own cost or produce positive cash flow. The value generated includes:

- carbon-negative biogas (energy)
- algae-derived materials
- duckweed
- ornamental aquatic beings such as aquarium fish and shrimp
- crops for food and materials
- pure revitalized water

The value of these products must be considered in light of substantial carbon reduction, resource preservation, and other environmental impacts. We should try to assign a value to the nutrients contributed by an average adult.

8 Wastewater Recirculation Through "PolyBio" to Attain Sustainability

The proprietary "PolyBio" system (Fig. 4) has been developed by Climate Survival Solutions at its Tataouine habitats which is intended to create a circular "Waste-Energy-Nutrient-Food-Water" system from the initial input of domestic wastewaters. The domestic or septic wastewater undergoes the primary anaerobic digestion treatment mediated by the microbial degradation of the organic matter to generate gaseous, solid, and liquid byproducts. The production of gaseous byproducts mainly in the form of methane (up to 60%), CO_2 (30–40%), and other trace gases is called biomethanation. The solid and liquid components consist of nutrient-rich slurry which is further processed by downstream processes. As human feces contain nitrogen (N), potassium (K), and phosphorus (P), it is tapped as a valuable nutrient resource after sequential septic treatment processes. The effluent from anaerobic and aerobic digesters still containing essential NPK nutrients are directed to the algal photobioreactors to simultaneously attain the growth and nutrient-rich biomass. The algal biomass undergoes downstream processing for various kinds of resource recovery.

The aquaponics system integrates the two separate systems of growing plants (1) Hydroponics—cultivating the plants in the water and (2) Aquaculture—cultivating and breeding fish. The wastewater of the habitat undergoes several primary treatments before entering the aquaponics system. The basic principle of aquaponics essentially involves the recycling of the water within the hydroponics and aquaponics systems. The waste generated by the fish gets effectively consumed by the plants for their growth and development. This results in the purification and oxygenation of the water which facilitates the cultivation of fish in the system. The vital role is played by the naturally occurring microbes in the system or the ones present in the plant's roots. The beneficial microbiota is specialized in converting the fish waste (ammonia) present in the water into nitrate (plant's food) by following the process of nitrification. The successive absorption of nitrates by the plants supports their growth and health. Fish, plants, and bacteria make the crucial components of the system which create a mutually beneficial environment for their survival in the system. The overall system of a functional aquaponics system is illustrated in Fig. 5.

The bacteria residing in the system process the fish waste, which is an inevitable stage in aquaponics as the accumulation of high levels of ammonia in fish waste becomes toxic to the fish which can lead to the system collapse. Certain type of bacteria by the process of nitrification converts the ammonia (from fish waste) into nitrate, which serves as the plant's fertilizer. The primary species of bacteria involved in this conversion include ammonia-oxidizing bacteria (AOB), like *Nitrosococcus, Nitrosospira,* and Nitrosomonas, and nitrite-oxidizing bacteria (NOB), like *Nitrobacter, Nitrospira* (Rurangwa & Verdegem, 2015), *Nitrococcus, and Nitrospina.* Similarly, archaea such as

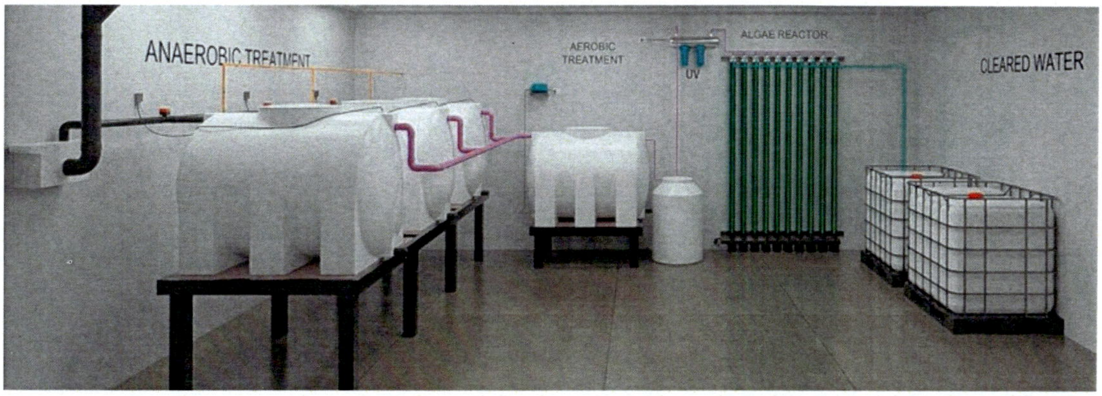

Fig. 4 A schematic presentation of the in-house "PolyBio" system developed by the Climate Survival Solutions at Hermosillo, Mexico site

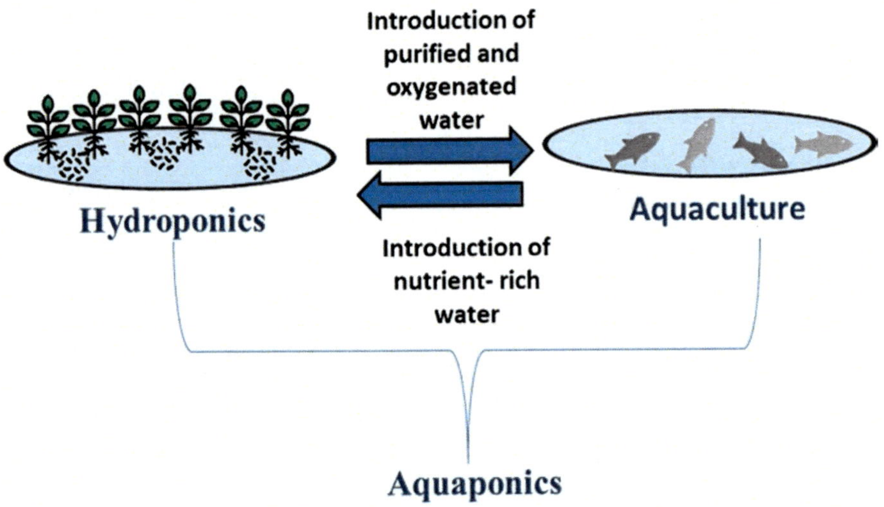

Fig. 5 Aquaponics represents sustainable food production systems wherein the resources such as water and nutrients are exchanged and reused

Thaumarchaeota can be also involved in the ammonia-oxidizing process (Bartelme et al., 2017). The nitrification reaction converts the toxin ammonia into usable nitrates for the plants. In the process of nitrification, certain autotrophic bacteria such as *Nitrosomonas* oxidize ammonia to nitrite. This nitrite is further oxidized to nitrate majorly by *Nitrobacter*.

Additionally, microorganisms are also involved in other vital processes such as extracting the various macro- and micronutrients from the feed leftovers and particles of solid feces. The process flow begins with the introduction of treated wastewater into the system, feeding the fish, followed by the release of waste matter in the form of excreta by fish, recirculation of the water containing the ammonia (fish waste) to the hydroponic system, conversion of ammonia to nitrate by the bacteria present in the grow bed (substrate to support the plant, majorly, LECA—Light expanded clay aggregate) or occurring on the plant's roots and re-introduction of the purified water into the fish tank. Aquaponics supports the cultivation of plants and culturing of the fish for food purposes. However, the major focus in the Tataouine Habitat will be on plant-based foods to improve and safeguard the life of its inhabitants.

The most commonly grown plants in the aquaponic system include leafy green herbs, such as Lettuce (Walsworth, 2016). These plants can grow well in high-nitrogen and low micronutrients, making them the best choice for aquaponics. Lettuce can be highly productive and rapid growers. However, these plants require a slightly higher nutrient as compared to the leafy green herbs. Similarly, even the fruiting plant needs higher amounts of nutrients and therefore, grows well in well-established and well-stocked aquaponics units (Walsworth, 2016). There could be a need to additionally incorporate potassium and calcium in aquaponics while dealing with fruiting plants if deficiencies are indicated. Large fruiting vegetables such as tomatoes, peppers, eggplant, cucumbers, peas, and beans are suited for aquaponics (Somerville et al., 2014). A few of the shallow-rooted crops such as radish and beets can perform well in aquaponics, however, carrots, potatoes, and other deep-rooted crops as well as large fruiting plants such as melons and pumpkins are not suited for the aquaponics system. The habitat shall also harbor the AI mechanism to sense the trouble maintaining/achieving environmental parameters of the system and rectify it immediately without any major damage to the system. The aquaponics offers a multitude of benefits which is the reason, it is being incorporated into the climate-proof habitat (Fig. 6).

9 Biosolids Utilization and Mushroom Cultivation

Mushrooms are the most widely utilized fungi for dietary needs with high medicinal and nutritive values. The protein content (on a dry weight basis) of some of the most commonly used mushrooms, namely, *Agaricus bisporus, Pleurotus* spp., *Volvariella volvacea,* and *Lentinula edodes* is 29.14%, 19.59%, 38.10%, and 18.85%, respectively (Gupta et al., 2018). The species *Ganoderma lucidum* has a proven record of improving the health and longevity of people in China, Japan, and other Asian countries (Wachtel-Galor et al., 2011). The global market for mushroom is exponentially growing owing to their beneficial uses and ease of cultivation. Mycoprotein obtained from fermented mushrooms forms a great meat-free environment-friendly alternative to proteins. As mushroom cultivation requires less water and inorganic fertilizers, it has a lower carbon footprint which thus creates a more sustainable food system.

Mushrooms are capable of growing on simple substrates and wastes of several kinds. For example, oyster

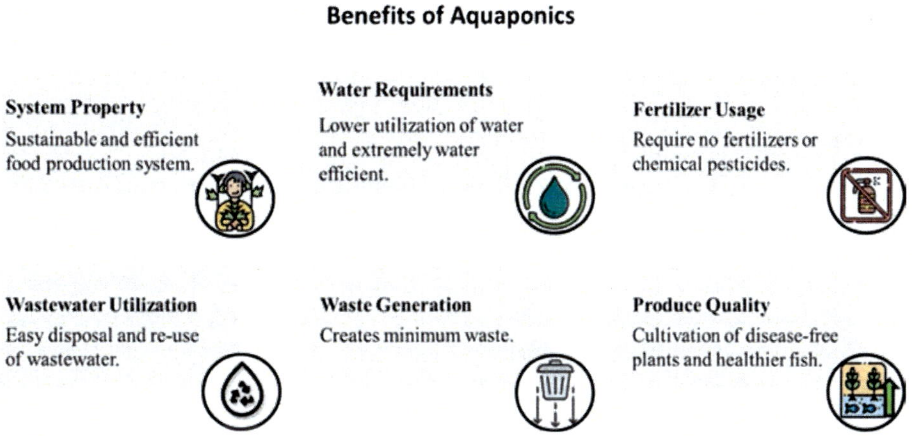

Fig. 6 Multitude of benefits offered by aquaponics. Illustration by author, after (Somerville et al., 2014)

mushrooms and straw mushrooms (*Volvariella volvacea*) are readily grown on cotton waste from textile industries, while several *Pleurotus* spp. utilize the lignocellulosic agricultural waste as their growth substrate (Moore & Chiu, 2001). The "spent compost" after mushroom harvesting provides a suitable source for animal feed. At Tataouine habitats, mushrooms are cultivated on biosolids obtained from the anaerobic digester which performs the primary treatment of the wastewater. Over time, the biosolids get accumulated on the inner surface of the anaerobic digester and require scrapping. The biosolids have a high nutritive value and thus, make an optimal substrate for growing mushrooms. The biosolids are dried and composted before being utilized as culture media for growing mushrooms. The spent mushroom culture is further processed and utilized as fertilizer (Imamura et al., 1984) and as animal feed. The general steps involved in growing mushrooms include (a) Composting—preparation of the substrate by utilization of biosolids in combination with other agro-waste such as straw. (b) Spawn-Run—mixing the spawn (fungal mycelium) with the compost (c) Casing—layering the spawn with an inert material to promote sporophyte production (d) Cropping and Harvest—induction of mushroom to fruit in well-defined flushes or breaks and picking up the mushrooms by gentle twisting and pulling it up very softly (Munshi et al., n.d.).

Tataouine has incorporated an efficient mushrooms cultivation system which has multiple advantages (Fig. 7). Mushroom cultivation requires minimal energy inputs which are also economical, such as sawdust bags, bottles,

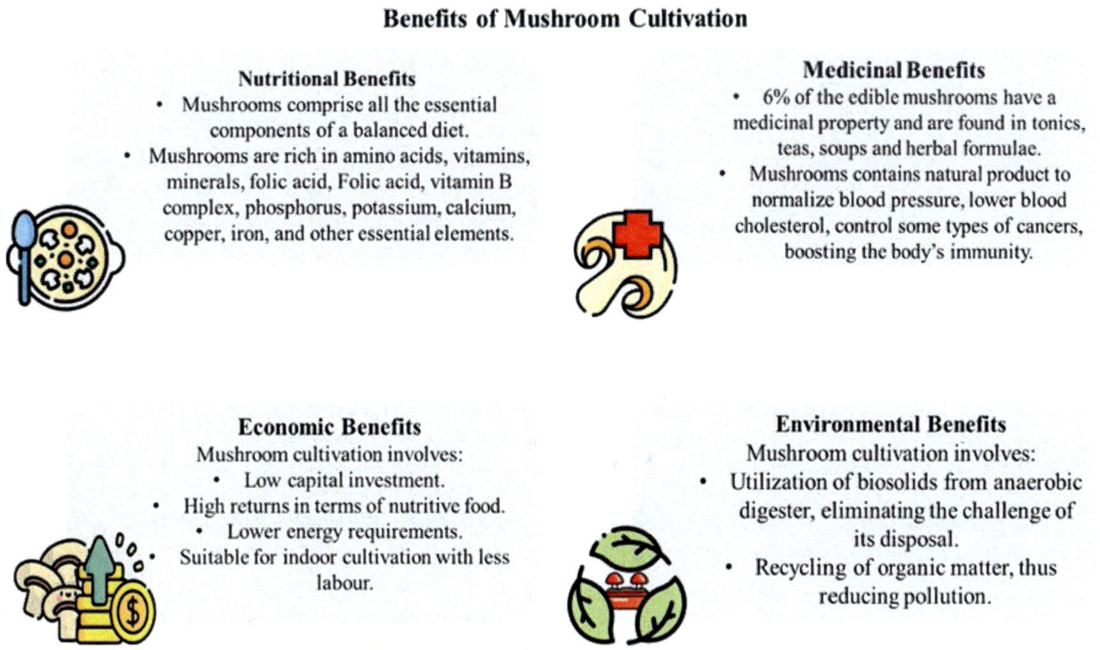

Fig. 7 Benefits of mushroom cultivation. Illustration by author, after (Ferdousi et al., 2019)

shelves, and logs. The infrastructure required for mushroom cultivation includes a well-sterilized and enclosed room with regulated environmental parameters such as (humidity, temperature, suitable substrate moisture levels, and/or light for the formation of fruiting bodies). The treated water and biogas generated through the "PolyBio" recirculation system would meet the energy and water requirements.

10 Wastewater Management and Its Revitalization at Tataouine

All the wastewater generated in the Tataouine Habitat is a part of the water reuse circuit providing a win–win situation for all. The end-users save money on water bills and wastewater disposal and simultaneously address the issues of environmental degradation due to excessive nutrient run-offs from sewage. Tataouine is primarily focusing on developing a Circular Waste-Energy-Nutrient-Food-Water system, which involves the recycling of all the wastewater generated in the household and additionally obtaining nutrients for food cultivation.

In summary, the process of water management starts with the treatment of black and greywater through a series of anaerobic and aerobic digesters. The blackwater effluent upon further dilution and sterilization is utilized in algal reactors to degrade any remaining potential toxins and to restore remaining nutrients, particularly phosphates. The solids from the digester are utilized for the cultivation of mushrooms and as plant fertilizers. Blackwater for Arizona refers to the wastewater from the kitchen, bathrooms, and toilets (having fecal matter and urine). Human waste released into the environment leads to environmental degradation and poses potent public health risks. Therefore, its suitable and sustainable management has become a serious issue and an area of concern. However, at the Tataouine community, we strive to develop a circular waste management system, wherein the human waste is efficiently converted into energy, which is further utilized for heating and power generation purposes. The individual processes involved in Tataouine's household sewage water treatment such as nutrient resource recovery, recycling, and reuse are proven methods (Fig. 8). Overall, it is more cost effective, sustainable, and resilient when done on a micro or neighbourhood level.

The greywater from the bathroom and laundry is collected into a sump which flows through a dual media filter to remove the suspended solids that lowers the turbidity and trap some pathogens. The water filtered through a dual media filter is disinfected and reused for toilet flushing, surface irrigation of non-food crops, or landscaping irrigation (in accordance with ADEQ best management practices for greywater reuse). Alternatively, the filtered and disinfected water stream could be fortified with additional nutrients to cultivate microalgae species for high-value nutritional bioproducts (Fig. 9).

We have isolated different species of microalgae from various aquatic habitats (Fig. 10A) to study their bioremediation and nutrient recovery potential for use in the "PolyBio" wastewater recirculation system. The isolated microalgae belong to various species of cyanobacteria or blue-green algae, diatoms, and green microalgae, namely, *Scenedesmus* spp. and *Botryococcus* sp. (Fig. 10B). The cultivation of these various species of microalgae isolates is being tested in different prototypes of tank and tubular photobioreactors (Fig. 11).

The effluents from the algal photobioreactor shall then be oxygenated by an artificial stream to improve the dissolved oxygen content of the water and revitalized to support the diverse pond ecosystem. This revitalization process shall improve the quality of water and make it suitable to support the aquatic plants. The water is treated further by aquaponics, a final algal reactor, reverse osmosis, and UV radiation to form a potential source of potable water. Food and agricultural wastes shall be disposed off, by converting them into vermicompost and utilizing it for supplying nutrients to plants and enriching the soil. The Tataouine Habitats with "PolyBio" wastewater recirculation systems are 'no-waste' with a complete recycling policy to adequately support the environment and sustainable living.

11 Conclusions

The global heat waves and extreme rise in land and air temperatures around different corners of the world are not unprecedented as climate scientists had long warned the policymakers and governments about the anthropogenic crisis that humanity is currently facing. In many places, the wet bulb temperatures have already crossed the limits of endurance for human bodies. Under such extreme weather conditions, the demand and supply sides of resources including water, food, and biodiversity (flora and fauna) have put extreme pressures on our existence and persistence. The utilization of natural resources has hitherto been unsustainable, as we have crossed the planetary boundaries for nutrients (phosphorous, nitrogen), climate change, land use, and loss of genetic biodiversity. The motto of the "Tataouine Community and Research Center" in Arizona, USA is intended to build a model climate-proof, self-sufficient, and sustainable habitat, to not just survive the climate crisis, but to live in a new and better environment that nurtures the human spirit and biodiversity. This would be achieved through the conservation and transformation of waste into energy and nutrients to feed algae, plants, and humans. The natural processes of water rejuvenation would be mimicked to obtain ultra-pure and energized water at

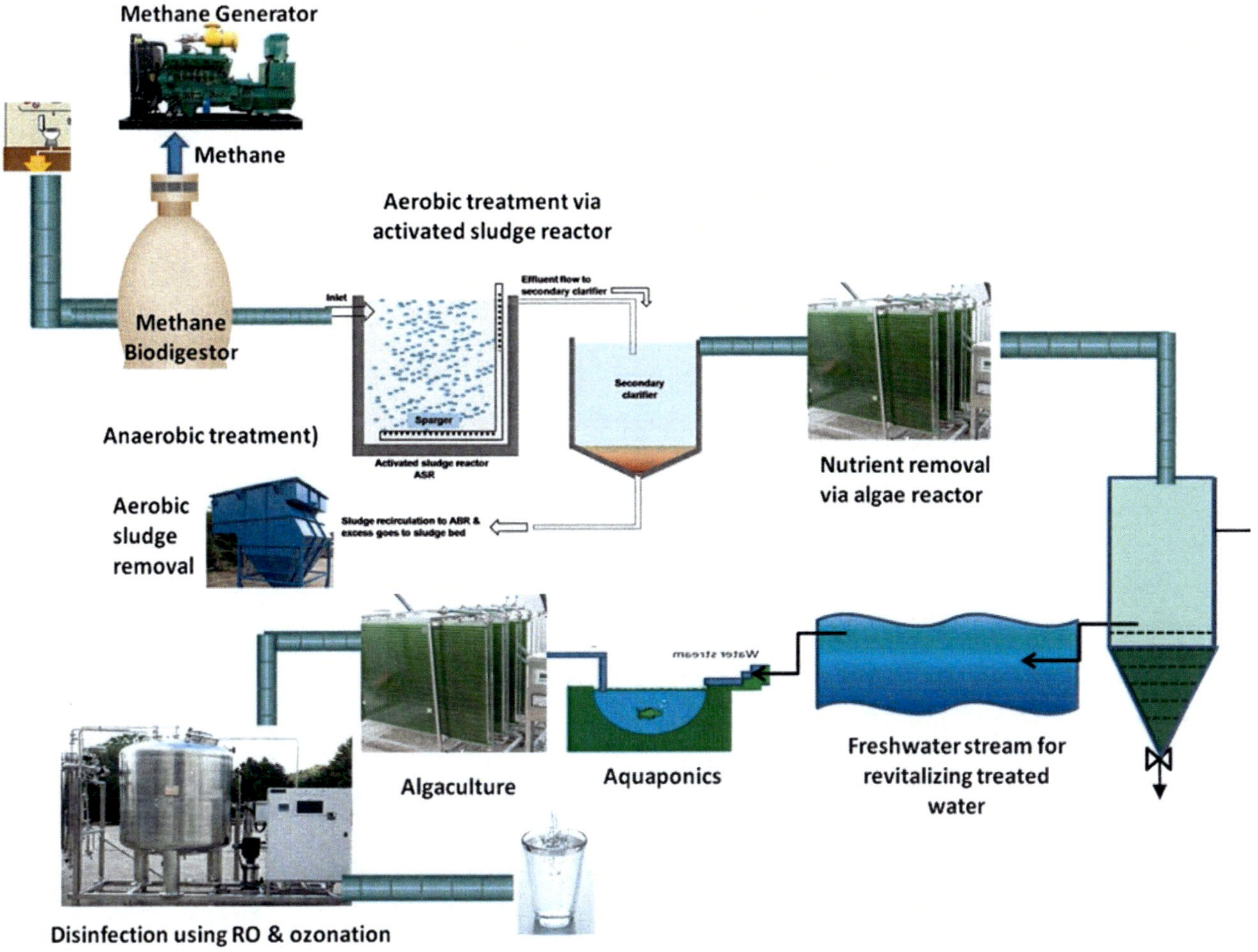

Fig. 8 Overview of methods and processes adopted for establishing circular waste-energy-nutrient-food-potable water system at Tataouine

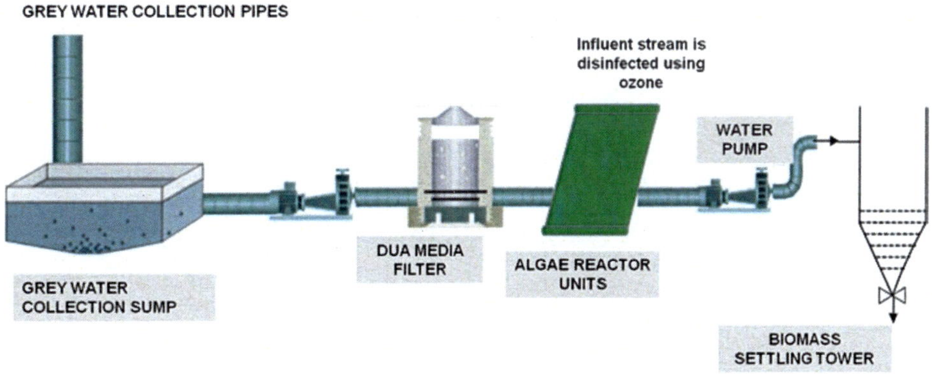

Fig. 9 Diagrammatic sketch of an integrated greywater treatment and recovery system for obtaining valuable products from microalgae

the end of carefully managed natural circular processes. Both black and greywater would be recycled and reused. The potential of certain strictly anaerobic microbes would be utilized to primarily treat the black water via anaerobic digestion and the matter would be transformed into both nutrients and biogas. The generated biogas shall be made available for the production of electricity and heat for several in-house needs. The byproducts of the anaerobic digestion shall be utilized as fertilizers to improve soil quality and for livestock bedding. The genetic biodiversity of other lower groups of organisms such as algae and fungi would be explored to recycle the nutrients and to make nutritious food in a sustainable way. The objective of the study is to highlight the most sustainable drought response by adopting a decentralized wastewater management system to recover nutrients, energy, and water.

Fig. 10 Microalgae cultures used in the "PolyBio" system. **A** Various sites from where the water samples were collected for microalgae isolation. **B** Microalgae isolated and tested for "PolyBio" wastewater recirculation system belong to different groups, namely, **a-b** cyanobacteria, **c** diatoms, **d-e** Scenedesmus spp., and **f** Botryococcus sp.

Fig. 11 Prototypes of tank and tubular types photobioreactors for microalgae cultivation under natural and artificial illumination

References

Bartelme, R. P., McLellan, S. L., & Newton, R. J. (2017). Freshwater recirculating aquaculture system operations drive biofilter bacterial community shifts around a stable nitrifying consortium of ammonia-oxidizing archaea and comammox Nitrospira. *Frontiers in Microbiology, 8*, 101.

Bartzas, G., Tinivella, F., Medini, L., Zaharaki, D., & Komnitsas, K. (2015). Assessment of groundwater contamination risk in an agricultural area in north Italy. *Information Processing in Agriculture, 2*(2), 109–129. https://doi.org/10.1016/j.inpa.2015.06.004

BBC. (2022). River Thames: More than 2bn litres of raw sewage discharged over two days. *BBC News*. Retrieved June 30, 2022. [Online]. Available: https://www.bbc.com/news/uk-england-london-60046320

Beardall, J., Stojkovic, S., & Larsen, S. (2009). Living in a high CO2 world: Impacts of global climate change on marine phytoplankton. *Plant Ecology and Diversity, 2*(2), 191–205. https://doi.org/10.1080/17550870903271363

Casadevall, A. (2022). *Microbes and climate change – science, people & impacts: Report on an American Academy of Microbiology virtual colloquium held on November 5, 2021*. American Society for Microbiology, Washington (DC), 2022. Retrieved March 03, 2023. [Online]. Available: http://www.ncbi.nlm.nih.gov/books/NBK580166/

Chagas, V. B. P., Chaffe, P. L. B., & Blöschl, G. (2022). Climate and land management accelerate the Brazilian water cycle. *Nature Communications, 13*(1). https://doi.org/10.1038/s41467-022-32580-x

Chakraborty, S., Maity, I., Dadashpoor, H., Novotný, J., & Banerji, S. (2022). Building in or out? Examining urban expansion patterns and land use efficiency across the global sample of 466 cities with million+ inhabitants. *Habitat International, 120*, 102503. https://doi.org/10.1016/j.habitatint.2021.102503

Cremades, R., et al. (2021). Guiding cities under increased droughts: The limits to sustainable urban futures. *Ecological Economics, 189*, 107140. https://doi.org/10.1016/j.ecolecon.2021.107140

Čuček, L., Klemeš, J. J., & Kravanja, Z. (2015). Chapter 5 - Overview of environmental footprints. In J. J. Klemeš (Ed.), *Assessing and measuring environmental impact and sustainability* (pp. 131–193). Butterworth-Heinemann. https://doi.org/10.1016/B978-0-12-799968-5.00005-1

Deutsch, C. A., et al. (2008). Impacts of climate warming on terrestrial ectotherms across latitude. *Proceedings of the National Academy of Sciences, 105*(18), 6668–6672. https://doi.org/10.1073/pnas.0709472105

Dilling, L., et al. (2019). Drought in urban water systems: Learning lessons for climate adaptive capacity. Elsevier Enhanced Reader. *23*, 32–42. https://doi.org/10.1016/j.crm.2018.11.001

Ferdousi, J., Al Riyadh, Z., Hossain, M. I., Saha, S. R., & Zakaria, M. (2019). Mushroom production benefits, status, challenges and opportunities in Bangladesh: A review. *Annual Research & Review in Biology*, 1–13.

Gupta, S., Summuna, B., Gupta, M., & Annepu, S. K. (2018). Edible mushrooms: Cultivation, bioactive molecules, and health benefits. *Bioactive Molecules in Food, 1*, 1–33.

Hoekstra, A. Y., & Mekonnen, M. M. (2012). The water footprint of humanity. *Proceedings of the National Academy of Sciences, 109*(9), 3232–3237. https://doi.org/10.1073/pnas.1109936109

Human Development Report. (2006). *UNDP, 2006 Coping with water scarcity*. Challenge of the twenty-first century. UN-Water, FAO, 2007

Imamura, Y., Ito, H., Sato, Y., & Nakamori, H. (1984). Mushroom growing by utilizing sewage sludge compost and re-utilization of the used compost as livestock feed.

Ionita, M., Dima, M., Nagavciuc, V., Scholz, P., & Lohmann, G. (2021). Past megadroughts in central Europe were longer, more severe and less warm than modern droughts. *Communications Earth & Environment, 2*(1). https://doi.org/10.1038/s43247-021-00130-w

Jafari, H., Raeisi, E., Hoehn, E., & Zare, M. (2012). Hydrochemical characteristics of irrigation return flow in semi-arid regions of Iran. *Hydrological Sciences Journal, 57*(1), 173–185. https://doi.org/10.1080/02626667.2011.636365

Kim, Y., Kim, S., Jeong, H., & An, H. (2022). Quantitatively defining megadrought based on drought events in central Chile. *Geomatics, Natural Hazards and Risk, 13*(1), 975–992. https://doi.org/10.1080/19475705.2022.2060763

Kollmann, R., et al. (2016). Renewable energy from wastewater—Practical aspects of integrating a wastewater treatment plant into local energy supply concepts. *Journal of Cleaner Production*. https://doi.org/10.1016/j.jclepro.2016.08.168

Kumar, S., Getirana, A., Libonati, R., Hain, C., Mahanama, S., & Andela, N. (2022). Changes in land use enhance the sensitivity of tropical ecosystems to fire-climate extremes. *Scientific Reports, 12*(1). https://doi.org/10.1038/s41598-022-05130-0

Libonati, R., et al. (2022). Assessing the role of compound drought and heatwave events on unprecedented 2020 wildfires in the Pantanal. *Environmental Research Letters, 17*(1), 015005. https://doi.org/10.1088/1748-9326/ac462e

Marengo, J. A., et al. (2021). Extreme drought in the Brazilian Pantanal in 2019–2020: Characterization, causes, and impacts. *Frontiers in Water, 3*, Retrieved June 28, 2022. [Online]. Available: https://www.frontiersin.org/article/https://doi.org/10.3389/frwa.2021.639204

Moore, D., & Chiu, S. W. (2001). Fungal products as food. In *Bio-exploit. Filamentous fungi fungal divers* (pp. 223–251). Press Hong Kong.

Munshi, N. A., Dar, G. H., Ghani, M. Y., Kauser, S., & Mughal, N. (n.d.). Button mushroom cultivation, p. 33.

Puppim de Oliveira, J. A., Bellezoni, R. A., Shih, W., & Bayulken, B. (2022). Innovations in Urban Green and Blue Infrastructure: Tackling local and global challenges in cities. *Journal of Cleaner Production, 362*, 132355. https://doi.org/10.1016/j.jclepro.2022.132355

Qin, P., Chen, S., Tan-Soo, J.-S., & Zhang, X.-B. (2022). Urban household water usage in adaptation to climate change: Evidence from China. *Environmental Science & Policy, 136*, 486–496. https://doi.org/10.1016/j.envsci.2022.07.019

Reddin, C. J., Aberhan, M., Raja, N. B., & Kocsis, Á. T. (2022). Global warming generates predictable extinctions of warm- and cold-water marine benthic invertebrates via thermal habitat loss. *Global Change Biology, 28*(19), 5793–5807. https://doi.org/10.1111/gcb.16333

Richardson, D., et al. (2022). Global increase in wildfire potential from compound fire weather and drought. *NPJ Climate and Atmospheric Science, 5*(23). https://doi.org/10.1038/s41612-022-00248-4

Rotiroti, M., et al. (2019). The effects of irrigation on groundwater quality and quantity in a human-modified hydro-system: The Oglio River basin, Po Plain, northern Italy. *Science of the Total Environment, 672*, 342–356. https://doi.org/10.1016/j.scitotenv.2019.03.427

Van Ruijven, J., & Berendse, F. (2010). Diversity enhances community recovery, but not resistance, after drought. *Journal of Ecology, 98*(1), 81–86. https://doi.org/10.1111/j.1365-2745.2009.01603.x

Rummer, J. L., et al. (2014). Life on the edge: Thermal optima for aerobic scope of equatorial reef fishes are close to current day temperatures. *Global Change Biology, 20*(4), 1055–1066. https://doi.org/10.1111/gcb.12455

Rurangwa, E., & Verdegem, M. C. (2015). Microorganisms in recirculating aquaculture systems and their management. *Reviews in Aquaculture, 7*(2), 117–130.

Shahid, S. (2012). Vulnerability of the power sector of Bangladesh to climate change and extreme weather events. *Regional Environmental Change, 12*(3), 595–606. https://doi.org/10.1007/s10113-011-0276-z

Shi, C., Wu, C., Zhang, J., Zhang, C., & Xiao, Q. (2022). Impact of urban and rural food consumption on water demand in China—From the perspective of water footprint. *Sustainable Production and Consumption, 34*, 148–162. https://doi.org/10.1016/j.spc.2022.09.006

Somerville, C., Cohen, M., Pantanella, E., Stankus, A., & Lovatelli, A. (2014). Small-scale aquaponic food production: Integrated fish and plant farming. *FAO Fisheries and Aquaculture Technical Paper, 589*, I.

Uwimbabazi, J., Jing, Y., Iyakaremye, V., Ullah, I., & Ayugi, B. (2022). Observed changes in meteorological drought events during 1981–2020 over Rwanda, East Africa. *Sustainability, 14*(3). https://doi.org/10.3390/su14031519

Voiland, A., & N. E. Observatory (2022). A human fingerprint on the Pantanal Inferno – Human activities primed the region to burn. *SciTechDaily*. https://scitechdaily.com/a-human-fingerprint-on-the-pantanal-inferno-human-activities-primed-the-region-to-burn/ (Retrieved June 28, 2022).

Wachtel-Galor, S., Yuen, J., Buswell, J. A., & Benzie, I. F. (2011). Ganoderma lucidum (Lingzhi or Reishi). In *Herb. Med. Biomol. Clin. Asp* (2nd ed.).

Walsworth, W. (2016). *Everything you need to know to start an expert DIY aquaponic system from home* (1st ed.). BPI.

Williams, A. P., Cook, B. I., & Smerdon, J. E. (2022). Rapid intensification of the emerging southwestern North American megadrought in 2020–2021. *Nature Climate Change, 12*(3). https://doi.org/10.1038/s41558-022-01290-z

Xi, N., et al. (2022). Drought soil legacy alters drivers of plant diversity-productivity relationships in oldfield systems. *Science Advances, 8*(18), eabn3368. https://doi.org/10.1126/sciadv.abn3368

Zarei, M. (2020). Wastewater resources management for energy recovery from circular economy perspective. *Water-Energy Nexus, 3*, 170–185. https://doi.org/10.1016/j.wen.2020.11.001

Zipper, S. C., Smith, K. H., Breyer, B., Qiu, J., Kung, A., & Herrmann, D. (2017). Socio-environmental drought response in a mixed urban-agricultural setting: synthesizing biophysical and governance responses in the Platte River Watershed, Nebraska, USA. *Ecology and Society, 22*(4). Retrieved July 26, 2022. [Online]. Available: https://www.jstor.org/stable/26799014

Non-intrusive Load Monitoring System for Decarbonization of Built Environment Based on Deep Learning Algorithm

Liu Yang, Wenbo Wu and Xiao Liu

Abstract

Load monitoring is a significant part of building energy-consumption management, but traditional electricity meters need to be connected to separate monitors on each device, a process with the problems of high cost, complex installation, and complicated maintenance. For decarbonization and the sustainable development of the built environment, the authors of this paper propose a nonintrusive load monitoring algorithm and platform based on a deep neural network model. With an intelligent electricity meter, centralized data collection, the cloud server processing of electricity data, and the analysis and identification of their building's electrical situation, users can check the electricity situation of their home or office building in real time through web or mobile clients. Users can also remotely control the switches of electrical appliances through these clients. By collecting, processing, and feeding back electricity-consumption details, the proposed algorithm can provide energy-saving suggestions and energy-consumption predictions, as well as monitor power redundancy, to reduce carbon emissions and costs.

Keywords

Building energy efficiency · NILM · Deep learning · Algorithm · Decarbonization · Sustainable development

L. Yang
Tsinghua University, Beijing, China
e-mail: yl-19@mails.tsinghua.edu.cn

W. Wu
Beijing Institute for General Artificial Intelligence, Beijing, China
e-mail: wuwb@bigai.ai

X. Liu (✉)
South China University of Technology, Guangzhou, China
e-mail: xiaoliu@scut.edu.cn

1 Introduction

Load monitoring is one of the key aspects of decarbonization management in the built environment. However, the traditional approach of connecting individual monitors to each device has many problems, such as: such as high monitoring hardware costs, monitoring hardware maintenance difficulties, and a static and closed evaluation algorithm system with poor mobility (Zeifman and Roth, 2011).

In order to better apply smart technologies and promote sustainable operations, we will discuss the following issues: How to improve the efficiency and accuracy of identification? How can real-time energy-consumption data be fed back to the user to adjust power usage manually or automatically?

The significance of this research is that users can control switches and appliances remotely through the app or web. By collecting, processing, and providing feedback on electricity-consumption details, the algorithm can provide energy-saving suggestions, energy-consumption prediction, and monitor electricity redundancy to reduce carbon emissions and costs.

2 Related Work

The carbon footprint can be reduced by reducing energy consumption and integrating low-carbon sources into power structures (Gershenfeld et al., 2010). Deep learning can reduce energy consumption by allowing devices and systems to adapt to usage patterns. In addition, buildings can respond to signals from the grid, providing flexibility for grid operators and reducing costs for consumers. The efficiency of many key systems within a building can be radically improved. The same is true of small appliances such as refrigerators, light bulbs, and HVAC systems, because they are inefficient and account for more than half of a building's energy consumption (Afroz et al., 2018). There

B.-J. He et al. (eds.), *Resilient Horizons: Building Sustainable Environments for Climate Adaptation and Health*, Advances in Science, Technology & Innovation, https://doi.org/10.1007/978-3-031-46109-5_3

are several promising ways to improve HVAC performance using deep learning: predicting the temperature required for the entire system, achieving better control to maintain these temperatures, and fault detection. Traditional temperature predictions, such as through building energy-consumption modeling, are accomplished by using detailed physical models of related systems. However, using deep learning can improve accuracy with less computational effort (Huber et al., 2021). In terms of control, Hussain et al. used deep reinforcement learning to achieve a scalable 20% reduction in energy with only three sensors: air temperature, water temperature, and energy use (Kazmi et al., 2018). Finally, deep learning automates building diagnosis and maintenance through fault detection. Wang et al. treated HVAC fault detection as a classification problem using only temperature readings for prediction. Automatic encoders can simplify information about machine operations so that deep neural networks can more easily predict multiple failures (Jia et al., 2016).

3 Research Objects

3.1 NILM Introduction

Non-Intrusive Load Monitoring (NILM) technology was developed in the 1980s by MIT Hart G. W. (Hart, 1985, 1992). The main purpose is to reduce the hardware cost of load monitoring, enabling power companies to obtain more specific power data at a lower cost. In other words, by detecting the voltage, current, and other power signals of the total load, NILM can detect what electrical appliances a user is using and realize the decomposition of the total power data to the consumed power of a single electrical appliance (Bonfigli et al., 2015; Nalmpantis & Vrakas, 2019; Pereira & Nunes, 2018; Zoha et al., 2012). The individual characteristics of each electrical device are unique in each particular electrical environment, and these characteristics are called load imprinting. Load imprinting can reflect the unique information of power consumption in the operation of an electrical device, such as voltage, active power waveform, and starting current. These characteristics are determined by the working conditions of the electrical equipment, so load imprinting can be divided into steady state, transient state, and operation mode. The operation mode is determined by the operation control policy of the device. During the operation of equipment, these load marks repeatedly appear, and they can be used to identify each electrical appliance when combined with a deep learning algorithm. Non-intrusive load decomposition is the basis of energy conservation. Strengthening energy-consumption monitoring, especially the monitoring of power consumption, is an important practical step in improving energy efficiency, realizing the sustainable development of energy, and building a conservation-oriented society.

3.2 Basic Framework of NILM System

The implementation of NILM is divided into five major steps: data measurement, data processing, event detection, feature extraction, and load identification.

The purpose of data measurement is to obtain the transient and steady-state signals of the total load. Data processing mainly includes denoising, active power, and other electrical quantity calculations and standardization. The main purpose of standardization is to deal with the disturbance caused by power quality fluctuation. Data measurement and processing comprises the preparation of NILM, and event detection, feature extraction, and load identification are the key tasks.

Event detection detects whether new events occur according to the changes in load imprinting over a period of time. For example, assume that the threshold value of the difference between the active power at two moments is set. If the change is greater than the threshold value, new events will be generated. After the event is detected, feature extraction can be carried out, that is, a series of different load imprinting features for load use can be extracted from the data before and after the event. By comparing the extracted features with those in the load bank, the corresponding electrical equipment can be identified. Of course, the establishment of the feature database is critical, and there are two main methods at present: one is to record the load imprinting characteristics of the electrical equipment with manual assistance, and the other is automatic classification through a deep learning algorithm. Obviously, method 1 has a low efficiency and is prone to human error, so method 2, namely, NILM combined with a deep learning algorithm, was used in this study to perform energy prediction and decomposition.

4 Methodology

4.1 System Architecture

In this section, we propose a monitoring system architecture based on NILM and deep learning (Fig. 1) that can be used to monitor the power consumption of each appliance, the visual data of which are transmitted to a user or energy system manager from the cloud in real time in order to achieve the user's goal of manually or automatically adjusting their energy consumption. Energy appliances in buildings are

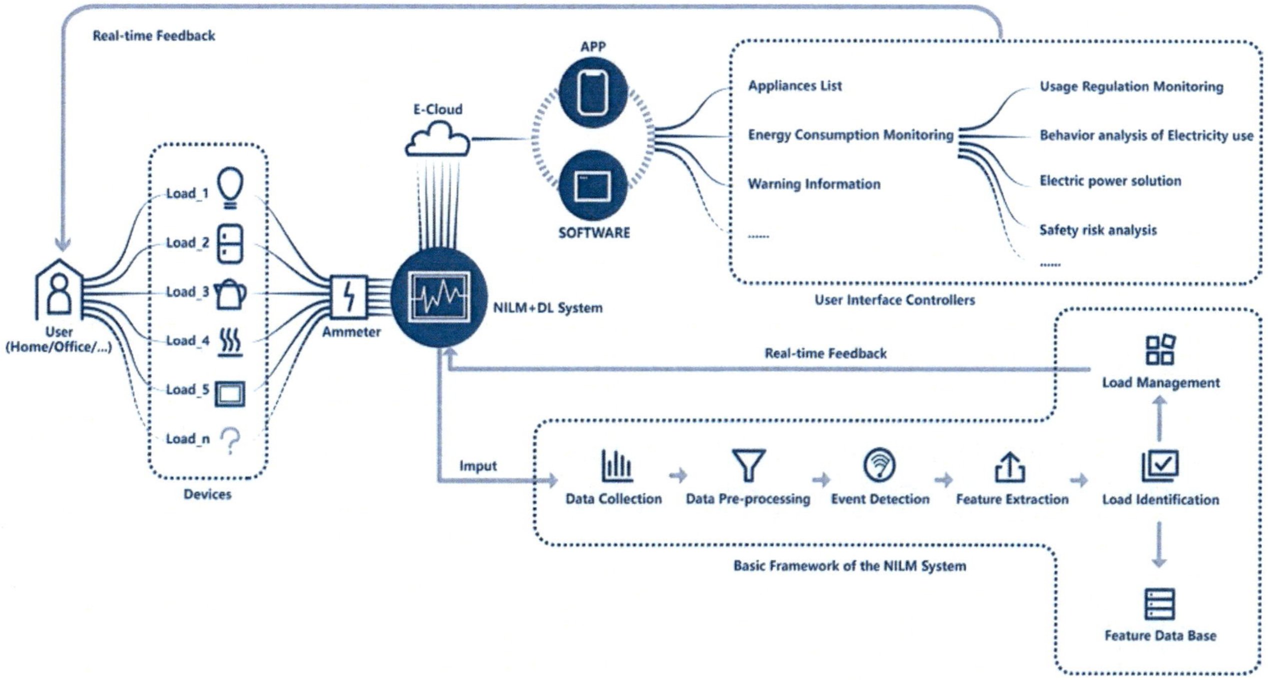

Fig. 1 Structural design of a smart house

complex and diversified. Data show that about 90% of a building's energy consumption is used for air-conditioning, so we will use an air-conditioning system as an example. Due to the irregular use time and users of office and conference spaces, the long-term use of air-conditioning results in energy waste—a conflict between energy saving and user experience. If the location of each user can be known in real time, the air-conditioning system can be turned off where there are no users or quickly adjusted to a comfortable temperature where there are users, therefore, significantly saving energy and enhancing the experience. In addition, this monitoring system can combine visual recognition and positioning technology to accurately locate a user's location and link the local air-conditioning, lighting, and background music systems to realize the integration of energy saving, safety monitoring, and consumption monitoring.

4.2 The Structure of Neural Network

According to the test and practice results of Kelly & Knottenbelt (2015), the accuracy and decomposition quality of deep neural networks are better than those of traditional probabilistic statistical methods, such as a Factor Hidden Markov Model (FHMM) (Kim et al., 2011; Kolter & Jaakkola, 2012; Makonin et al., 2016; Parson et al., 2014). In this paper, a load decomposition model was developed based on the results of various depth model experiments.

Considering the periodicity of electrical appliance use, the continuity of electrical energy consumption, and the accuracy of decomposition, we solved the load decomposition problem in steps using a combination mode of convolutional neural networks (CNNs), long short-term memory (LSTM), and a fully connected neural network (dense) (LeCun et al., 2015; Mauch & Yang, 2015; Tongta and Chooruang, 2020).

The network consists of a three-level model that processes the total data of the input and the results of the previous level model (Fig. 2).

4.2.1 First Level

The first network adopts a combination of a one-dimensional CNN and LSTM. The total load is input, and the open state of all kinds of home appliances is output. This level adopts the sequential processing method and outputs the result based on the time rule and the load fluctuation.

4.2.2 Second Level

The second network adopts a combination of a one-dimensional CNN and a fully connected DNN. The input is the output of the first-level network, that is, the open state of each electrical appliance. The output is the continuous open state of each appliance. The function of this model is to filter out the short-term irregular fluctuations in the actual electrical operation process in order to ensure the continuity and stability of the open state of electrical equipment.

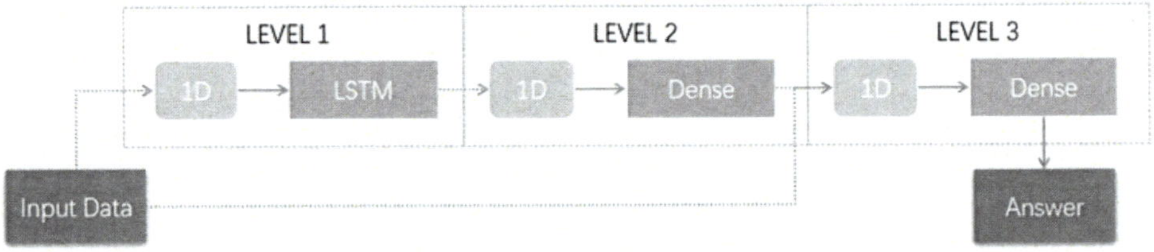

Fig. 2 The network consists of a three-level model

4.2.3 Third Level

The third network adopts a combination of a one-dimensional CNN and a fully connected DNN. The input is the output and total load of the two-stage network, that is, the optimized open state of the electrical appliances. The function of this model is to decompose the load according to the type of open point electrical appliances and the total load, which enables users to obtain the number of appliances and the load used for each type of appliance.

4.3 Choice of Dataset

UK Domestic Appliance Level Electricity (UK-DALE), which consists of five UK households, records the electricity demand of household appliances and the whole household approximately every six years (Kelly & Knottenbelt, 2015). Full household data for three families were also recorded at 1 s and 16 kHz.

We used five target appliances in all our experiments: the fridge, TV, kettle, microwave, and washing machine. These five appliances consume a significant proportion of energy and represent a range of different power "signatures" from the simple on/off of the kettle to the complex pattern shown by the washing machine.

For small devices, such as game console and mobile phone, the power consumption of a single unit is small and the fluctuation of power consumption usually disappears in the noise, but there is usually a large quantity of small loaders, and the total power consumption may be huge. We plan to develop a network for low-power devices in the future.

We extracted the activation energy information by finding strictly continuous samples of power above certain thresholds. We then discarded any activation energy below the threshold duration (to ignore the false peak) because the power demand for more complex devices, such as washing machines, may be below the threshold for a short period of time within a cycle.

5 Experiment and Results

5.1 Prediction and Decomposition Result

The line chart reflects the real-time detection of data household power consumption, analyzes the current electrical switch prediction through the DL-NILM model, and has a good accuracy rate (Fig. 3). In the case of high-power consumption and low-power consumption, the use of different electrical appliances can be subdivided and separated. The model has good and smooth switching state prediction for simultaneous electrical operation.

The pie chart uses 20 days of household electricity-consumption data information to sum and seek the proportion, reflecting the proportion of household electricity consumption (Fig. 4). The algorithm can accurately and effectively make a specific electricity-consumption analysis report (Fig. 5), which helps families to understand their electricity consumption, and also helps power grid companies and some power consumers to deploy and make decisions about electric energy and household appliances.

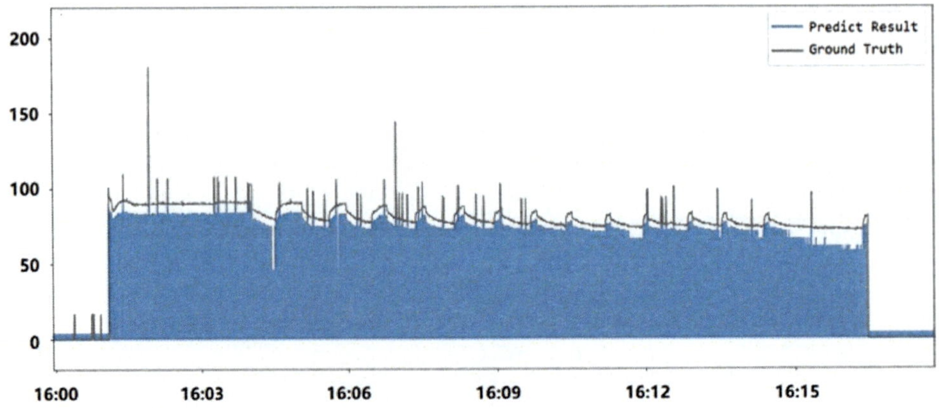

Fig. 3 Comparing the ground truth fridge signals to that of our DL–NILM algorithm

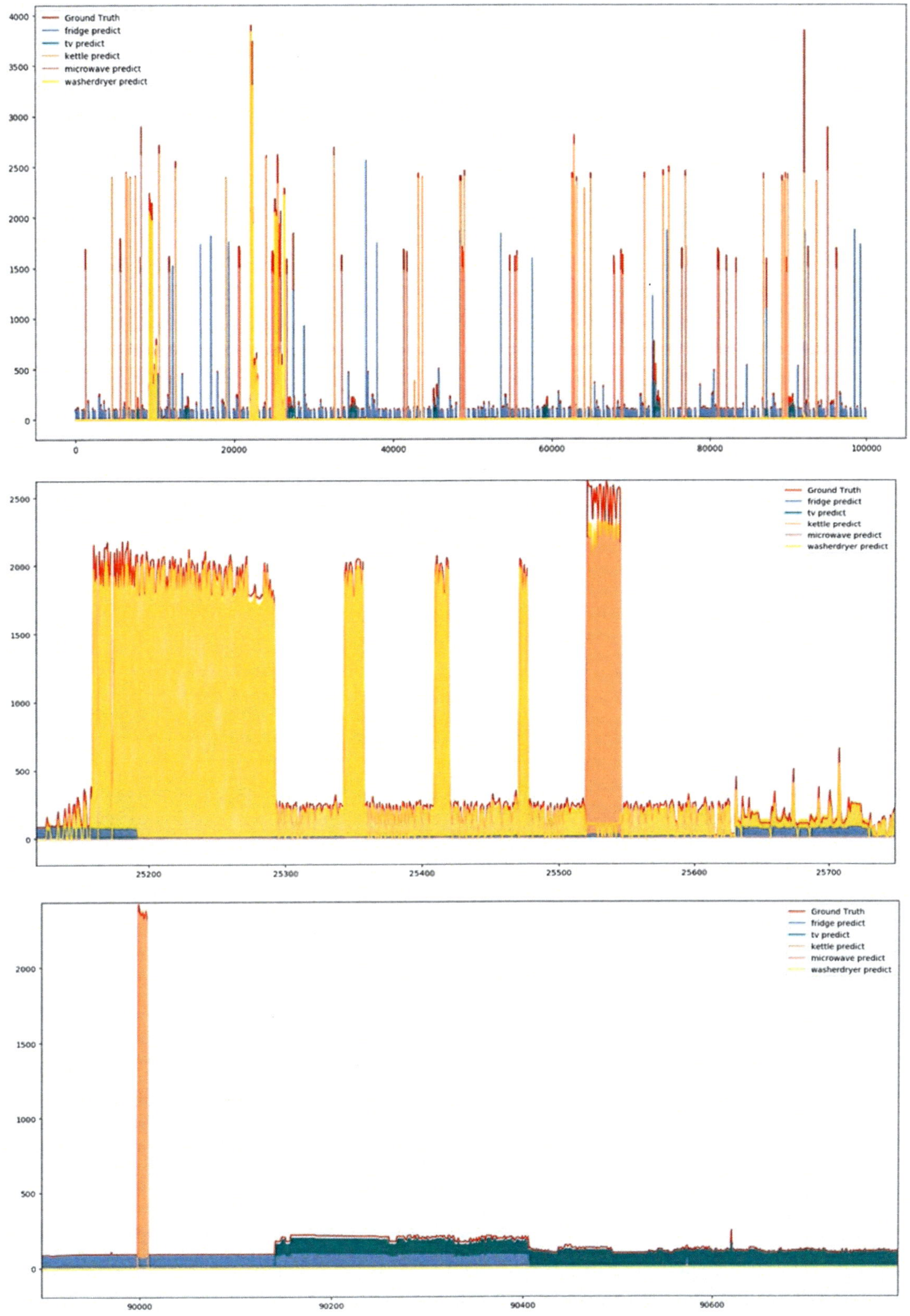

Fig. 4 Decomposition results by DL–NILM in 2 weeks: ground truth aggregate compared to the disaggregated aggregate (top), showing 98.7% accuracy in tracking; parts of the sequence (middle and bottom)

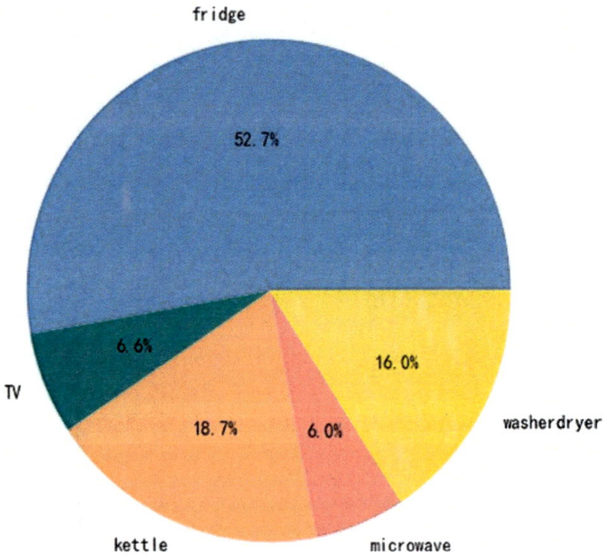

Fig. 5 Five applications' total power cost in 20 days decomposed by DL–NILM

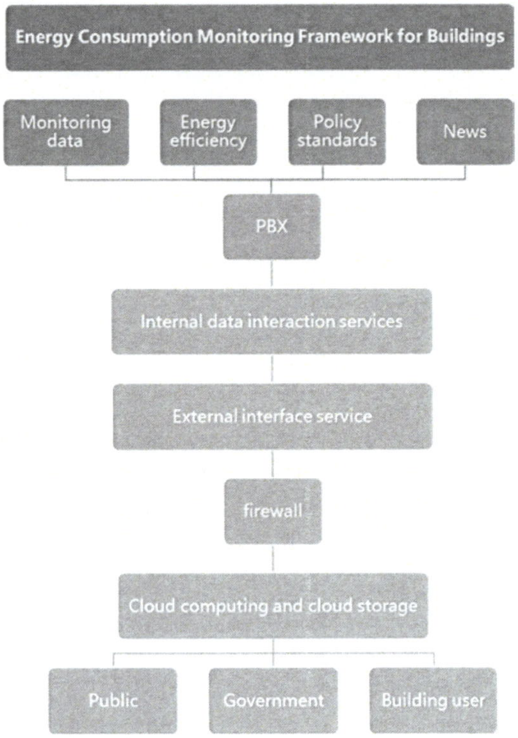

Fig. 6 Network design architecture of website portal

5.2 Energy-Consumption Monitoring Application Architecture Design

5.2.1 Portal Website Architecture Design

Energy-consumption monitoring requires the establishment of a real-time energy-consumption data-visualization website. Based on the energy-consumption monitoring platform, a series of internet products can be developed to enhance the publicity of building energy conservation, facilitate the management of the competent department, and promote the energy-saving operation of owners. Government and large public building energy-monitoring centers can promote energy conservation and carbon reduction through the establishment of a website. The website is used to carry out policies, regulations, standards, and specifications related to the monitoring of energy consumption in public buildings, energy-conservation demonstration projects, information publicity, news trends, and other information queries and browsing, as well as to download various data reports on municipal platforms, therefore providing a working information service window for energy-conservation workers and owners. The main functional modules are as follows.

The website architecture adopts double-layer security protection. The first layer comprises a front hardware firewall between the internet network and the energy-consumption monitoring center. The second layer comprises a process in which all services invoked by the portal website are kept in an external interface service before database access, so the internal data service needs to be connected before and after the program; this achieves the effects of internal and external data isolation, as shown in Fig. 6.

5.2.2 Mobile Application Architecture Design

For private residential users, mobile applications can also be built based on a previous system. The app aims to provide better services for building owners or property managers, enhance the interactive experience, increase the convenience and efficiency of building energy management, and ultimately assist users' energy management (Haugen, 2009; Popa et al., 2019). Considering the privacy of building energy-consumption data, the app is divided into registered and authenticated users. Registered users are open to the public, who can see the data of a demonstration building (for which the names and detailed information of residents are hidden), so that all users can experience the main functions of the app while maintaining data privacy. Verified users connect their buildings to the energy-consumption monitoring platform. After the application is downloaded to their mobile phone, a user can log into the app and see the data of the building, thus providing guidance for building operation. The main functions of this app are as follows (Fig. 7).

Following the establishment of the first version of the app, the energy-monitoring center can also launch an "Energy-Consumption Online" app, aimed at the municipal competent department, which is in charge of the district;

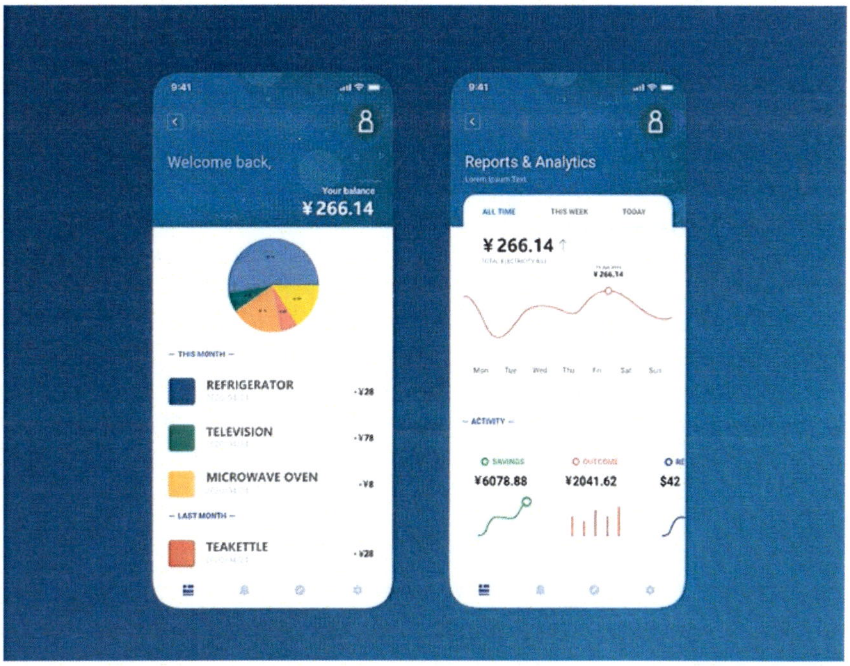

Fig. 7 Energy-consumption online (building version) user interface

this version of the app has different functions that allows administrative departments to conduct macro data analysis at any time, eliminate information islands, and realize information sharing between different administrative departments in construction and energy-conservation work.

Both apps use hybrid app development technology and the IONIC framework. IONIC is a powerful HTML5 hybrid app development framework that is seamlessly integrated with AngularJS, uses the MVVM model, and offers significant productivity improvements, native styling, aesthetically appealing UI components, and strong performance. IONIC can be integrated with Cordova for hardware access; it can be integrated with Crosswalk to significantly improve the running speed of an app. Crosswalk is an open-source WebView engine based on the open-source Chromium kernel of Google, with advanced features such as Web GL, Web RTC, IndexedDB, Web Sockets, Flexbox, and other HTML5 and CSS3 technologies, that can be used to solve the Android fragmentation compatibility problem and significantly improve performance.

6 Discussion

Our circulation network can be used to easily decompose high-power electrical appliances and multi-state electrical appliance loads. However, there is currently no way to distinguish low-power appliances from noise with a circulation network without a denoising autoencoder. Further problems with our proposal are that the presented network cannot detect unknown appliances and that we used the most current UK-DALE dataset, which only comprised data for five houses. A machine learning algorithm can only be generalized if there is sufficient diversity in the training set.

Our proposed application can be used in the general home environment and the office. The total power consumption and use time of each appliance can be predicted by feedback in real time. The application can also be used in the electrical energy detection of some instruments using an energy-consumption diagram. It is not applicable to environments with many low-power appliances that exist, because they will interfere with decomposition.

The overall system relies on trained electrical equipment and cannot identify appliances that have not been identified. There are also obvious differences in power consumption for many of the same kinds of electrical appliances in the market. The system can transmit low-frequency data to a cloud server through a smart electricity meter, and web and app clients can be used for real-time monitoring. The system is currently most suitable for homes and office buildings, and its recognition accuracy can be further improved through pre-training in specific environments.

7 Conclusions and Future Work

The results of this study show that the use of deep learning in NILM is efficient, low-cost, and accurate. Here, only 20 days' worth of data were extracted from a room for training, but the algorithm was shown to be a predictor of electricity usage. The next step is to expand the data range to several different buildings and the time range of the training data in order to enhance performance. The model discussed in this paper is divided into three levels according to the function, which enables the convenient adjustment of parameters and the optimization of each module's function. A further step will be to try to separately train the model for each type of appliance, and it may also be interesting to identify specific appliances.

It would also be useful to set up a server that could obtain data in real time through Wi-Fi, calculate current electricity consumption, and record history for enhanced training. Currently, the server can be deployed locally or in the cloud for developer design and user selection. Web and mobile clients also provide a convenient and intuitive experience for users. Additionally, our design for a new house or old house renovation requires few resources and costs. There is no need to change circuits, no need to enter the home, and no need for users to pay high additional fees. With this system, users can see monthly electricity-consumption details at any time, thus providing basic and necessary support to companies and residents in energy conservation.

Funding This research is supported by Guangdong Basic and Applied Basic Research Foundation (Grant No. 2023A1515011137); the National Natural Science Foundation of China (Grant No. 52108011); Guangzhou Philosophy and Social Science Planning 2022 Annual Project (Grant No. 2022GZQN14); the 2022 Guangdong Philosophy and Social Science Foundation (Grant No. GD22XGL02); Fundamental Research Funds for the Central Universities (Grant No. QNMS202211); Department of Education of Guangdong Province (Grant No. 2021KTSCX004); Department of Housing and Urban–Rural Development of Guangdong Province (Grant No. 2021-K2-305243).

References

Afroz, Z., Shafiullah, G. M., Urmee, T., & Higgins, G. (2018). Modeling techniques used in building HVAC control systems: A review. *Renewable and Sustainable Energy Reviews*, *83*, 64–84.

Bonfigli, R., Squartini, S., Fagiani, M., & Piazza, F. (2015). Unsupervised algorithms for non-intrusive load monitoring: An up-to-date overview. In *Proceedings of the 2015 IEEE 15th International Conference on Environment and Electrical Engineering (EEEIC)* (pp. 1175–1180), 10–13 June 2015.

Gershenfeld, N., Samouhos, S., & Nordman, B. (2010). Intelligent infrastructure for energy efficiency. *Science*, *327*(5969), 1086–1088.

Hart, G. W. (1985). *Prototype nonintrusive appliance load monitor*; Technical report 2. MIT Energy Laboratory and Electric Power Research Institute.

Hart, G. W. (1992). Nonintrusive appliance load monitoring. *Proceedings of the IEEE*, *80*, 1870–1891.

Haugen, C. (2009). *Technology and business models for smart homes*. Technical report. Norwegian University of Science and Technology. 1.4, 2.1, 2.2.

Huber, P., Calatroni, A., Rumsch, A., & Paice, A. (2021). Review on deep neural networks applied to low-frequency NILM. *Energies, 14*(9), 2390.

Jia, F., Lei, Y., Lin, J., Zhou, X., & Lu, N. (2016). Deep neural networks: A promising tool for fault characteristic mining and intelligent diagnosis of rotating machinery with massive data. *Mechanical Systems and Signal Processing*, *72*, 303–315.

Kazmi, H., Mehmood, F., Lodeweyckx, S., & Driesen, J. (2018). Gigawatt-hour scale savings on abudget of zero: Deep reinforcement learning based optimal control of hot water systems. *Energy, 144*, 159–168.

Kelly, J., & Knottenbelt, W. (2015). Neural nilm: Deep neural networks applied to energy disaggregation. In *Proceedings of the 2nd ACM international conference on embedded systems for energy-efficient built environments* (pp. 55–64). https://arxiv.org/pdf/1507.06594.pdf

Kim, H., Marwah, M., Arlitt, M., Lyon, G., & Han, J. (2011). Unsupervised disaggregation of low frequency power measurements. In B. Liu, H. Liu, C. Clifton, T. Washio, & C. Kamath (Eds.), *Proceedings of the 2011 SIAM International Conference on Data Mining* (pp. 747–758). Society for Industrial and Applied Mathematics.

Kolter, J. Z., & Jaakkola, T. S. (2012). Approximate inference in additive factorial HMMs with application to energy disaggregation. *AISTATS, 22*, 1472–1482.

LeCun, Y., Bengio, Y., & Hinton, G. (2015). Deep learning. *Nature, 521*, 436–444.

Makonin, S., Popowich, F., Bajić, I. V., Gill, B., & Bartram, L. (2016). Exploiting HMM sparsity to perform online real-time nonintrusive load monitoring. *IEEE Transactions on Smart Grid, 7*, 2575–2585.

Mauch, L., & Yang, B. (2015). A new approach for supervised power disaggregation by using a deep recurrent LSTM Network. In *Proceedings of the 2015 IEEE Global Conference on Signal and Information Processing (GlobalSIP), Orlando, FL, USA, 14–16 December 2015* (pp. 63–67). IEEE.

Nalmpantis, C., & Vrakas, D. (2019). Machine learning approaches for non-intrusive load monitoring: From qualitative to quantitative comparation. *Artificial Intelligence Review, 52*, 217–243.

Parson, O., Ghosh, S., Weal, M., & Rogers, A. (2014). An unsupervised training method for non-intrusive appliance load monitoring. *Artificial Intelligence, 217*, 1–19.

Pereira, L., & Nunes, N. (2018). Performance evaluation in non-intrusive load monitoring: datasets, metrics, and tools—A review. *Wiley Interdisciplinary Reviews: Data Mining and Knowledge Discovery*, *8*, e1265.

Popa, D., Pop, F., Serbanescu, C., & Castiglione, A. (2019). Deep learning model for home automation and energy reduction in a smart home environment platform. *Neural Computing and Applications, 31*, 1317–1337.

Tongta, A., & Chooruang, K. (2020). Long short-term memory (LSTM) neural networks applied to energy disaggregation. In *Proceedings of the 2020 8th International Electrical Engineering Congress (iEECON)* (pp. 1–4), 4–6 March 2020.

Zeifman, M., & Roth, K. (2011). Nonintrusive appliance load monitoring: Review and outlook. *IEEE Transactions on Consumer Electronics, 57*, 76–84.

Zoha, A., Gluhak, A., Imran, M. A., & Rajasegarar, S. (2012). Non-intrusive load monitoring approaches for disaggregated energy sensing: A survey. *Sensors, 12*, 16838–16866.

Sustainable Renovation on Aosta Residential Building for Carbon Neutrality

Bin Li, Yuqing Zhang, Xiao Liu, Luca Caneparo, Weihong Guo and Qinglin Meng

Abstract

Under the effects of global warming and further environmental issues, sustainable development has gained worldwide relevance. The United Nations (UN) proposed Sustainable Development Goals (SDGs) for sustainable development, including carbon neutrality. This research focuses on decreasing the building energy use and carbon emissions for carbon neutrality, especially in the building operation phase. It is based on the current status of building stock retrofit in Aosta, Italy. Aosta social housing is used as an example of sustainable renovation research. The research methods include onsite investigation and energy simulation. The onsite investigation was made on May 30, 2022, by interviewing the users and onsite measurement to analyse the building status quo in subjective and objective aspects. The used software are Revit, Ecotect, and Phoenics to simulate the outdoor and indoor environments and to analyse the building's behaviour under a range of conditions. Glare and overheating during the day are considered. The design deploys outdoor shading, according to the traditional local Walser haylofts wisdom. Up-to-date technologies, such as high-performance windows profiles and triple glazing, are considered. Through sustainable renovation, indoor problems are addressed. Therefore, the sustainable retrofit improves the living environment and positively reduces energy consumption. The contribution of the measures to carbon neutrality through sustainable renovation is assessed.

Keywords

Carbon neutrality · Sustainable development · Renovation · Residential building · Aosta

B. Li · Y. Zhang · X. Liu (✉) · W. Guo · Q. Meng
School of Architecture, South China University of Technology, Guangzhou 510641, China
e-mail: xiaoliu@scut.edu.cn

B. Li
e-mail: bin.li@polito.it

Y. Zhang
e-mail: yuqing.zhang@polito.it

W. Guo
e-mail: whguo@scut.edu.cn

Q. Meng
e-mail: arqlmeng@scut.edu.cn

B. Li · Y. Zhang · L. Caneparo
Department of Architecture and Design, Politecnico Di Torino, 10125 Torino, Italy
e-mail: luca.caneparo@polito.it

B. Li · X. Liu · W. Guo
Energy Saving Technology Research Institute, South China University of Technology, Guangzhou 510641, China

B. Li · Y. Zhang · X. Liu · W. Guo · Q. Meng
State Key Laboratory of Subtropical Building and Urban Science, South China University of Technology, Guangzhou 510641, China

B. Li · X. Liu · W. Guo
Architectural Design & Research Institute Co., Ltd., South China University of Technology, Guangzhou 510641, China

1 Introduction

1.1 Background

The problem of global warming should be a top priority (Li et al., 2022; Liu et al., 2023). In 1972, the first world conference on the environment, United Nations Conference on the Human Environment (UNCHE), held in Stockholm, recommended attention to the climate change problem (United Nations, 1972). In 1982, the Nairobi Declaration was published to commemorate the tenth anniversary of the UNCHE. Importantly, this declaration further emphasized the global warming problem for sustainable development (Nairobi Declaration, 1982). In 1987, Our Common Future,

B.-J. He et al. (eds.), *Resilient Horizons: Building Sustainable Environments for Climate Adaptation and Health*, Advances in Science, Technology & Innovation, https://doi.org/10.1007/978-3-031-46109-5_4

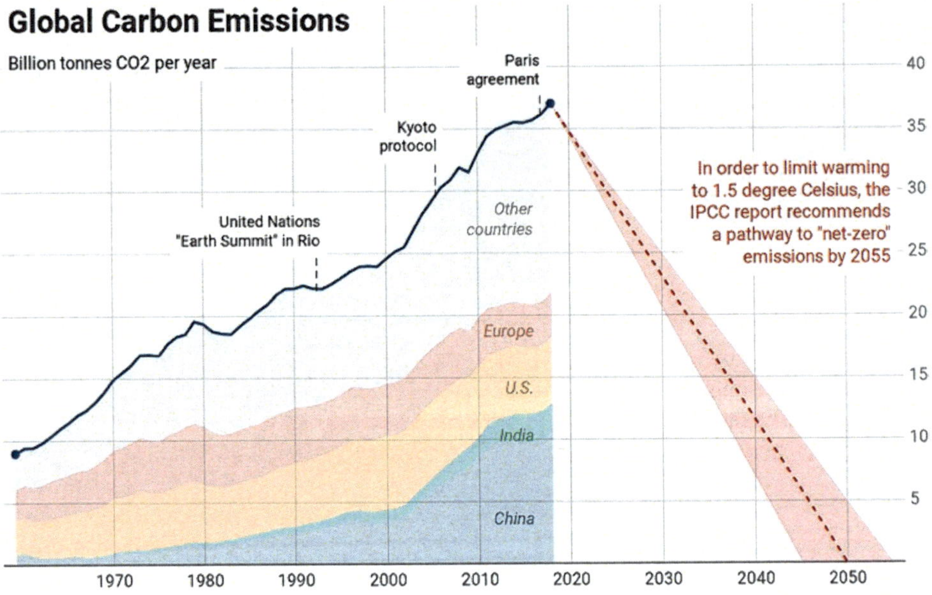

Fig. 1 Global carbon emissions (https://www.reddit.com/r/dataisbeautiful/comments/ca5bku/comment/et5x13f/)

also known as the Brundtland Report, mentioned the carbon dioxide build-up leading to global warming (World Commission on Environment and Development, 1987). In 1992, the United Nations Conference on Environment and Development, also called Earth Summit, was held in Rio de Janeiro. This conference resulted in the agreement of the United Nations Framework Convention on Climate Change (UNFCCC), the Rio Declaration on Environment and Development, and the Agenda 21 for sustainable development. In 1997, the Kyoto Protocol, based on the UNFCCC, was adopted to promote global energy conservation and emissions reduction for sustainable development, and this entered into force in 2005 (Li et al., 2021). In 2002, the World Summit on Sustainable Development was held in South Africa to promote sustainable development issues following the Rio Earth Summit in 1992 and the Millennium Declaration of 2000 (Schnoor, 2002). In 2007, Bali Road Map was published at the 2007 United Nations Climate Change Conference to mention that humans must reduce emissions to decrease the risks of climate change (Bali Road Map, 2007). In 2015, the United Nations Climate Change Conference was held in Paris, France. Negotiation resulted in the adoption of the Paris Agreement, which entered into force in 2016 with one of the aims to reach net zero by the middle of the twenty-first century (Gunfaus & Waisman, 2021). So, carbon neutrality is more and more relevant to addressing global warming, to achieve sustainable development. However, global carbon emissions are still increasing, as Fig. 1 shows. Moreover, according to the Intergovernmental Panel on Climate Change (IPCC)'s Summary for Policymakers in 2021, the global surface temperature has also increased rapidly, especially in the last 150 years with observation (Fig. 2). Global warming,

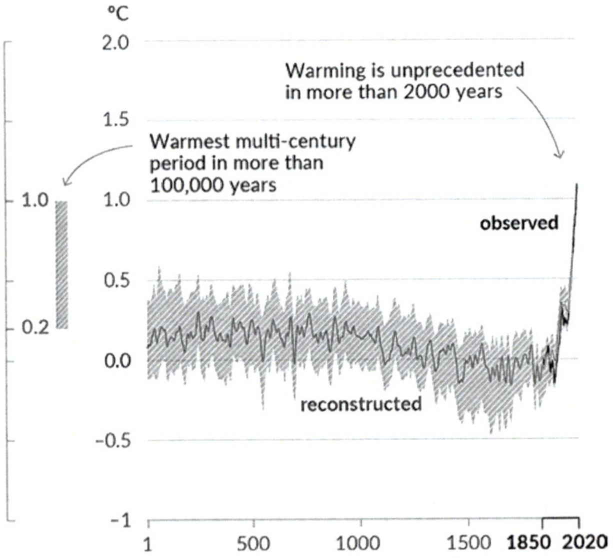

Fig. 2 Changes in global surface temperature (https://www.ipcc.ch/report/ar6/wg1/figures/summary-for-policymakers/figure-spm-1)

energy crisis, urban heat islands, air pollution, bushfire, flooding, COVID-19, and the Russian-Ukrainian War challenge are all urged for sustainable development.

1.2 Target

Sustainable development to address the above climate issues has gained momentum worldwide. The most cited sustainable development definition is "meets the needs of the present without compromising the ability of future generations to meet their own needs" from the 1987 Brundtland

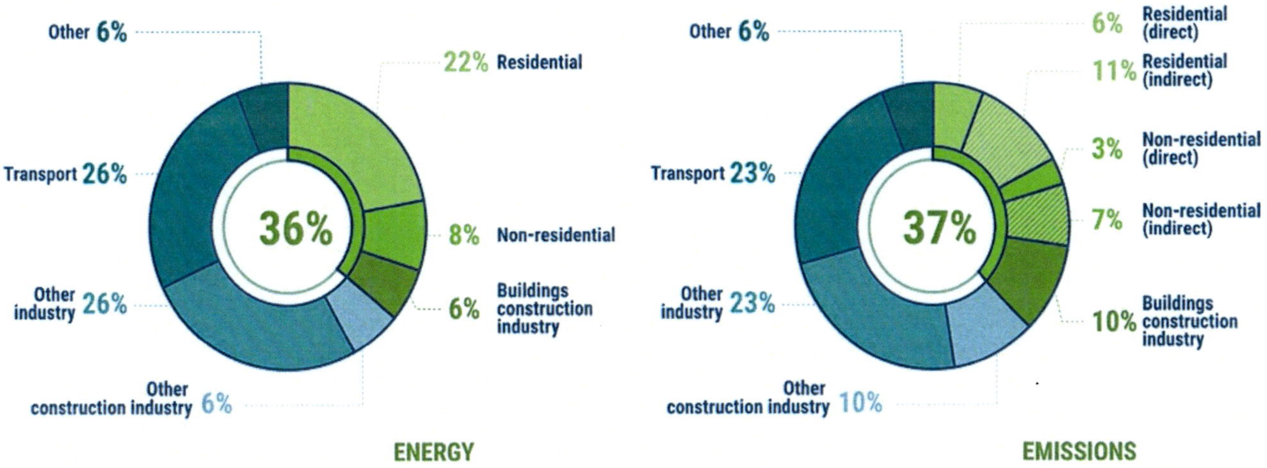

Fig. 3 Buildings and construction's share of global final energy and energy-related CO_2 emissions, 2020 (https://globalabc.org/resources/publications/2021-global-status-report-buildings-and-construction/)

Report (Roostaie et al., 2019). In 2015, United Nations (UN) formulated the 17 Sustainable Development Goals (SDGs) in the 2030 Agenda, which promoted a universal agreement to address climate change through carbon neutrality (Kawakubo et al., 2018). Therefore, carbon neutrality is recognized as a requirement to achieve sustainable development.

Carbon neutrality is offsetting the generated carbon dioxide (CO_2) through carbon capture, storage, and conversion within a certain period to achieve zero emission (Wu et al., 2022). This concept started in the 1990s, especially in Denmark's competition for a carbon–neutral community (Mundaca et al., 2018). In 2018, the glossary in the IPCC's Working Group Report defined it as "net zero CO_2 emissions" (Global Warming of 1.5 °C, 2013).

However, in the buildings and construction sector, carbon neutrality has not been achieved. The 2021 Global Status Report for Buildings and Construction, which was published by the UN Environment Programme (UNEP), indicates that buildings accounted for 36 percent of global energy demand and 37 percent of energy-related CO_2 emissions in 2020 (Fig. 3), even though the emissions were 10 percent lower than 2019 reaching lows not seen since 2007 due to COVID-19 and other factors (United Nations Environment Programme, 2021). Nevertheless, research should be done on decreasing the building energy use and carbon emissions for carbon neutrality, especially in the building operation phase of the programme of stock building renovation.

1.3 Significance

The European Commission 2021 Work Programme announced a 'Fit for 55' package to reduce greenhouse gas (GHG) emissions. In the proposal for a Directive of the European Council and Parliament on the energy performance of buildings (recast COM/2021/802 final), "The main objectives of this revision are reducing buildings' GHG emissions and final energy consumption by 2030 and setting a long-term vision for buildings towards EU-wide climate neutrality in 2050. In order to meet them, the initiative is grounded in several specific objectives: to increase the rate and depth of buildings' renovations, to improve information on energy performance and sustainability of buildings, and to ensure that all buildings will be in line with the 2050 climate neutrality requirements. Strengthened financial support and modernization and system integration are levers to deliver on these objectives."

To this aim, retrofitting the existing building is really relevant, because the residential building stock in Italy counts about 12,200,000 buildings, and about 60% of them were built after 1991, when the initial Italian energy law (number 10/1991) was issued. Furthermore, the yearly rate of building substitution is low, below 1%.

Italian laws on energy implement European Union Directives (EPBD, 2010), namely, the requirements for nearly Zero Energy Buildings (nZEB), stated in the Legislative Decree of June 26, 2015. In Europe nZEB definition was set by European Directive 2010/31/EU and was embraced beyond European boundaries. The Directive has

Table 1 nZEB requirements further to the Legislative Decree of June 26, 2015

Parameter	Criteria
H'_T [W/m²K]	$H'_T < H'_{T,limit}$ (depending on the building compactness ratio and the climatic zone)
$A_{sol,sum}/A_f$ [−]	$A_{sol,sum}/A_f < (A_{sol,sum}/A_f)_{limit}$ (depending on the building category)
$_H$ [−]	$_H > _{H,limit}$ (reference building)
$_W$ [−]	$_W > _{W,limit}$ (reference building)
$_C$ [−]	$_C > _{C,limit}$ (reference building)
$EP_{H,nd}$ [kWh/m²]	$EP_{H,nd} < EP_{H,nd,limit(2019/21)}$ (reference building)
$EP_{C,nd}$ [kWh/m²]	$EP_{C,nd} < EP_{C,nd,limit(2019/21)}$ (reference building)
$EP_{gl,tot}$ [kWh/m²] For residential buildings: $EP_{gl,tot} = EP_H + EP_W + EP_V + EP_C$	$EP_{gl,tot} < EP_{gl,tot,limit(2019/21)}$

led researchers and practitioners to focus their attention on the cost-optimized energy performance and has led to major advancements in terms of innovative energy saving technologies and innovative design methods. The leading requirements are outlined in Table 1.

2 Research Object

2.1 Aosta Social Housing

The case building studied is in the Cogne Neighbourhood in Aosta. The Neighbourhood was settled to accommodate the workers of the steel company Cogne Acciai Speciali. The Neighbourhood, over a century of life span, gradually deteriorated; because of the incremental effect of occasional changes, of the inability to match the changing societal needs, and of increasing sustainability targets. Thus, sustainable retrofit nowadays is an assumption for both the research and the planned interventions.

Notwithstanding its location in the city of Aosta, which has a fundamental relationship with the Alpine environment, the Cogne Neighbourhood presents several characteristics which make it similar, in some aspects, to numerous other districts of the same type and era, and these characteristics are also present in environments which are not strictly Alpine. So, the case study of the Aosta residential building renovation can have a meaning and applicability wider than the local context.

The case study takes one building in the Cogne Neighbourhood, the Gazzera Buildings, highlighted in Fig. 4. The Gazzera in-line houses constitute the expansion towards the West during the period of post-war reconstruction, on the model of multi-storey building with balcony access of workers' houses, in parallel lines with a north–south orientation.

Fig. 4 Location of the chosen building

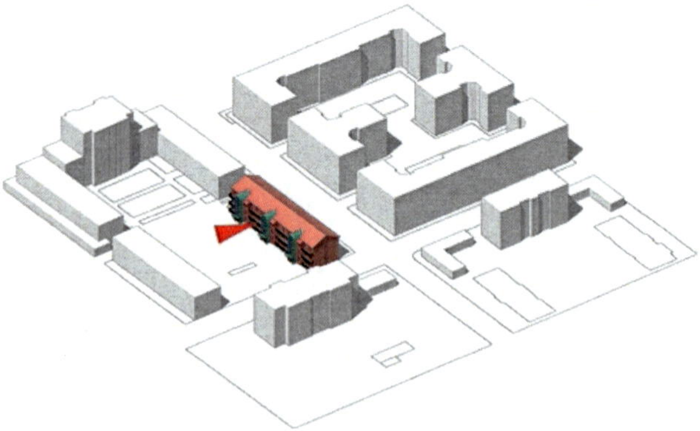

Fig. 5 Researched apartment

In this building, measurements have been taken in an apartment on the second floor (Fig. 5). This apartment has been chosen because it has a double frontage and offers a common typology and layout for the buildings in this area (Fig. 6).

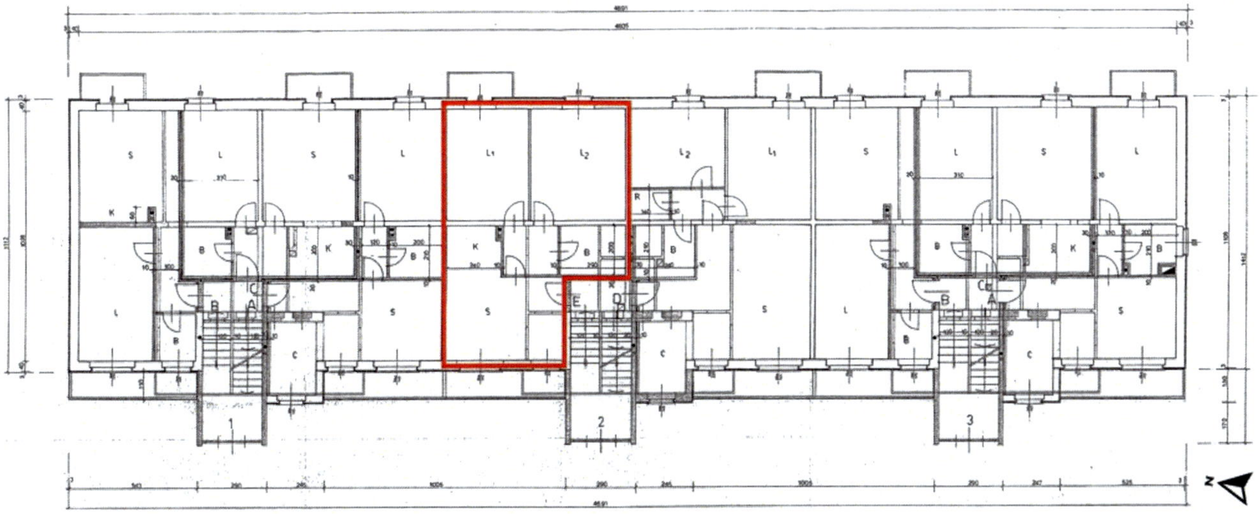

Fig. 6 Standard plan

2.2 Research Framework

Building Energy Modelling (BEM) has been extensively applied during the past decades to support the design of retrofit interventions. The nZEB targets require the designers to match and fulfil increasing levels of energy saving and integration with renewable energies, with the aim of improved overall efficiency and sustainability. These needs are supported by improved energy modelling.

The research addresses data collection at a neighbourhood scale from a weather station, outdoor and indoor measurements, and modelling with simulation.

Finally, the discussion and conclusion are put forward for carbon neutrality in a sustainable renovation.

3 Research Methods

3.1 Measurements

The data survey can achieve different levels of detail, according to the time-span and time-interval of the observations and of the data collected. A widely accepted differentiation is defined by Reddy who sets five possible levels of survey (Reddy, 2006) (Table 2).

The energy bills data are available at the scale of buildings and are not reliable at the scale of apartments. The measurements fitted the Level 4 since they spanned a full week-long.

The local weather data has been collected by the meteorological station in Piazza Plouves, Aosta. The station includes barometer, wind direction, hygrometer, rain gauge, total solar radiation, solarimeter, thermometer, and wind speed. The measurements are taken at an hourly time scale.

For onsite investigation, the research team went to the Aosta residential building on May 30, 2022. The test apartment was visited for the interviews with the residents and the acoustic and thermal measurements by the instruments. For the acoustic measurements, the outdoor balcony and the indoor bedroom were monitored (Fig. 7). As for the thermal measurement, the living room's temperature and humidity were recorded (Fig. 8). In fact, the indoor natural lighting environment was monitored too. Unfortunately, due to a logging problem of the device, the data were not recorded properly (Fig. 9). Figure 10 shows the measurement instruments' position in the visited apartment.

Table 2 Survey levels

Levels	Building input data available					
	Utility bills	As-built data	Inspection	Detailed audit	Short-term monitoring	Long-term monitoring
Level 1	x	x				
Level 2	x	x	x			
Level 3	x	x	x	x		
Level 4	x	x	x	x	x	
Level 5	x	x	x	x	x	x

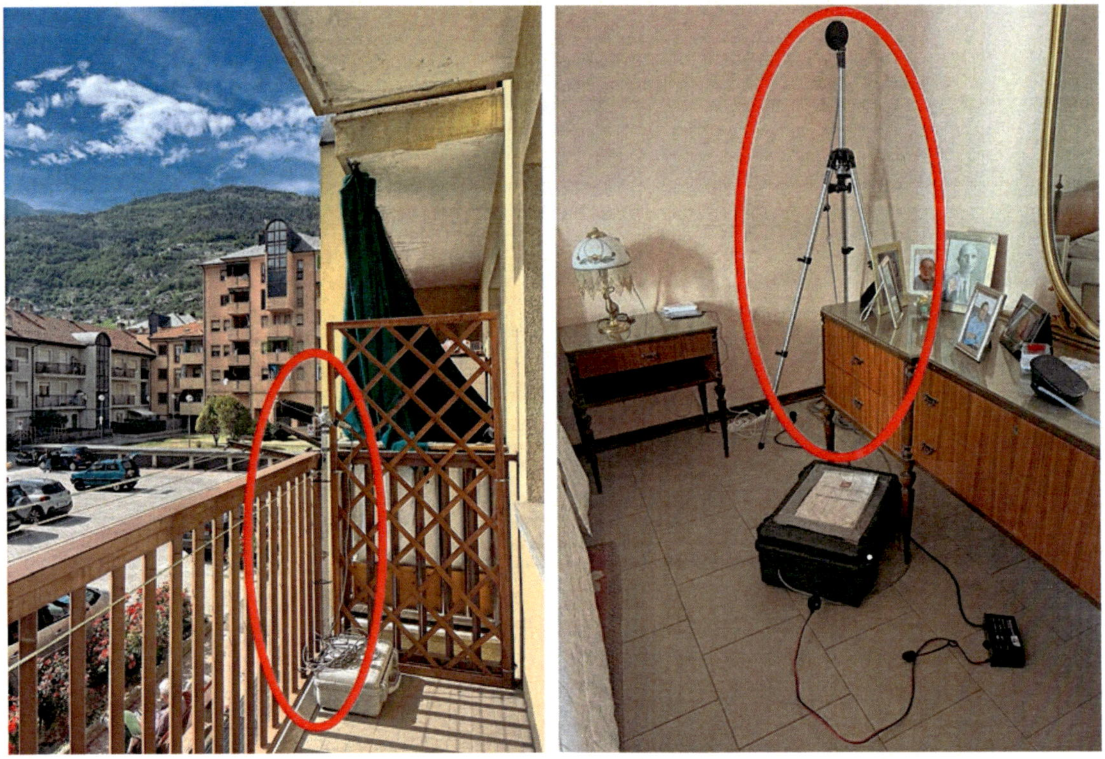

Fig. 7 Acoustic monitoring (outdoor and indoor)

Fig. 8 Thermal monitoring

3.2 Software Simulation

BEM is used as a complementary method. Revit software with the Insight Solar Analysis plugin simulated the outdoor solar insolation during the whole summer. Ecotect software simulated the indoor average daily radiation on the summer solstice. Moreover, Phoenics software simulated the indoor dry resultant temperature and Predicted Mean Vote (PMV). The software simulation method has been integrated with the onsite investigation method, to study the lighting, acoustic, and thermal environments all together. The research outcome managed both subjective and objective aspects.

Using the software simulation method combined with the onsite investigation method, overall, the lighting, acoustic, and thermal environments were analysed. Then the initial research results were obtained in subjective and objective aspects.

4 Results

4.1 Subjective Results

The Gazzera Buildings are in-line houses with a north–south orientation. Therefore, the sunshine directly shoots

Fig. 9 Lighting monitoring

indoors, especially during the afternoon due to west sunlight, which causes the glare issue. After the interview with the apartment residents, this issue has been confirmed. The data on the status quo of the west facade of the residential building testify to the glare issue. In Fig. 11, the case apartment used curtains on the west-facing balcony. And for others, almost all the west-facing windows added sunblinds in different ways except for the communal stairwell windows.

4.2 Objective Results

4.2.1 Onsite Measurement

The data from the onsite measurements, the software simulations, and the interviews aim to assist the design from the preliminary stage onwards.

The onsite measurement was taken on May 30, 2022. Both the acoustic and thermal environments were monitored. For the acoustic environment, two class 1 phonometers compliant with IEC-61672 were installed, respectively on the balcony and in the bedroom, for surveying the outdoor and indoor noise levels. The survey period started on Monday (May 30, 2022) and ended on the next Monday (June 6, 2022), nearly 7 days. Figure 12 shows the acoustic levels of outdoors and indoors side-by-side. Outdoor noise shows some regularity patterns, sometimes reaching to peak possibly due to outdoor parking noise. The indoor noise is hardly greater than 20dBA, which proves the sound insulation effect of doors and windows to a certain extent. The sporadic peaks show correlations with the outdoor conditions. According to the WELL, LEED, and series acoustic standards requirement, the acoustic level of a bedroom in a residential building should be lower than 35dBA (Torresin et al., 2019). Consequently, the acoustic environment here is nearly satisfactory and does not demand major interventions.

For the thermal environment, two AirBeam 2 and one multisensor Honeywell HPMA115C0-003 were installed, respectively in the living room and in the bedroom, the survey period overlapping the monitoring of the acoustic environment. Figure 13 shows the temperature and humidity over the period. In May and June, in Aosta the climate is temperate. The survey data shows that the trend in the temperature is slightly increasing: it almost reached a max of 26 °C indoors. The humidity is increasing from 37.5 to 72.4% until the noon on June 5, 2022. Then, the humidity is decreasing. According to the historical record of weather data in Aosta, the changing of humidity is clearly affected by the showery weather on that day. Moreover, the detection position in the living room is close to the window, so it is influential.

Regarding the indoor temperature during summer, in the weather history record for 2022, on July 19 outdoor temperature reached its highest peak in Aosta, 35 °C. In fact, this year's summer was hotter than in the past two centuries at least. Due to global warming, the indoor overheating issue has to be addressed by the design, while the simulation and the analysis have to aid it.

4.2.2 Software Simulation of the Current Status

Revit software with the Insight Solar Analysis plugin has been used to simulate the solar energy on the envelope. The cumulative insolation simulation of the whole summer from June 21 to September 21 has been performed. Figure 14 shows for the Gazzera that the cumulative insolation for the west facade is much more than for the east facade.

Ecotect has been used to simulate the apartment. The Aosta historical weather data, from 2001 to 2021, has been used. The simulation day is the summer solstice (June 21). Although this day is not the hottest day of the whole summer, it sets the beginning of the summer and has the longest hours of sunshine in the northern hemisphere. Figure 15 shows the average daily radiation situation with the material optical properties based on wall, ceiling, floor, and window (Fig. 16). The west balcony and the communal stairwell got the most radiation on this day. The simulation highlights these areas as the hottest in the whole apartment.

Fig. 10 Instrument location in the case-apartment

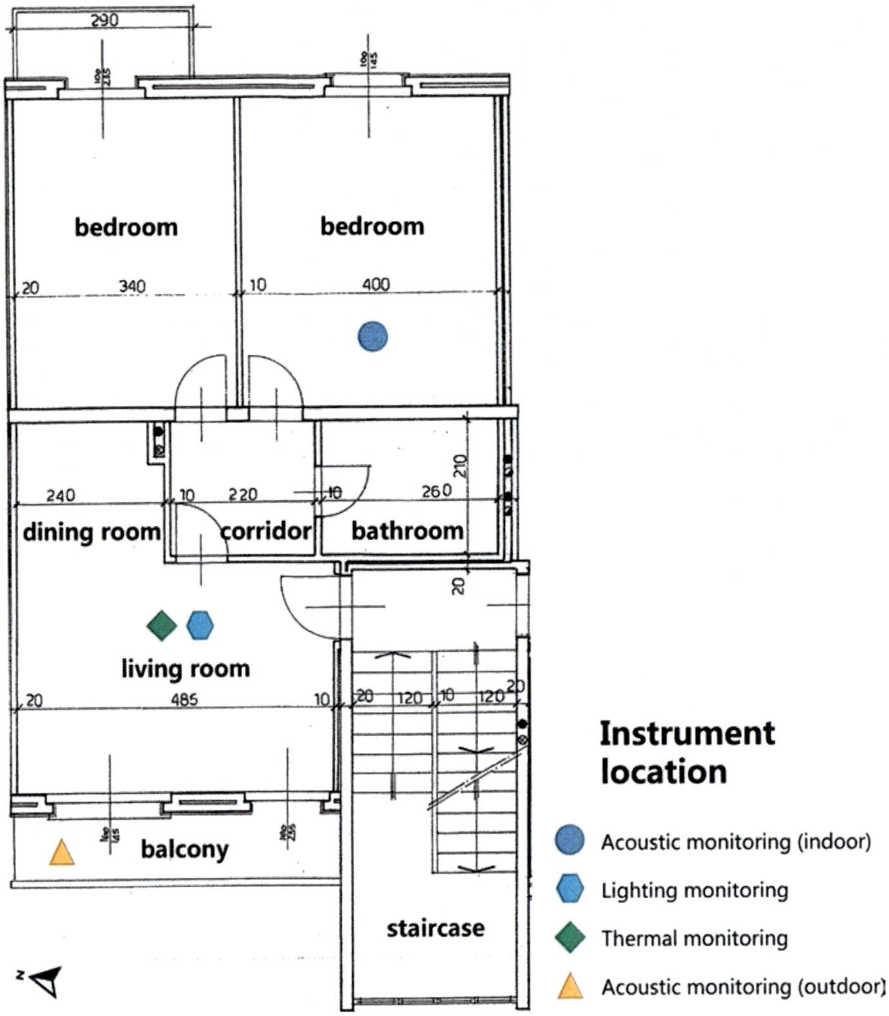

Phoenics software was used to quantify the heat level. For natural ventilation indoor environments, PMV is generally accepted for assessing the thermal comfort of users (Ole Fanger & Toftum, 2002). Here, PMV on the summer solstice at 14:00 was simulated. At the same time, Fig. 17 shows the simulation outcome for the indoor dry resultant temperature. The east wind blows from the east window to the west window. Even the east side bedroom which has relatively low solar insolation and radiation, also presented a relatively high temperature. Even though the temperature on the summer solstice is not the hottest compared with the overall summer, the bedroom's dry resultant temperature

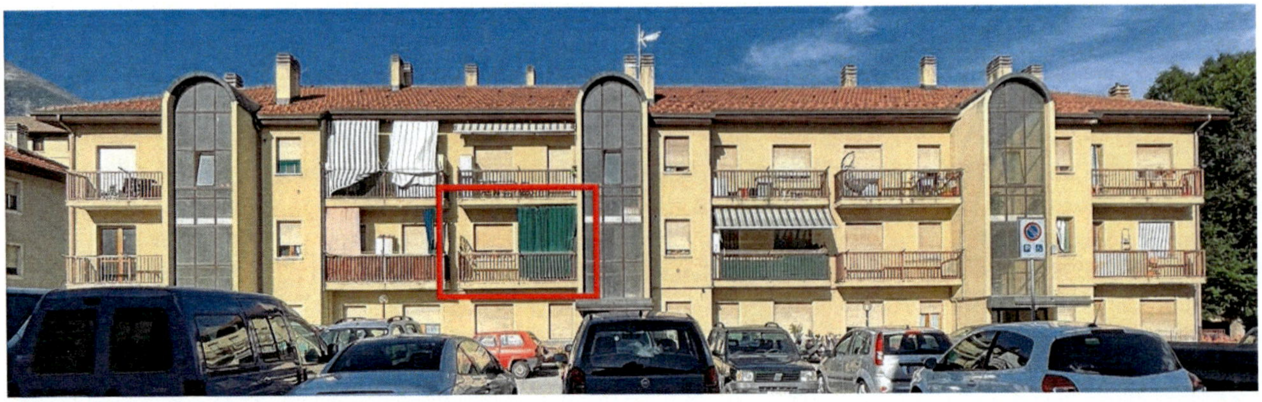

Fig. 11 Glare issue

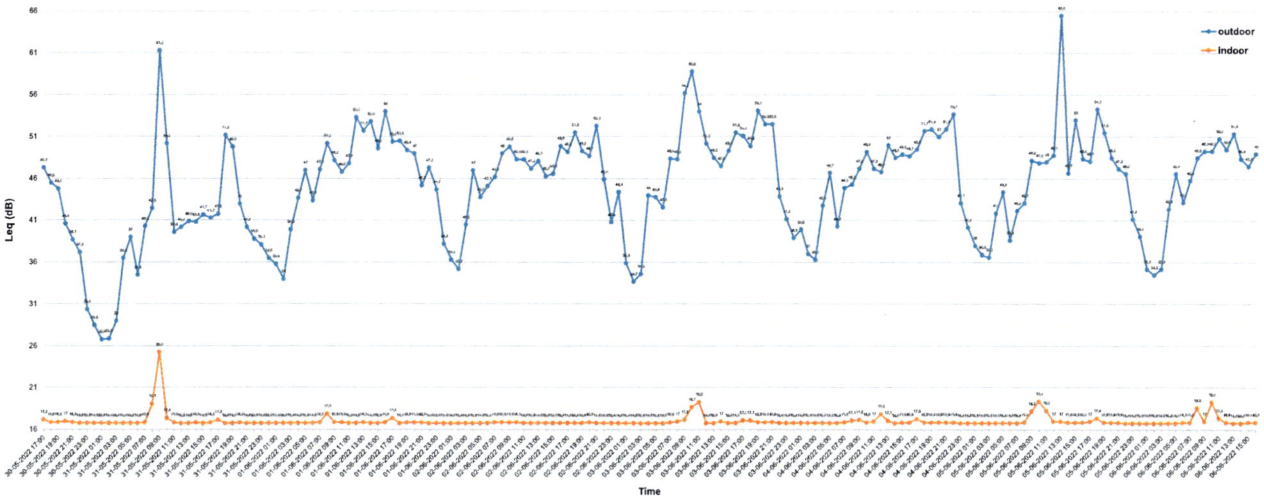

Fig. 12 Acoustic environment results

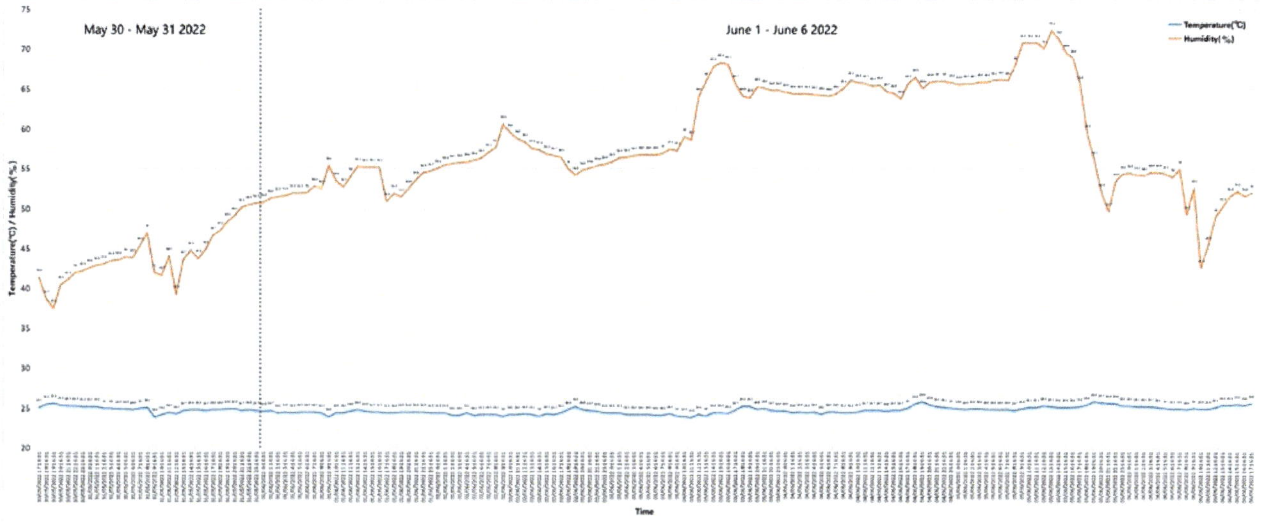

Fig. 13 Thermal environment results

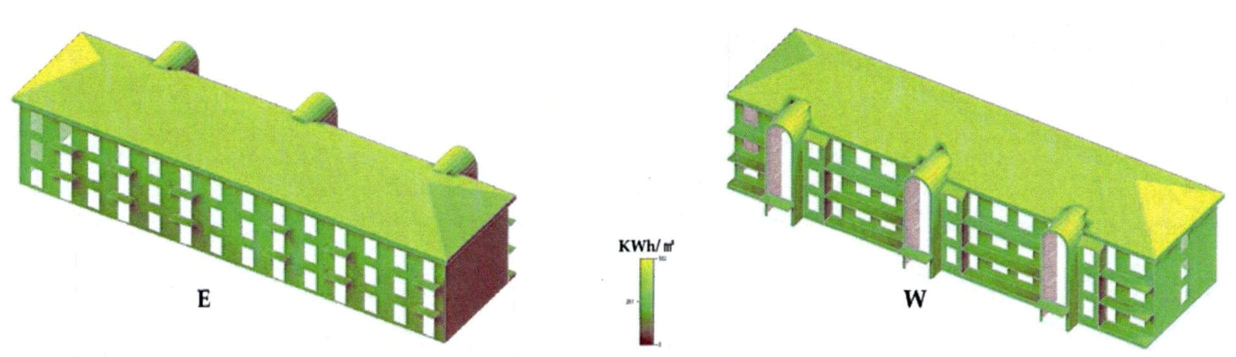

Fig. 14 Cumulative insolation simulation results

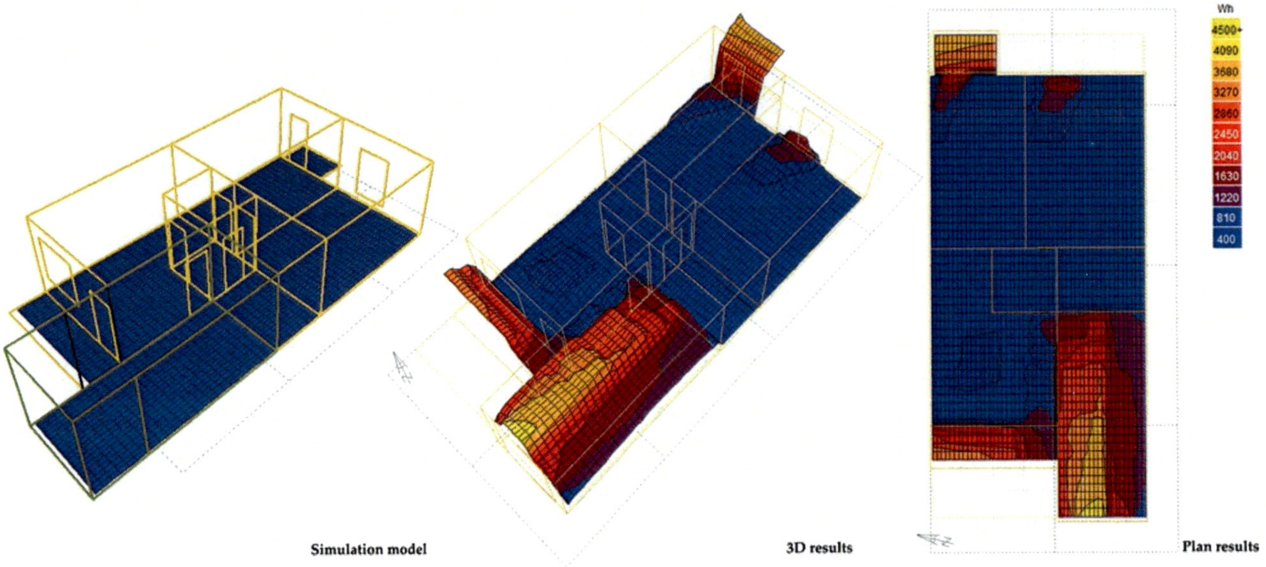

Fig. 15 Average daily radiation simulation results

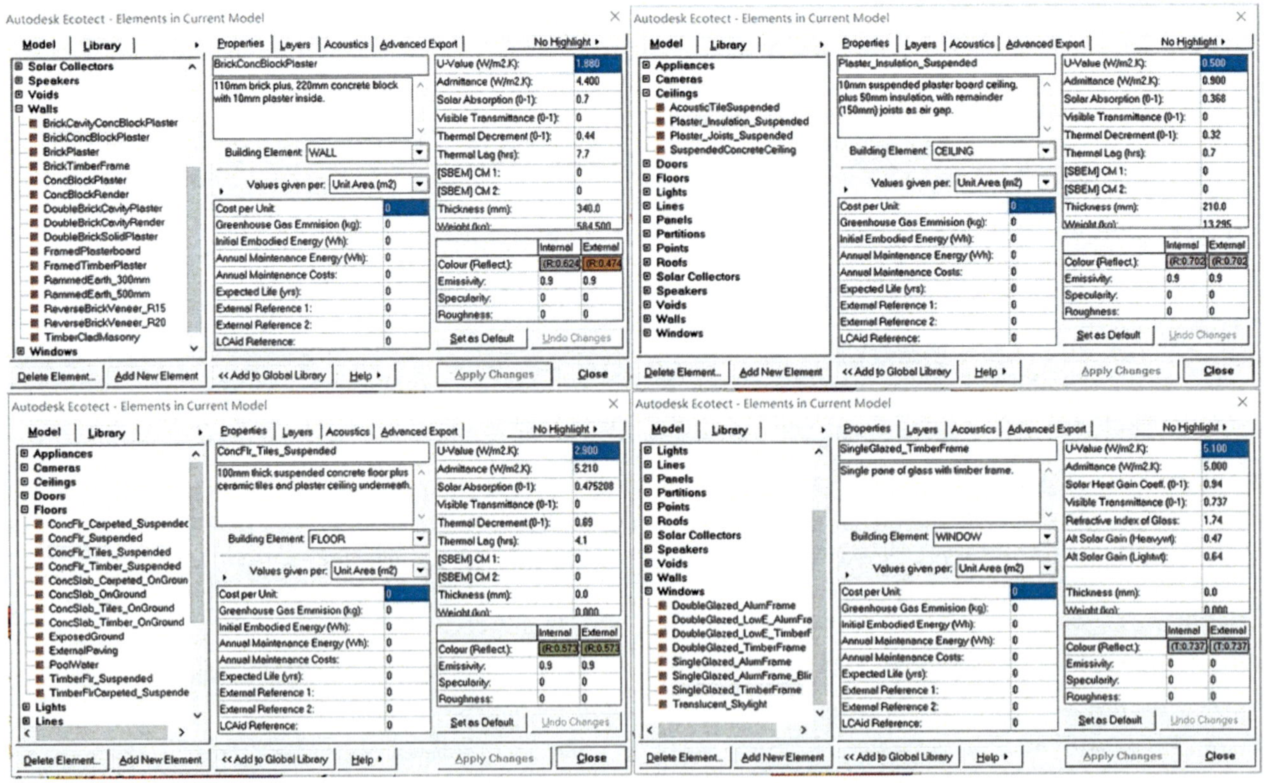

Fig. 16 Material optical properties of wall, ceiling, floor, and window

reached about 27 °C. Figure 18 shows the indoor PMV, the values are more than 2.34, which means it is close to hot thermal perception. Correspondingly, the predicted percentage of dissatisfied (PPD) associated with PMV represents the thermal dissatisfaction rate. Figure 19 shows that indoor PPD is close to 90%. The data confirm that indoors is out of the comfort zone. It is hot indoors from the beginning of the summer (summer solstice).

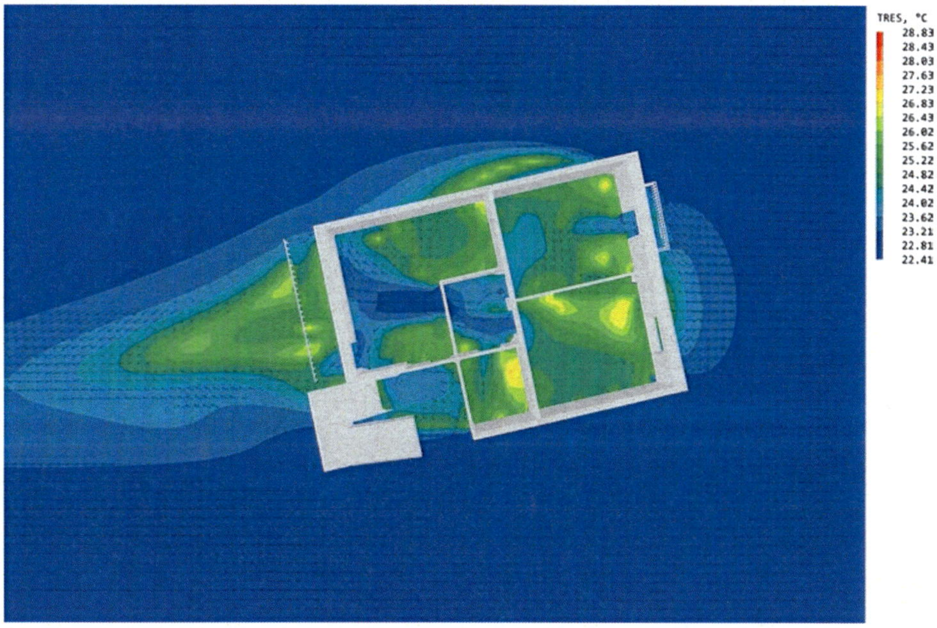

Fig. 17 Dry resultant temperature simulation

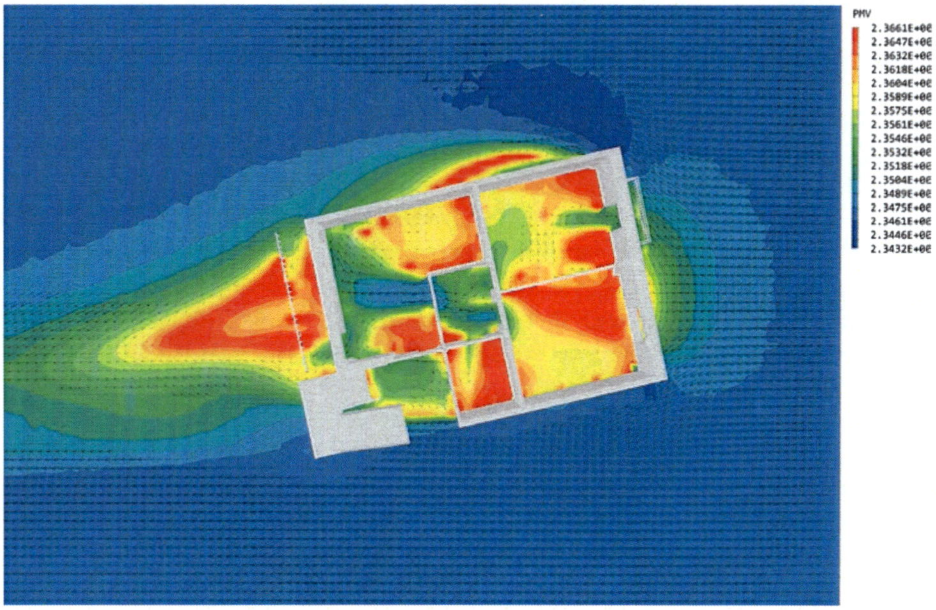

Fig. 18 PMV simulation

5 Discussion

5.1 Sustainable Renovation Analysis

Based on the outcomes of the analyses and simulations of the subjective and objective aspects, some design issues in Aosta residential buildings can be sorted out. The glaring problem is assessed especially on the west facades. In fact, the angle of incidence of the sun in summer is higher than in winter. The glare effect is manifest not just during the summer, but in winter too. The glaring issue affects the indoor light environment during the whole year. It also contributes to indoor overheating.

For the thermal environment, further to the onsite measurements taken in May, although the outdoor temperature was not exceedingly high. The analyses and the simulations show that the indoor environment tends to be uncomfortable, because of overheating during the afternoon, by the start of the summer (June 21).

To address the glare and overheating issues, shading devices are integrated with the envelope design. The residents tend to install sunblinds outside their apartments

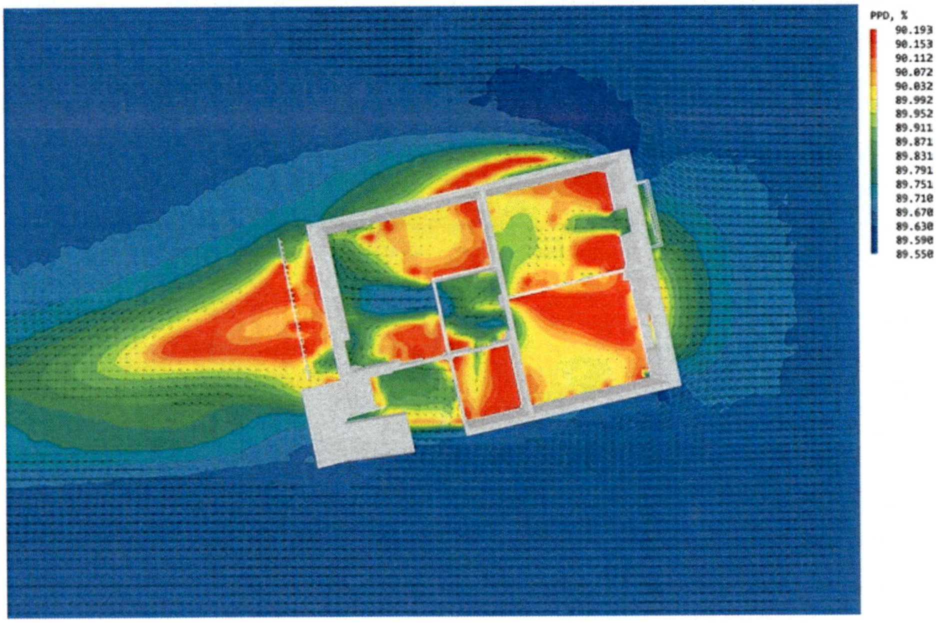

Fig. 19 PPD simulation

autonomously. The residents' self-construction testifies of real issues, but at the same time, it raises further issues such as the façade design and safety, because of the dangers arising from severe weather events, e.g. storms.

Because of the extreme climatic conditions of Aosta, with cold winters and hot summers, the integration with passive design measures has been considered. According to European Union Directives (EPBD, 2010), the design process deploys an integrative approach aiming at (1) Integrating passive cooling strategies; (2) Optimizing the envelope; (3) Reducing energy needs for heating and cooling.

To achieve passive cooling, sun shading is provided through steel louvre blades. They refer to the wooden balconies of heritage Walser haylofts, where they are used to increase the useful volume and drying capacity for forage storage (Fig. 20).

During the summer months, to increase the shading capacity of louvre blades, they support espalier plants, in which the lateral branches are made to grow horizontally from the central trunk. Vigorously growing apple trees will be planted directly into the soil, to reduce the need for watering and care.

In Val d'Aosta, and also generally in regions with a harsh climate, there is a widespread practice of leaning fruit trees against a wall which protects them and assists the ripening process. The espalier cultivation offers the residents an effective collection of fruits.

Alongside the traditional local architecture wisdom, advanced systems are being considered for this sustainable building renovation. Currently, the transparent envelope includes outdated low performing wooden window frames. The renovation measures consider the removal of the old windows and the installation of certificated new windows (UNI 11673), consisting of wooden frames with high quality profiles and triple glazing.

To increase the performance of the envelope, the walls and the roof will be insulated by means of the External Thermal Insulation Composite System (ETICS), a compact multilayer insulation solution, designed to enhance energy efficiency.

Fig. 20 Walser haylofts's wisdom (https://www.alagna.it/en/routes-walks-trekking-monterosa/the-antique-hamlets/)

These sustainable renovation strategies contribute to passive measures and to carbon neutrality. They will improve the indoor environments and to savings on energy consumptions and bills over their life span (Zhang et al., 2023).

5.2 Limitations

Due to the access restrictions to the dwellings, the onsite investigation was taken in May, which is not the hottest period in Aosta. Anyhow, the glare problem was evident. For the problem of indoor overheating, a set of software simulations have been performed, based on local meteorological data. Further simulations, to assess a wider range of passive technologies, would provide a larger set of scenarios for decision. In previous researches (Lush et al., 1999; Visscher et al., 2009), the overheating phenomena for nZEB in hot climates are frequently ascribed to the sum of insulation, thermal mass, and air tightness. We are committed to the appraisal of the contribution from further aspects, i.e. solar shading or solar protections, contribution from available passive cooling sources, including means to control indoor air exchanges and speed, outdoor air temperature during the day and night, sky and soil vault temperature. As for sustainable renovation, more testing is to be performed. Currently, the paper has addressed the observed issues only and is advancing possible measures. Further onsite measurements and dynamic thermal and lighting simulations are required to test the design proposals.

6 Conclusions

It is a current open challenge to design practices to balance active and passive measures with the aim of improving the sustainable design, namely, the concepts of solar shading, night ventilation, and thermal mass. Summer comfort is challenging to be assessed quantitatively, because of its dynamic and subjective features. The present study aims to set some measurements and evaluation tools to support the design with assessment and methodologies of active and passive design.

Based on the present research, several considerations can be drawn. (1) Global warming and other problems are increasingly affecting living environments. So, sustainable development has become a common goal worldwide. (2) Carbon neutrality is a necessary way to achieve sustainable development, especially in the building and construction sector. (3) The research into Aosta residential buildings offers an instance, to try to appraise and analyse site-specific issues. For example, glare and indoor overheating problems are matched. (4) Based on the onsite investigation by interviewing residents and onsite measurement on May 30, 2022, the acoustic environment was demonstrated as satisfactory. (5) The software simulation by Revit, Ecotect, and Phoenics has assessed indoor overheating during summer. (6) For addressing these issues, to achieve sustainable renovation, the heritage local Walser Haylofts wisdom has been deployed for shading. At the same time, high-performance window profiles and triple glazing have been proposed. (7) The present research is trying to increase the contributions from passive and active measures to enhance sustainability, especially with the aim of carbon neutrality. The renovation design has addressed indoor and outdoor comfort, pursuing an architectural expressiveness respecting the specific Alpine context and environment.

Acknowledgements This work was supported in part by the scholarship from the China Scholarship Council (CSC) under the CSC grant nos. 202006150053 and 202106150080.

We are also grateful for the collaboration to Azienda Regionale Edilizia Residenziale Aosta, manager of the social housing of Cogne Neighbourhood, and to Agenzia Regionale per la Protezione dell'Ambiente for the measurement activities.

Funding This research was funded by the National Key R&D Program of China (grant no. 2021YFC2009400); National Natural Science Foundation of China (grant no. 52108011); Guangdong Basic and Applied Basic Research Foundation (grant no. 2023A1515011137); Department of Housing and Urban–Rural Development of Guangdong Province (grant no. 2021-K2-305243); Department of Education of Guangdong Province (grant no. 2021KTSCX004); Guangzhou Philosophy and Social Science Planning 2022 Annual Project (grant no. 2022GZQN14); the Fundamental Research Funds for the Central Universities (grant no. QNMS202211); the 2022 Guangdong Philosophy and Social Science Foundation (grant no. GD22XGL02); Science and Technology Program of Guangzhou, China (grant no. 202102020302); China Postdoctoral Science Foundation (grant no. 2021M701249); and State Key Laboratory of Subtropical Building Science, South China University of Technology (grant no. 2021ZB16).

References

Bali Road Map. (2007). Available online: https://en.wikipedia.org/wiki/Bali_Road_Map (Retrieved July 29, 2022).

EPBD. (2010). *Directive 2010/31/EU of the European Parliament and of the Council of 19 May 2010 on the energy performance of buildings (recast)*. Official Journal of the European Union.

Global Warming of 1.5 °C. (2013). Available online: https://www.ipcc.ch/sr15/ (Retrieved July 31, 2022).

Gunfaus, M. T., & Waisman, H. (2021). Assessing the adequacy of the global response to the Paris agreement: Toward a full appraisal of climate ambition and action. *Earth System Governance, 8*, 100102. https://doi.org/10.1016/j.esg.2021.100102

Kawakubo, S., Murakami, S., Ikaga, T., & Asami, Y. (2018). Sustainability assessment of cities: SDGs and GHG emissions. *Building Research & Information, 46*(5), 528–539. https://doi.org/10.1080/09613218.2017.1356120

Li, B., Guo, W., Liu, X., Zhang, Y., & Caneparo, L. (2022). The third Solar Decathlon China buildings for achieving carbon neutrality. *Buildings, 12*(8), 1094. https://www.mdpi.com/2075-5309/12/8/1094

Li, B., Guo, W., Liu, X., Zhang, Y., Russell, P. J., & Schnabel, M. A. (2021). Sustainable passive design for building performance of healthy built environment in the Lingnan area. *Sustainability, 13*, 9115. https://doi.org/10.3390/su13169115

Liu, X., He, J. T., Xiong, K., Liu, S., & He, B. J. (2023). Identification of factors affecting public willingness to pay for heat mitigation and adaptation: Evidence from Guangzhou, China. *Urban Climate, 48*, Article 101405. https://doi.org/10.1016/j.uclim.2022.101405

Lush, D., Butcher, K., & Appleby, P. (1999). Environmental design: CIBSE Guide A.

Mundaca, L., Busch, H., & Schwer, S. (2018). 'Successful' low-carbon energy transitions at the community level? An energy justice perspective. *Applied Energy, 218*, 292–303. https://doi.org/10.1016/j.apenergy.2018.02.146

Nairobi Declaration. (1982). Available online: http://www.un-documents.net/nair-dec.htm (Retrieved July 29, 2022).

Ole Fanger, P., & Toftum, J. (2002). Extension of the PMV model to non-air-conditioned buildings in warm climates. *Energy and Buildings, 34*(6), 533–536. https://doi.org/10.1016/S0378-7788(02)00003-8

Reddy, T. A. (2006). Literature review on calibration of building energy simulation programs: Uses, problems, procedures, uncertainty, and tools. *ASHRAE Transactions, 112*(1).

Schnoor J. (2002). Examining the world summit on sustainable development. *Environmental Science & Technology, 36*(21), 429A-430A. https://doi.org/10.1021/es0224611

Roostaie, S., Nawari, N., & Kibert, C. J. (2019). Sustainability and resilience: A review of definitions, relationships, and their integration into a combined building assessment framework. *Building and Environment, 154*, 132–144. https://doi.org/10.1016/j.buildenv.2019.02.042

Torresin, S., Albatici, R., Aletta, F., Babich, F., Oberman, T., & Kang, J. (2019). Acoustic design criteria in naturally ventilated residential buildings: New research perspectives by applying the indoor soundscape approach. *Applied Sciences, 9*(24).

United Nations (1972). *Report of the United Nations conference on the human environment - Stockholm, 5–16 June 1972*. https://wedocs.unep.org/20.500.11822/30829

United Nations Environment Programme. (2021). 2021 *Global status report for buildings and construction: Towards a zero-emission, efficient and resilient buildings and construction sector*.

Visscher, H. J., Mlecnik, E., & Meijer, F. M. (2009). The impact of energy performance regulations on systems of building control. In RCIS COBRA 2009, *The construction and building research conference of the Royal Institution of Chartered Surveyors, Cape Town* (pp. 1356–1367), 10–11 September, 2009. RCIS.

World Commission on Environment and Development. (1987). *Our common future*. Oxford University Press.

Wu, X., Tian, Z., & Guo, J. (2022). A review of the theoretical research and practical progress of carbon neutrality. *Sustainable Operations and Computers, 3*, 54–66. https://doi.org/10.1016/j.susoc.2021.10.001

Zhang, Y., Li, B., Caneparo, L., Meng, Q., Guo, W., & Liu, X. (2023). Physical Environment Study on Social Housing Stock in Italian Western Alps for Healthy and Sustainable Communities. Land, 12(7), 1468. http://dx.doi.org/10.3390/land12071468

Features of ESG Risk Management in the Implementation of Megaprojects the Belt and Road Initiative

Olesya Dobryakhina, Anastasia Brazhnikova and Vadim Beketov

Abstract

The article is devoted to the problems of ESG risk management (Environmental, Social, and Corporate Governance) in implementing megaprojects. The essence of megaprojects and their features are considered. The essence of the ESG concept is analyzed, and the definition of ESG risks is given. Shareholders and management investigate the features of ESG risk identification and management. In the practical part of the study, the main trends in implementing China's water management megaprojects under the Belt and Road Initiative (OECD (2018) are analyzed, ESG risks of these projects are investigated, and recommendations on risk management are given.

Keywords

ESG risk management · Megaprojects · China's water management

1 Introduction

Large-scale socio-economic development projects (megaprojects) are constantly implemented in various countries (Flyvbjerg, 2014). It is not a secret that megaprojects are often characterized by adverse environmental and social effects (impact on transport infrastructure, pollution, and others).

O. Dobryakhina (✉)
Said Business School, Oxford University, Oxford, UK
e-mail: dobryakhina@gmail.com

A. Brazhnikova
Lomonosov Moscow State University, Moscow, Russia
e-mail: brazhnikovanastasia@yandex.ru

V. Beketov
Novosibirsk State Technical University, Novosibirsk, Russia

The problem of rational nature management during the implementation of megaprojects, the assessment of environmental and social damage, and the problem of social responsibility of project initiators are the subject of many studies. The necessity of investigating the impact of megaprojects on the "green" (ecological) economy and the further interconnection with the concept of sustainable development (United Nations, 1987) are admitted by society. Despite the comprehensive study of these problems, some features of implementing megaprojects are not always considered. In particular, when justifying and implementing such projects, it is of great importance to create a mechanism for compensating for damage to public health and the environment due to environmentally hazardous activities, environmental audits, and the development of ethnological expertise procedures during the industrial development of territories.

Ignoring significant environmental and social factors in the financial models of megaprojects means that investors may not consider the full range of risks and, in some cases, benefits associated with investments. This situation may cause severe errors in the justification of the required rate of return and the calculation of the model parameters, which ultimately will not allow the project initiators to estimate the fair value of investments. The most critical risks affecting many megaprojects' critical parameters are ESG risks (environmental, social, governance). As the number of megaprojects grows, the need for comprehensive and reliable ways to understand and assess how these risks affect the socio-economic effects of investment decisions increases. This aspect is essential for both private and public investments. At the same time, it should be borne in mind that megaprojects, as a rule, are implemented on the terms of public–private partnerships. Sustainability considerations and ESG risks are becoming particularly important for institutional investors.

Megaprojects are an adequate basis for the country's sustainable development policy. The megaprojects also

B.-J. He et al. (eds.), *Resilient Horizons: Building Sustainable Environments for Climate Adaptation and Health*, Advances in Science, Technology & Innovation, https://doi.org/10.1007/978-3-031-46109-5_5

influence the sustainable development of regions and the country's economy (Bagaynicov, 2017). According to Bagaynicov (2017, p. 1), "The implementation of megaprojects, due to their scale, significantly changes the economic landscape of vast territories, forming a qualitatively different institutional environment and contributing to their more effective global geo-economic positioning." Simultaneously, the sustainable development of territories is an essential factor in the success of megaprojects.

This article discusses the most important theoretical and practical aspects of ESG risk management in implementing megaprojects. This work relies on empirical data on China's water management megaprojects implemented within the Belt and Road Initiative (OECD 2018) framework.

2 Theoretical and Methodological Aspects of ESG Risk Management in the Implementation of Megaprojects

2.1 The Essence of Megaprojects

Megaprojects are large-scale and complex projects that meet several criteria:

- the cost of $1 billion and above, development and construction over many years,
- the unification of public and private stakeholders, and
- the impact on millions of people.

Such projects are created to change the structure of society, which is why the same project management methodologies and metrics cannot be applied to them as in conventional projects. The term itself appeared in 1976.

In the generally accepted terminology, "mega" means a million, and the cost of the first megaprojects was measured in millions. The stakes began to rise from the beginning of World War II: the Manhattan (the program to create the first atomic bomb) (Jones, 1985) and Apollo (the lunar program) projects were already estimated at billions of dollars (Wild, 2019).

Megaprojects are large not only in terms of the number of resources spent but also in terms of the intensity of work. So, from 2015 to 2018, China built as many kilometers of railways for high-speed trains as Europe in 20 years (Economist.Com, 2019).

The growth of competition and the need to ensure the sustainable development of the regions require megaproject implementation. According to Bagaynicov (2017, p. 1), megaprojects always amaze society with their scale that, during implementation, can "significantly change the economic landscape of vast territories." Further, the change in the economic landscape can "contribute to more effective global geo-economic positioning" (Bagaynicov, 2017, p. 1) of the territories. Thus, as we can see, the implementation of megaprojects affects huge territories and "contribute to the formation of balanced economic complexes in the regions" (Bagaynicov, 2017).

2.2 ESG Risks of Megaprojects

The success of megaprojects largely depends on the degree of protection from various risks generated by fluctuations in the external environment and the projects' complexity. According to Algin (1991, p. 31), "risk is the activities of economic entities associated with overcoming the uncertainty in the situation of inevitable choice, during which the possibility of a desired result, failure and deviation from the goal contained in the alternatives can be assessed."

As a matter of fact, the uncertainty is the result of incomplete knowledge and understanding of social, nature, and governance features during the megaproject implementations. The pre-project investigation and preparation can help reasonably estimate the uncertainty factor and protects the project in the future (Flyvbjerg et al., 2004).

The uniqueness of each megaproject affects the occurrence of typical and specific risks, including macroeconomic, political, financial, organizational, legal, competence, reputational, and others (Flyvbjerg et al., 2004). Special attention has recently been paid to ESG risks when implementing megaprojects.

ESG (Environmental, Social, and Corporate Governance) is a set of company management characteristics in which this company's involvement in solving environmental, social, and managerial problems is achieved.

ESG is a set of company performance standards that "socially conscious investors use to verify potential investments" (Investopedia, 2023). Thus, the environmental element influences how the company takes care of nature. The social element is responsible for interaction with employees, suppliers, customers, and society. Corporate governance elements influence company management, internal control, procurement process, and shareholders' rights (Chen, 2021).

Thus, ESG is three elements according to which companies provide sustainable development management. Accordingly, ESG risks are environmental, social, and corporate governance risks.

Research in the field of ESG risk management began to develop as a response to environmental degradation, global warming, and increasing economic inequality between rich and developing countries, which created additional ground for the emergence and development of new financial instruments and criteria. Modern society has begun imposing additional requirements on companies' activities and evaluating them from the point of view of investing in sustainable development, not only financial statements.

The core reason for society's close attention to the environmental risks of megaprojects is that the increased concentration of greenhouse gases is already increasing the global average temperature of the Earth's surface. The use of fossil fuels, together with agriculture and waste, generated 56% of methane emissions into the atmosphere in 2021 and made the main contribution to the imbalance in the Earth's ability to absorb it (Olivers & Schulz, 2022). Although there is much less methane in the atmosphere than carbon dioxide, it absorbs thermal infrared radiation much more efficiently. As a result, methane has a global warming potential 86 times stronger per unit mass than CO_2 (Jackson et al., 2020).

The public currently pays special attention to various social aspects of megaproject implementation: working conditions, the use of child labor, compliance with workplace safety, harmful working conditions, respect for human rights, discrimination in the workplace (when hiring), responsibility to society, and others.

To be successful, a company must demonstrate steady growth in its financial indicators and a particular reputation. Effective ESG risk management allows the company to improve its reputation and increase brand awareness (Van Burren et al., 2016).

The most significant number of ESG risk studies are devoted to the following aspects (Fig. 1):

- ESG risk and financial performance of projects relationship analyses (Barnett & Salomon, 2006 Cheng et al., 2014);
- identification and quantification of the financial significance of various ESG risks in specific industries (Eccles et al., 2013);
- analysis of requirements for completeness of the information disclosed by companies about ESG risks (Busch et al., 2016; Khan et al., 2015);
- development of non-financial reporting standards considering the investment community's requirements.

Different researches show that information about ESG risks can significantly positively or negatively impact investment decision-making. Thus, the works of Goss and Roberts (2011) concluded that a high ESG rating and a high level of corporate governance could reduce the cost of debt financing.

ESG risks can have a different impact on the financial return on investment. Friede et al. (2015) found a correlation between ESG investment and corporate financial indicators. This correlation is at least non-negative in about 90% of cases. Amel-Zadeh and Serafeim (2017) analyzed the data from their global survey of investment community representatives. The study showed that almost 97% of investors recognize the reputational role of ESG-related data. According to this study, most respondents believe stocks with good ESG indicators have a lower systematic risk.

Certain aspects of E, S, and G may affect a project's financial efficiency in different ways in terms of the scale of investments, their industry affiliation, and the corresponding time horizon. Dhaliwal et al. (2011) concluded that corporate governance's effectiveness significantly impacts financial performance more than social and environmental efficiency. It happens because the impact of changes in corporate governance is more noticeable than the impact of environmental and social changes. In addition, the events of the COVID-19 pandemic have shown that companies with a high level of corporate governance and adaptability of the business model could most successfully withstand environmental and social risks and shocks.

According to Barnett and Salomon (2006), the interrelation between ESG elements and financial factors is not linear. The company's social and environmental costs are related to the present period's influence, while the ESG elements are long-term fiches (Barnett & Salomon, 2006). This finding can show us that leading and lagging indicators play a significant role, which should be taken into account

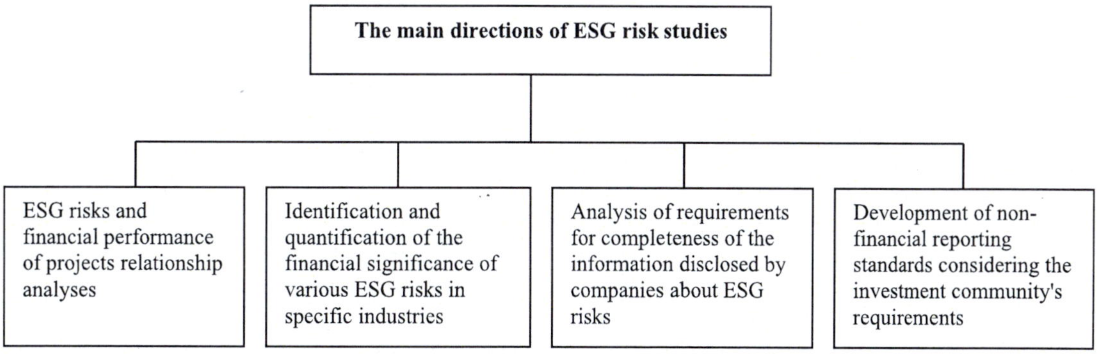

Fig. 1 The main directions of ESG risk studies

"in particular, when developing regression models to find out the dependence of financial indicators on sustainability indicators" (Efimova, 2018, p. 667).

If the idea of the need to account for ESG risks is supported by both the investment and academic community, then accounting methods for them in investment analysis vary greatly. According to the work of Eccles and Kastrapeli (2017), as well as a study by Hayat (2015), the most significant obstacle to integrating ESG into investment decisions is the lack of common standards for uniform analysis and assessment of the impact of significant ESG risks and, as a result, the lack of comparability of data disclosed by companies. Due to these and some other problems, accounting for ESG risks in investment analysis, in most cases, transforms into a format of qualitative input data used in the form of a system of criteria and restrictions along with traditional financial criteria (NPV, IRR, and others).

Of particular interest is the work of Schramade (2016), the author of which suggests using the approach of adjusting financial cost drivers in the project financial model, taking into account the impact of ESG risks on the business model and competitive positions.

As a rule, the approaches of investors and management of the companies initiating megaprojects to the assessment of ESG risks differ. The management of companies proceeds from the need to increase revenues, reduce costs, and increase profits in the short term. The remuneration of top managers depends on these indicators. For investors, the prospect of stock growth and an increase in business scale in the broadest time horizon is essential. Accordingly, the decisions made by investors and management differ. First, they may contradict each other in terms of performance over time. This contradiction was shown in detail in a specialized study by Shu-Lin Yang (Yang, 2016).

The author notes that implementing sustainable development measures does not always have an immediate positive effect (Yang, 2016). Moreover, with the high involvement of the company in social and environmental activity in the short term, the current corporate efficiency may decrease. However, it may increase in the long term. With a low involvement of the company in work to avoid ESG risks, a positive effect arises due to the rejection of the related costs. Meanwhile, in the long-term perspective, the efficiency of the company's activities decreases sharply (Yang, 2016).

Megaproject managers usually try to find a compromise in this situation, using important societal factors to increase the project's reputation. The attention of target audiences is focused on the megaproject's social significance and environmental safety. However, management focuses primarily on actions that can reduce costs and increase revenue.

On the other hand, investors' growing attention to environmental protection and social problems makes it possible to raise the value of the company-initiator of the megaproject extremely quickly and highly.

Companies interpret ESG risks as something that causes financial and/or reputational losses, including the risk of business interruption/closure. The view of state environmental protection agencies on such risks focuses on the likelihood of harm to human health and the environment from implementing specific hazards. Companies need to comprehensively and adequately assess the possibilities of such losses using various qualitative and quantitative analysis tools. This analysis is based on the possibility of identifying ESG risks, which is not easy, especially when there is no prohibiting legislation. This situation often happens because it takes time for society and the authorities to realize the dangers. From this point of view, the main thing is to identify the compliance of potential risk with the ethical principles of ESG (no harm to the environment, social and corporate relations).

Maximalism in the analysis of ESG risks (this process includes an assessment of the probability of risks and the importance and degree of their impact on the company) should be welcomed since the stage of formulating the company's attitude to each risk should be preceded by a thorough threat assessment. The attitude to risk can manifest in acceptance, rejection, sharing it with a third party, or transferring responsibility for it to a third party. However, suppose the risk is not identified. In that case, the probability of losses due to its implementation and failure to take measures to reduce or eliminate the probability of risk occurrence is exceptionally high.

When taking a significant ESG risk (in whole or part), "its management is not autonomous; it must be fully integrated with other business issues" (Vivian et al., 2003). Work to reduce or eliminate the impact of risk has a crucial influence on the company's supply, production, sales, and finances (project).

The significance of the same ESG risks varies in different industries. Moreover, this value varies over time. As noted by the most significant researchers in the ESG—Freiberg et al. (2019, p. 0.32), "the dynamic nature of risk materiality will manifest itself at different times and with different intensity in different companies and industries. The discrepancy between corporate behavior and social needs is a critical initial condition for materiality. With early attention to certain risk catalysts, it is possible that emerging problems can be solved before they become tangible, which is the best possible outcome for all stakeholders."

2.3 ESG Risk Assessment and Management

The importance of ESG risk assessment is also determined by the fact that companies' adherence to the principles of sustainable development (United Nations, 2022) will be the leading force of prosperity and value creation in the current decades. This thesis was voiced by experts at the World Economic Forum in Davos in 2020 (Eltobgy & Brown, 2020). Effective management of ESG risks is crucial for business owners because the buyers are willing to pay a higher price than competitors to a manufacturer of a product of equal consumer qualities only because it is produced, taking into account the principles of sustainable development (United Nations, 2022). In a global survey, 57% of consumer goods buyers said this in 2020 (Falk et al., 2020).

The importance of ESG risk assessment is also determined by the fact that companies adhere to the principles of sustainable development (United Nations, 2022) will be the leading force of prosperity and value creation in the current decades. This thesis was voiced by experts at the World Economic Forum in Davos in 2020 (Eltobgy & Brown, 2020). Effective management of ESG risks is crucial for business owners because the buyers are willing to pay a higher price than competitors to a manufacturer of a product of equal consumer qualities only because it is produced, taking into account the principles of sustainable development (United Nations, 2022). In a global survey, 57% of consumer goods buyers said this in 2020 (Falk et al., 2020).

When justifying megaprojects, ensuring an optimal sequence of actions for a particular megaproject is crucial when assessing ESG risks and evaluating the project's effectiveness. In a broad sense, the effectiveness of a megaproject is an indicator that characterizes the ratio of total benefits and losses from the project. At the same time, the external environmental effects and the project's associated social and economic consequences, which affect the interests of the population and future generations, are considered.

Environmental and social benefits and costs (damage) from implementing a megaproject are environmental effects expressed in positive or negative consequences of the project for the environment and the population. The benefits include additional tax receipts and payments to budgets, an increase in household income, a decrease in unemployment, the implementation of related social programs, and others. Environmental costs (environmental costs) can include damage from the loss of natural resources, degradation of environmental components, and the health of the population in the territory.

ESG risk assessment of megaprojects is most often given by specialized rating agencies. Ratings are mainly used by investors, less often by lenders, to supplement their analysis. To do this, they use the final assessment of the rating agency and the ESG data about the company provided as part of the assessment. In addition, some use several ESG ratings simultaneously to get the complete picture. This practice is widespread, among other things, due to the problem of standardization of ratings. Different agencies can assign very different ratings to the same company.

To analyze the ESG risks the methods shown in Fig. 2 (Hubbard, 2009) were used.

It should be noted that some practical experience in managing ESG risks has been accumulated and obtained during the implementation of large-scale infrastructure projects. Specialists are actively involved in decision-making regarding project implementation's environmental and social components. However, at the same time, using ESG risk reduction tools that ensure the implementation of the principles of sustainable development (United Nations, 2022) remains an important task (Fig. 3).

As we can see, compensation and guarantee obligations play the most crucial role in the system of ecological and economic regulation of the implementation of megaprojects.

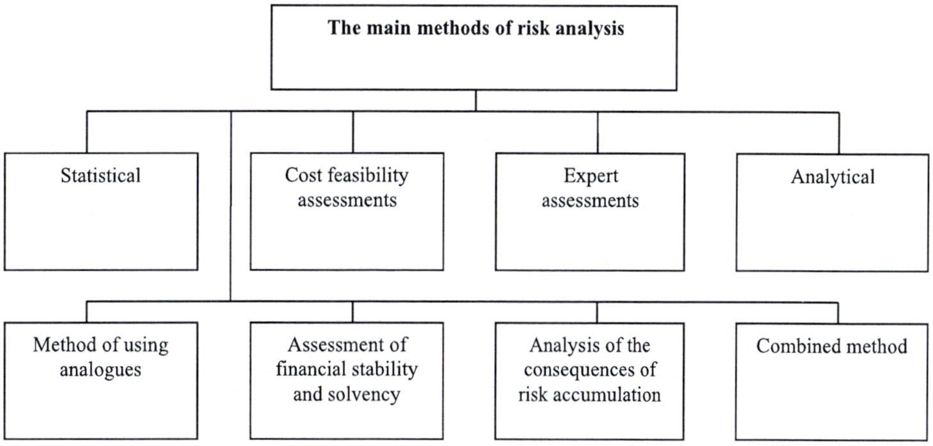

Fig. 2 The main methods of risk analysis Resource: Hubbard (2009)

2.4 Features of ESG Risks of the Chinese Belt and Road Initiative

Within the framework of this article, research on the problems of ESG risk management in the implementation of Chinese megaprojects related to the Belt and Road Initiative (OECD 2018) is of particular interest.

In 2013, China launched the Belt and Road Initiative (B&R, BRI) (OECD 2018), consisting of the Silk Road Economic Belt and the Maritime Silk Road of the twenty-first century (The World Bank, 2018). This initiative has significantly impacted many aspects of Eurasia's development and use of resources.

This activity was associated not only with creating a single new Eurasian transport and energy corridor but, first of all, with strengthening the Eurasian economic and cultural interaction. The Belt and Road Initiative (OECD 2018) is a crucial element of the 13th Five-year plan (OECD 2018) and the main focus of China's foreign policy from 2015 to 2025.

In the frame of the Silk Road Economic Belt project, all environmental and social risks are already known. As we can see, strategic environmental assessment tools (CCICED, 2018) have become essential for developing the "joint construction of the Green Silk Road" in China. It is, therefore, imperative that the environmental risks of investment and the environmental aspects of the behavior of Chinese companies outside China are also under Chinese control (CCICED, 2018).

Since 2006, the Chinese government has been actively issuing recommendations for companies related to environmental protection when investing abroad. Several documents were implemented in order for Chinese enterprises to take care of the environment and comply with local laws when implementing their projects (Davis, 2013), such as "Administrative Regulations for Foreign Projects on a Contractual Basis" (State Council, 2008), "Nine Principles of the State Council for Stimulating and Regulating China's Foreign Investments" (Davis, 2013). The Ministry of Commerce and the Ministry of the Environment also issued a "Guide to Environmental Protection in the Implementation of Foreign Investment Projects and Cooperation" later in 2013 (MOFCOM & MEP, 2013).

3 ESG Risks of Implementing China's Water Management Megaprojects

3.1 The Main Megaprojects and Their Specifics

The promotion of the B&R Initiative (OECD 2018) was caused by both the socio-economic and environmental situation inside China and the revision of the country's foreign policy priorities. Among the foreign policy reasons is China's desire to lead regional economic cooperation and establish close relations with the region's states in the economic, political, and humanitarian spheres. This aspiration

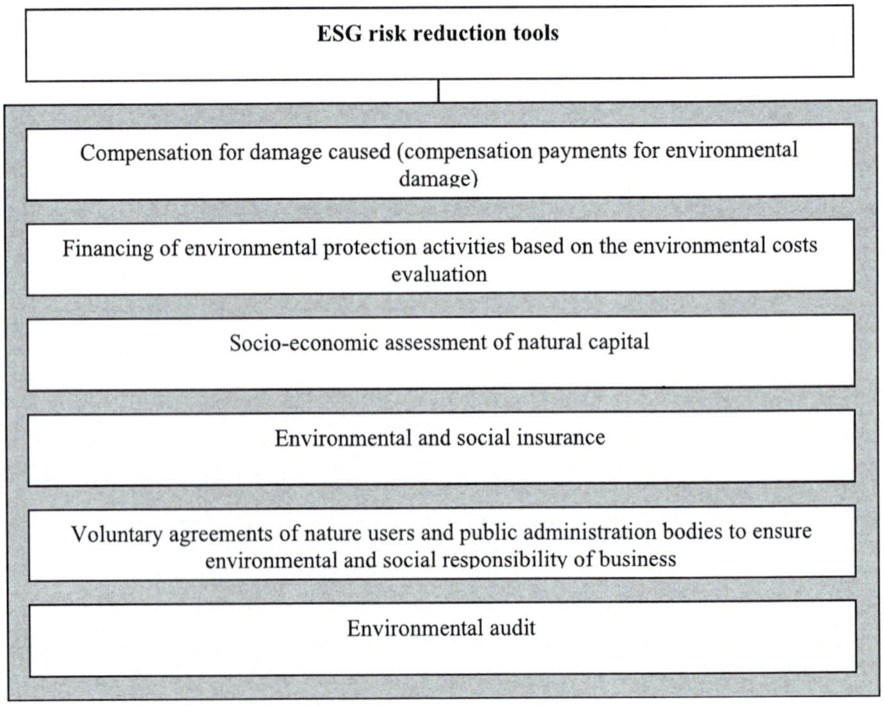

Fig. 3 ESG risk reduction tools

is sometimes interpreted as a desire to practice a "new model of globalization." From the point of view of internal problems, B&R (OECD 2018) helps China harmoniously develop all the provinces of the state and, at the same time, solve problems associated with an excess of industrial capacity and an increased burden on the country's environment by environmental degradation.

In recent years, the possibilities of creating large hydraulic structures in China have decreased due to the growth of socio-environmental costs and the already achieved surplus of mega-infrastructure. Thus, the hydropower development program in the 13th five-year plan (OECD 2018) has been cut in half compared to the last 12 five-year plans. Therefore, the conquest of the international hydro-engineering market is essential for the survival of such large state-owned companies in the Chinese construction industry as Gezhouba, Sinohydro (Power China), CMEC, etc.

In the water sector, the most popular areas of Chinese private and public investment in megaprojects abroad are hydropower and the creation of urban agglomerations and infrastructure, including the construction of flood protection systems, water supply, sanitation, and purification (Simonov, 2018a). In addition, China invests in agriculture, the development of river and sea transport routes, the water supply of industrial parks, and the creation of tourist complexes on the water.

For example, the following tasks are highlighted in the Program for the Creation of the China-Pakistan Economic Corridor (CPEC Authority, 2022), announced in December 2017 (CPEC Authority, 2017):

(a) the creation of large water-saving agricultural complexes, the development of drip irrigation, and other forms of water conservation;
(b) application of new concepts of urbanization in the creation of water supply and sanitation systems;
(c) cooperation in the field of integrated management of water resources and river basins, the creation of comprehensive plans to reduce the risks of floods and droughts;
(d) developing basin hydropower plans, preparatory work on HPP projects, and developing a network of high-voltage power lines.

The China–Pakistan Economic Corridor (CPEC Authority, 2017) represents several infrastructure projects implemented by the Government of the People's Republic of China in Pakistan (China Daily Information Co (CDIC), 2015). Initially, the cost of implementing the projects was estimated at 46 billion US dollars, but by 2017, it had increased to 62 billion US dollars (Anil Kumar Lal, 2020). In 2016, CPEC began partially functioning when cargo from China was delivered overland to the port of Gwadar for further sea transportation to Africa and West Asia (Saeed, 2015). Some extensive energy facilities were put into operation in 2017.

Thanks to the China–Pakistan Economic Corridor (CPEC Authority, 2022), an extensive network of roads and railways will be built throughout Pakistan. The Government of Pakistan believes that inefficiency caused mainly by the country's dilapidated transport network leads to a loss in the country's annual GDP by up to 3.55% (Kiani, 2017). Modern transport networks were built within the project of the China–Pakistan Economic Corridor (CPEC Authority, 2022). These transport networks help connect Karachi and Gwadar seaport with northern Pakistan and the cities in western China and Central Asia (Saeed, 2015). As a part of the China–Pakistan Economic Corridor megaproject (CPEC Authority, 2022), a 1,100-km highway will be built between Lahore and Karachi cities (Dawn Media Group, 2014).

Moreover, the Karakoram highway will be totally renovated and help to connect the territories from Hassan Abdal to the border with China (Dawn Media Group, 2014). The main railway line between the Karachi–Peshawar will also be renovated, allowing trains to travel at 160 km per hour (Alam, 2015). The expansion of the Pakistan Railway Network is also envisaged to connect with the South Xinjiang Railway in Kashgar (Zhen, 2015). Constantly, according to Rafi (2016), around $11 billion needed to upgrade transport networks will be provided through subsidized concessional loans.

The statistics provided by Kiani (2019) show that 20% of the China–Pakistan Economic Corridor is debt financing, while 80% is an investment in joint ventures between Pakistan and China. This approach allows the creation of more than 120,000 jobs for Pakistanis and Chinese (40,000 and 80,000 jobs, respectively) (Bennett-Jones, 2021). Relying on statistics, we can estimate the possibility of a return of $6 billion to $8 billion a year from taxes such as road and bridge tolls (CPEC Authority, 2022).

Now in Pakistan, the Chinese corporation "Three Gorges" is building the Karot hydroelectric power station (700 MW) (Fig. 4, Jamal, 2021) with the money of the Silk Road Fund and the filling of the reservoir of the Neelum–Jhelum hydroelectric power station (1 GW) (Fig. 5, Khan, 2012) has begun. The World Bank is financing the Dasu hydroelectric power station creation by Gezhouba on the Indus River.

In the Asian Infrastructure Investment Bank (AIIB) portfolio, where China dominates (26% of the capital), there are already 14 major water management projects (Simonov, 2018b). These projects include the reconstruction of hydroelectric power plants in Pakistan and Tajikistan, flood protection in the Philippines and Sri Lanka, repair of old dams in Indonesia, irrigation projects in India, and the Nenskra hydroelectric power plant in Georgia, sewage treatment

Fig. 4 Aerial view of the Karot Hydropower Station in Pakistan Resource: Jamal (2021), Image credited: China Three Gorges Corporation

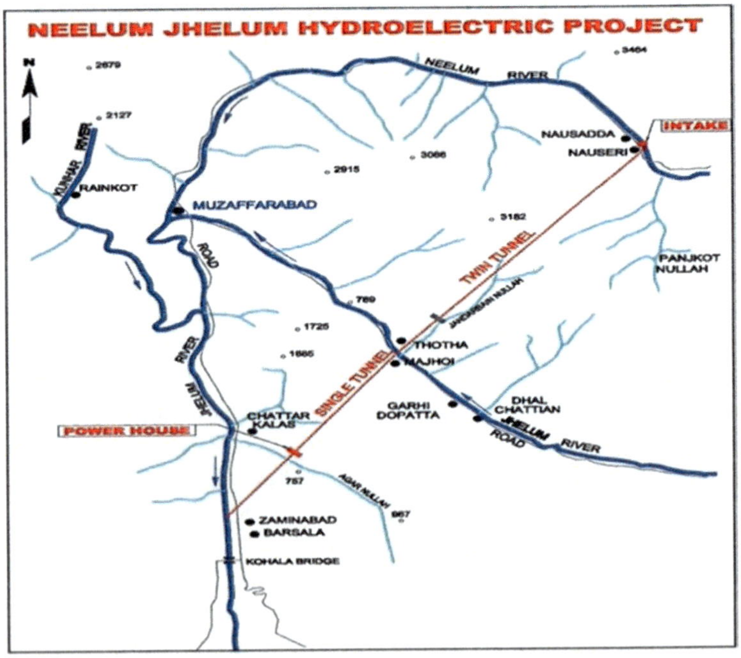

Fig. 5 Project layout of Neelum–Jhelum dam in Pakistan Resource: Khan (2012)

plants in many Asian cities. The AIIB's ability to manage socio-environmental risks is limited, and civil society actively criticizes its activities (Simonov, 2018b).

The scheduled commissioning of HPP capacities shows a significant growth in 2018–2919 (Fig. 6, Simonov, 2022). According to the NGO International Rivers (IR) (2021), from 2006 to 2019, Chinese companies signed contracts for the construction of at least 266 HPPs, with a total capacity of 130 GW, while 76 GW fell on contracts under which a Chinese company will own the station for the first decades of its existence. The share of HPP projects of Chinese companies carried out abroad is growing. In 2019, Chinese companies participated in 58% of all commissioned hydroelectric power plants worldwide and 77% of all investments in hydropower (Rivers Without Boundaries, 2018).

The Chinese megaproject "Global Energy Interconnection" (GEI) (Downie, 2020) is an example of systematic cooperation

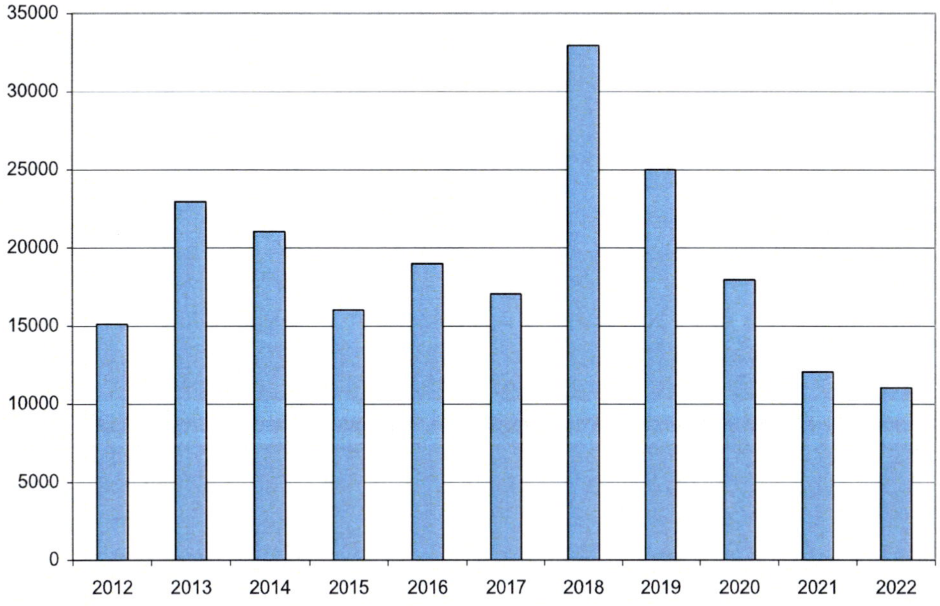

Fig. 6 Schedule of commissioning of HPP capacities (MW) for 2012–2022 Resource: Simonov (2022)

in technology and manufacturing exports in the frame of the B&R (OECD 2018) Initiative. The global network assumes uniform technical standards and the most decisive influence of the principal owners of the network on its formation (IEC, 2020). Large hydroelectric power plants are considered in GEI as environmentally friendly "energy bases." (Downie, 2020). The global network of transmission lines will create an opportunity to carry the places of electricity generation from the places of consumption for thousands of kilometers. So, consumers will be far from the generation's environmental and social costs. In particular, one of the "pilot projects" is a power line carrying energy to Brazilian cities from the super-large Belo Monte hydroelectric power plant (13GW) in the Amazon (NS Energy, 2022). The construction of Belo Monte led to substantial negative consequences for nature and aboriginal tribes. It was also connected with a corruption scandal that led to the impeachment of the country's president. The distrust of partners is holding back the development of GEI, the high price of power lines, and the availability of local alternatives to energy supply (Shkradyuk & Simonov, 2018).

Among other water management initiatives of Chinese companies, the most resonant is projects to ensure navigation. In 2017, the inevitable collapse of the most ambitious of them, the Nicaraguan Canal, announced as an alternative to the Panama Canal by a Chinese private company previously known for large-scale projects for developing the Crimean Peninsula, became evident (Stargardter, 2014). The fiasco of this megaproject was facilitated by unaccounted economic costs, geopolitical risks, the difficulty of taking land from the population, and massive damage to the tropical ecosystems of Central America. Of the several significant canals and river navigation projects announced with the participation of the Chinese capital, clear progress is being observed only in the design of deepening the channel and destroying rock rapids on the Mekong River for transporting goods from Yunnan Province to Thailand and Vietnam (Ball, 2017).

In recent years, China has been aggressively seizing the global market of water supply, sanitation, and water treatment systems in the cities of Eurasia: from Belgrade to Ulaanbaatar (Batchimeg, 2018). This direction is designated "environmental business" in the 13th five-year Plan (OECD 2018). Several guidance documents encourage exporting relevant technologies (Belt and Road portal, 2017).

3.2 ESG Risks and Risk Management Tools

Many water management megaprojects in China face serious ESG risks. For example, the pulp enterprise "Polarnaya," with 100% capital from the PRC, is now completing the construction of a reservoir on the first tributary of the Amur—the Amazar River (Simonov, 2017a). This situation is the first dam in the history of Russia (12–18 m high) created exclusively in the interests of foreign investors with gross violations and without environmental expertise. A reservoir unsuitable for drinking water supply has been created, and rare fish populations have been damaged. Due to the environmental community's active opposition and investors' distrust, the project has slowed down and may not be completed.

The most significant ESG risks of B&R are associated with HPP projects. Several examples of cancellation or freezing of Chinese hydroelectric power projects exist for various reasons. For example, with the help of the International Finance Corporation, Myanmar conducted a Strategic Environmental Assessment of the prospects for hydropower and stated that shortly it does not intend to return to joint projects with China of large hydroelectric power plants on the Irrawaddy River (Dasgupta & Pasricha, 2017). This decision was made under unprecedented pressure from the PRC. Today, there are at least 40 frozen Chinese hydroelectric projects in Myanmar, the construction of which was agreed upon during the military dictatorship. At the same time, in 2018, some projects on tributaries built by supposedly more environmentally responsible European contractors received approval (Kean, 2018).

In 2017, the new government of Nepal terminated the contract with the Chinese Gezhouba for the creation of the Budhi–Gandaki hydroelectric power plant (1200 MW) in the upper reaches of the Ganges (Rivers Without Boundaries, 2017), as the contract was concluded without competition on terms not favorable for the country. In addition, a proper environmental assessment was not carried out. Nepal soon received a "symmetrical response" from the Three Gorges Corporation, which stated that it would not build the Western Sethi hydroelectric power plant (700 MW) unless the Nepalese government changed the tariff policy to a more favorable one for the company.

The example of Myanmar and Nepal shows that developing countries are beginning to understand the vulnerability of Chinese construction companies to competition and are learning to bargain with them. At the same time, the creation of many hydroelectric power plants associated with maximum environmental risks continues, despite international protests. Thus, in Indonesia, on the island of Sumatra, Sinohydro is preparing to construct a hydroelectric power plant that will destroy the habitats of the only known population of the Tapanuli orangutan (Jong, 2020).

It is crucial to force the B&R institutions to form a responsible environmental policy. Thus, in 2017, when discussing the AIIB's Energy Strategy, non-governmental organizations prepared a detailed analysis of the opportunities and risks of financing hydroelectric power plants (Simonov, 2017b). As a result of the discussion of the report, the AIIB Strategy includes a recommendation to conduct a strategic assessment of basin-wide hydropower plans before making decisions on individual projects.

Concerns about transboundary basins are reflected in "China's Guide to investing abroad," issued in August 2017 by the State Committee for Development and Reforms. This document states that investments affecting transboundary water resources are the riskiest and require additional licensing procedures. This decision will have far-reaching consequences for all B&R initiatives related to transboundary water bodies.

A helpful tool is the SEA—strategic environmental assessment of development plans. From 2012 to 2015, such an assessment of all options for hydropower development in the Amur River basin was jointly carried out by the Russian companies Eurosibenergo and En+, the World Wildlife Fund, and the Wildlife Conservation Center. Russian companies and their Chinese partners were allowed to more soberly assess the prospects and risks of creating hydroelectric power plants in the basin. As a result, they abandoned the projects previously selected for development (Simonov et al., 2015).

In some cases, using the provisions of international law leads to eliminating unpromising projects. Thus, in Mongolia, the loan of the Russian Exim Bank to China in the amount of 1 billion US dollars previously intended for the construction of the Egiin Gol hydroelectric power station in the Baikal basin (the contractor is the Gezhouba company) was redistributed to other projects. These projects include creating higher educational institutions, an extensive strategic power line in the Gobi, and repairing the CHP in Erdenet (Rivers Without Boundaries, 2018). The status of Baikal facilitated this decision as a World Heritage Site and the active demands of the Heritage Committee, the Government of the Russian Federation, and international and national NGOs to look for alternative ways of energy supply.

Developing and promoting natural alternatives to China's megaprojects is critical for leveling ESG risks. Achievements in this direction are still modest, but the experience is gradually gained, for example, during the Mongolian hydroelectric power plants campaign in the Baikal basin. Moreover, the problem is not so much in the unwillingness of the Chinese to invest in the green economy abroad as in the complete unwillingness to build this economy in close cooperation with national authorities, public organizations, and businesses in the host countries.

4 Conclusions

The analysis of ESG risk management features using empirical data on China's water management megaprojects showed the following. The potential of risk management is not fully realized. We can be supposed that the main reasons for this situation are:

- administrative problems, namely the difficulty of controlling the use of funds to solve environmental and social problems;
- the duration of the procedures for approving decisions by the relevant ministries in the host countries (since environmental aspects may be contained in projects in the transport sector, housing, communal services, and others).

Insufficient involvement of regional and municipal authorities and weak motivation of civil servants in carrying out measures to minimize ESG risks remain a severe problem. In many host countries, there are no precise mechanisms for providing state (municipal) guarantees for investments in environmental protection and the social sphere. Environmental initiatives involve a significant amount of initial investment, which requires additional state support in tax incentives, subsidies, grants, soft loans, and state guarantees.

The following typical features of modern megaprojects can be distinguished:

- the “imperative” nature of such projects (the decision on their implementation is submitted as final and non-negotiable);
- state participation in one form or another (both in the form of direct financing and the form of participation of state-owned companies or tax benefits for private companies);
- "simplified" and selective application of legislation, extensive use of administrative resources, and mandatory propaganda support with the help of state media and customized PR campaigns.

These features are challenging for the public to oppose megaprojects, even if they are incredibly destructive to the environment and the social sphere. However, the main obstacle is the chronic passivity of the majority of the population in most countries.

The public must apply particular techniques and mechanisms to counteract ESG risks of megaprojects:

- "localize" large projects, bringing to residents accessible and understandable information about the expected projects’ negative consequences;
- conduct an active information campaign and counter-propaganda;
- counteract attempts by government agencies and businesses to “marginalize” public protest by actively attracting reputable experts to their side, developing and offering alternatives to a malicious project;
- use international legal mechanisms and bring the problem to the interstate level in cases where it is possible.

References

Alam, I. (2015). CPEC may get extra billion dollars, *The Nation*. June 22, 2015, https://www.nation.com.pk/22-Jun-2015/cpec-may-get-extra-billion-dollars

Algin, A. P. (1991). *Edges of economic risk* (p. 197). Knowledge.

Amel-Zadeh, A., & Serafeim, G. (2017). Why and how investors use ESG information: Evidence from a global survey. *Harvard Business School, Working Paper*, *74*(3), 87–103.

Anil Kumar Lal (2020), The CPEC challenge and the India-China standoff: An opportunity for war or peace? *The Times of India*. https://timesofindia.indiatimes.com

Bagaynicov, M. (2017). Megaproject as an institutional basis of establishing geo economic region. *Baikal Research Journal. 8*(2). https://doi.org/10.17150/2411-6262.2017.8(2)21

Ball, A. (2017). China to clear path on Mekong River for large ships. *Voanews*, January 23, 2017. https://learningenglish.voanews.com/a/chinese-project-will-attempt-to-clear-a-path-for-large-ships-on-the-mekong-river/3688494.html

Barnett, M., & Salomon, R. (2006). Beyond dichotomy: The curvilinear relationship between social responsibility and financial performance. *Strategic Management Journal, 27*(11), 1101–1122.

Batchimeg, B. (2018). Construction of new wastewater treatment plant of Ulaanbaatar city begins. *Oananews.org*. http://www.oananews.org/

Belt and Road portal. (2017). Guidelines for the promoting of the Green Belt and Road. https://eng.yidaiyilu.gov.cn/zchj/qwfb/12479.htm

Bennett-Jones, O. (2021). Who benefits more? *Dawn.Com.*, August 3, 2021. https://www.dawn.com/news/1638510

Busch, T., Bauer, R., & Orlitzky, M. (2016). Sustainable development and financial markets: Old paths and new avenues. *Business and Society, 55*(3), 303–329.

CCICED. (2018). Special policy study on Green Belt and Road and 2030 agenda for sustainable development. *China Council for International Cooperation on Environment and Development*, October 2018. https://www.iisd.org

Chen, J. (2021). What are environmental, social, and governance (ESG) criteria? *Investopedia*. https://www.investopedia.com/terms/e/environmental-social-and-governance-esg-criteria.asp

Cheng B., Ioannou I., Serafeim, G. (2014). Corporate social responsibility and access to finance. *Strategic Management Journal, 35*, 1–23.

China Daily Information Co (CDIC). (2015). Backgrounder: China-Pakistan economic corridor. *China Daily*. https://www.chinadaily.com.cn

CPEC Authority. (2017). CPEC’s Long Term Plan (LTP) 2017–2030. https://cpec.gov.pk

CPEC Authority. (2022). China-Pakistan economic corridor. cpecinfo.com

Dasgupta, S., Pasricha, A. (2017). Pakistan, Nepal, Myanmar back away from Chinese projects. *Voice of America*, No. 4. https://www.voanews.com/a/three-countries-withdraw-from-chinese-projects/4148094.html

Davis, K. (2013). China investment policy: An update. In *OECD working papers on international investment, 2013/01*. OECD Publishing. https://doi.org/10.1787/5k469l1hmvbt-en

Dawn Media Group. (2014). Karachi to Lahore motorway project approved. *Dawn.com*, July 4, 2014. https://www.dawn.com/news/1116948

Dhaliwal, D., Zhen Li Tsang, A., & Yang, Y. (2011). Voluntary nonfinancial disclosure and the cost of equity capital: The initiation of corporate social responsibility reporting. *The Accounting Review, 86*(1), 59–100.

Downie, E. (2020). Lessons from southeast Asia for China global energy interconnection initiatives. *Energypolicy.columbia.edu.*, April 2020, 53.

Eccles, R. G., Ioannou, I., & Serafeim, G. (2013). The impact of corporate sustainability on organizational processes and performance. *Management Science, 60*(11), 2835–2857.

Eccles, R., & Kastrapeli, G. (2017). Investing enlightenment: How principle and pragmatism can create sustainable value. http://www.statestreet.com/content/dam/statestreet/documents/Articles/The_Investing_Enlightenment.pdf

Economist.com. (2019). Building BRICs of growth (2019). https://www.economist.com/finance-and-economics/2019/06/05/building-brics-of-growth

Efimova, O. (2018). Integrating sustainability issues into investment decision evaluation. *Journal of Reviews on Global Economics, 7*, 668–681.

Eltobgy, M., & Brown, K. (2020). Embracing the new age of materiality: Harnessing the pace of change in ESG. *World Economic Forum*.

Falk, R., Wang, K., Morris, T., & Chan, E. (2020). Environmental impact and outperformance: A data-driven approach to integrating carbon footprinting into the investment process. *Arlington, VA: S&P Global Market Intelligence*. https://www.spglobal.com/marketintelligence/en/documents/environmental-impact-and-outperformance.pdf

Flyvbjerg, B. (2014). What you should know about megaprojects and why: An overview. *Project Management Journal*. https://doi.org/10.1002/pmj

Flyvbjerg, B., Skamris, M. K., & Buhl, S. L. (2004). What causes cost overrun in transport infrastructure projects. *Transport Reviews, 24*(1) (January), 3–18.

Freiberg, D., Rogers, J., & Serafeim G. (2019). Pathways to materiality: How sustainability issues become financially material to corporations and their investors. *Harvard Business School Working Paper*, 20–056, 1–34.

Friede, G., Busch, T., & Bassen, A. (2015). ESG and financial performance: Aggregated evidence from more than 2000 empirical studies. *Journal of Sustainable Finance and Investment, 5*(4), 210–233.

Goss, A., & Roberts, G. (2011). The impact of corporate social responsibility on the cost of bank loans. *Journal of Banking & Finance, Elsevier, 35*(7), 1794–1810.

Hayat, U. (2015). Environment, social and governance issues in investing. A Guide for Investment Professionals. *CFA Institute*. https://www.cfainstitute.org. http://www3.weforum.org/docs/WEF_Embracing_the_New_Age_of_Materiality_2020.pdf

Hubbard, D. (2009). *The failure of risk management: Why it's broken and how to fix it* (p. 46). John Wiley & Sons.

IEC. (2020). White paper on GEI. *International Electronical Commission*. http://www.iec.ch/

International Rivers (IR). (2021). NGO database (2021). https://www.internationalrivers.org

Investopedia. (2023). What is environmental, social, and governance (ESG) investing? March 22, 2023. https://www.investopedia.com/

Jackson, R. B., Saunois, M., Bousquet, P., Canadell, J. G., Poulter, B., Stavert, A. R., Bergamaschi, P., Niwa, Y., Segers, A., & Tsuruta, A. (2020). Increasing anthropogenic methane emissions arise equally from agricultural and fossil fuel sources. *Environmental Research Letters, 15*(7), 071002.

Jamal, S. (2021). Pakistan's first hydropower project under CPEC begins water storage. *GulfNews*. https://gulfnews.com/world/asia/pakistan/pakistans-first-hydropower-project-under-cpec-begins-water-storage-1.83860070

Jones,V. C. (1985). *Manhattan: The army and the atomic bomb*. Center of Military History United States Army.

Jong, H.N. (2020). Dam that threatens orangutan habitat faces three-year delay. *News Mongabay*, July 15, 2020. https://news.mongabay.com

Kean, T. (2018). Hydropower is back. *Frontier, 9*, https://frontiermyanmar.net/en/hydropower-is-back (Retrieved January 22, 2023).

Khan, A. K. (2012). Engineering of headrace tunnel for Neelum Jhelum hydropower project. *Pakistan Engineering Congress*. https://pecongress.org.pk/

Khan, M., Serafeim, G., & Yoon, A. (2015). Corporate sustainability: First evidence on materiality. *The Accounting Review, 91*(6), 1697–1724.

Kiani, K. (2017). Transport policy: need of the day. *Dawn.Com*, 13 February, 2017. https://www.dawn.com/news/1314522

Kiani, K. (2019). CPEC: Govt seeks more joint ventures. *Dawn*.Com. *The Business and Finance Weekly,* January 14, 2019. https://www.dawn.com/news/1457315

MOFCOM and MEP. (2013). Guide to environmental protection in the Implementation of foreign investment projects and Cooperation. *Ministry of Commerce People's Republic of China*. http://english.mofcom.gov.cn/

NS Energy. (2022). Belo Monte hydroelectric power. https://www.nsenergybusiness.com/projects/belo-monte-hydropower-project-brazil/

OECD. (2018). The Belt and Road Initiative in the global trade, investment and finance landscape. In *OECD business and finance Outlook 2018*. OECD Publishing. https://doi.org/10.1787/bus_fin_out-2018-6-en

Olivers Bredariol, T., & Schulz, R. (2022). Methane emissions from oil and gas operations. *IEA*. https://www.iea.org/reports/methane-emissions-from-oil-and-gas-operations

Rafi, S. (2016). A new 'East India Company' in the making? *Atimes*, October 23, 2016. www.atimes.com

Rivers Without Boundaries. (2017). Budhi Gandaki Hydro in Nepal: BRI project from the previous century. http://www.transrivers.org/2017/2116/

Rivers Without Boundaries. (2018). Statistics confirm: Chinese industry is the engine of the global hydropower boom. http://www.transrivers.org/2018/2211/

Saeed, S. (2015). China's Xi Jinping launches investment deal in Pakistan. *The Wall Street Journal*. https://www.wsj.com

Schramade, W. (2016). Integrating ESG into valuation models and investment decisions: The value driver adjustment approach. *Journal of Sustainable Finance & Investment, 6*(2), 95–111.

Shkradyuk, I., & Simonov, E. (2018). Notes on the global energy connection: The forks of the world energy and the Chinese network initiative (GEI). *Technical Report*. https://doi.org/10.13140/RG.2.2.21452.10885

Simonov, E. (2017a). *Report on the central processing plant "Polyarnaya"*. Technical Report, December, 2017. https://doi.org/10.13140/RG.2.2.17264.35846

Simonov, E. (2017b). Hydropower and related issues in the draft energy sector strategy of the Asian infrastructure investment bank (AIIB). *Technical Report*. https://doi.org/10.13140/RG.2.2.22887.34723

Simonov, E. (2018a). China is at home and on the Silk Road. *Newsletter of SIC ICWC*, 50.

Simonov, E. (2018b). The AIIB's road of disillusionment: From the ESF to the transport sector strategy. Notes on early relationships between the Asian infrastructure investment bank and civil society. *ResearchGate*. https://doi.org/10.13140/RG.2.2.25349.24809

Simonov, E. (2022). *Hydroelectric power station and climate: From COP26 to COP27*. Presentation, Conference: Towards COP27: Climate finance and water infrastructure, ResearchGate.

Simonov, E., Nikitina, O., & Egidarev, E. (2015). Basin-wide environmental assessment of hydropower development scenarios. Comprehensive environmental and socio-economic assessment of hydropower development in the Amur river basin. *Technical Report*. https://doi.org/10.13140/RG.2.2.17042.86729

Stargardter, G. (2014). Nicaragua announces start of China-led canal to rival Panama. *Reuters*, December 22, 2014. https://www.reuters.com/article/nicaragua-canal-idUSL1N0U60XN20141222

State Council. (2008). Administrative regulation on contracting foreign projects. http://www.lawinfochina.com/

The World Bank. (2018). Belt and Road initiative, March 29, 2018. https://www.worldbank.org/

United Nations. (1987). Report of the world commission on environment and development: Our common future.

United Nations. (2022). Transforming our world: The 2030 agenda for sustainable development. https://sdgs.un.org/

Van Burren, M., Plantinga, A., & Scholtens, B. (2016). ESG integration and the investment management process: Fundamental investing reinvented. *Journal of Business Ethics, 138*(3), 525–533.

Vivian, S., Hughes, A., Westwood, D., Pollard, S., & Pearce, H. (2003). Managing corporate environmental risk. A review of guidance and selected industry practice. *R&D Technical Report*, (E 2–056/TR). https://assets.publishing.service.gov.uk/government/uploads/system/uploads/attachment_data/file/290335/se2-056-tr-e-e.pdf

Wild, F. (2019). What was the Appolo program? *NASA Educational Technology Services*, July, 2019. https://www.nasa.gov/audience/forstudents/5-8/features/nasa-knows/what-was-apollo-program-58.html

Yang, S.-L. (2016). Corporate social responsibility and an enterprise's operational efficiency: Considering competitor's strategies and the perspectives of long-term engagement. *Quality & Quantity, 50*(6), 2553–2569.

Zhen, S. (2015). Chinese firm takes control of Gwadar Port free-trade zone in Pakistan. *South China Morning Post*, November, 2015. https://www.scmp.com

Sustainability and Post-COVID-19 Era Changing Housing Design and Models in the Cities of Istanbul

Hülya Coskun

Abstract

The main objective of this research is aimed at a trans-disciplinary investigation into the intersection of architectural design and urban planning in the post-pandemic era, with a focus on the future contribution specifically design process housing models and typologies in world cities and Istanbul. Paradoxically, while academic communities focus on climate issues since the 1990s, the COVID-19 pandemic has emerged as a significant factor in new city and housing planning, overshadowing the importance of climate-change. Nevertheless, climate-change had already led to fortitudes such as the 2016 Paris Agreement and UN declarations for more livable environments and housing, and the pandemic made this process relatively more visible. The recent pandemic has had a profound global and local impact, transforming cities, societies, and even communities in socio-spatial ways. COVID-19 has led to unpredictable changes in the habits and lifestyles of individuals, including remote working, and adaptions in housing models and typologies. The socio-spatial influences of pandemic have prompted significant changes and transformations in the cities made them into anti-social spaces also affected neighborhoods, public spaces, urban networks, roads, metro systems, and even information technologies. Major cities have experienced an exodus of people seeking to escape to nearby small towns and villages in pursuit of more spacious and sunlit abodes, resulting in a demographic shrink of the urban population from city centers to peripheries. The implementation of social distancing and urban lockdowns in response to the pandemic has significantly impacted the preferences and demands of individuals seeking housing. There has been a notable rise in the desire for living arrangements that provide greater access to green areas and outdoor areas, such as houses with garden or expansive terraces and balconies. This study examines the newly designed housing models developed in the post-pandemic era, with a particular emphasis on Istanbul as a paradigmatic global city. It was observed that the COVID-19 pandemic had a profound influence on the cities and housing markets in the world also in Istanbul and throughout Türkiye, leading to a reconfiguration of housing models and typologies, as well as heightened importance on sustainability and environmental issues has been witnessed in the post-pandemic era.

Keywords

Cities · COVID-19 · Green design · Housing models · Istanbul · Sustainable design

1 Introduction

The central aim of this research is to examine the intersection of architectural design and urban planning with a trans-disciplinary approach in the post-pandemic era, with a particular focus on housing models and typologies in world cities, notably Istanbul. The study serves two primary objectives: first to investigate sustainability and the second to examine the recent COVID-19 pandemic and its impact on housing models and typologies in urban areas. The primary goal of this research is to reconsider emerging new lifestyles, working norms and patterns, as well as the sustainability, and housing models that have gained traction, particularly over the past few years, due to COVID-19 pandemic. Additionally, this study concentrates on the critical issue of sustainable urban planning and the recent

H. Coskun (✉)
Faculty of Architecture, MSGSU, Mimar Sinan Fine Arts University, Istanbul, Türkiye
e-mail: her_222@yahoo.com

B.-J. He et al. (eds.), *Resilient Horizons: Building Sustainable Environments for Climate Adaptation and Health*, Advances in Science, Technology & Innovation, https://doi.org/10.1007/978-3-031-46109-5_6

significant impact of the COVID-19 pandemic on urban development. Initially, the profound consequences of a typical epidemic disease on society, living and working systems, and housing issues were unexpected.

Recently, we have witnessed COVID-19, initially considered only a humanitarian and some specific health issues has led to unforeseeable changes in both our daily lives and cities. However, the main issue emphasized in this study is that despite the world's attention to sustainability and resilience concerns for the last 30–40 years, COVID-19, which emerged abruptly in the last 2–3 years, has diverted attention from these health priorities. In addition to its relatively visible impacts on human health, the coronavirus had an unprecedented influence on cities, houses, as well as the urban planning, and architectural design with no analogous phenomenon in recent years. Especially when considering the past few decades, the intersection of urban planning, architectural design, and public health has focused less on infectious diseases and more on hazards and disasters affecting vulnerable populations. This priority is ambitious and broad, and achieving it requires support from multiple sectors, and the World Health Organization makes a plea to encourage other professions and disciplines to become more involved in the Healthy Cities agenda, recognizing their crucial contributions to health and well-being (WHO, 2008, p. 6) which is focuses specifically on cities achieving healthier outcomes, as architects are among the key professions to participate (Rice, 2020, p. 249). Thus, with the emergence of COVID-19, new disciplinary changes in urban and housing transformations design and planning have occurred. Coronavirus has not only altered people's lifestyle and working norms and habits with new hygiene and isolation regulations and restrictions concerns due to illness but also changed and transformed cities to reflect housing models and typologies.

This research aims to address a noticeable topic in the literature, specifically, the examination of sustainability, resiliency, and the post-COVID-19 era in Istanbul and other cities worldwide, thereby contributing to the existing knowledge base. The current pandemic has had a profound impact on the design world, and future research and practice must prioritize the development of healthy environments that are integral to architecture and urbanism. It is noteworthy, however, that health considerations have largely been absent from the education and practice of architecture and urban design and planning professions (Forsyth, 2020).

Despite these concerns, the uncontrolled growth of cities' demographic structures and populations, as well as their unplanned development and urban sprawl, persist as major challenges. This can be attributed to various factors, such as inevitable internal or transnational migrations worldwide, which tend to concentrate in specific regions due to economic or other reasons, resulting in people relocating between regions and even countries. These cities are often referred as "megacities" or "global cities" due to their large population exceeded over 10 million, which had led creation of some specific urban terms to identify them properly. There are approximately 291 cities with a population of over 10 million in the world today like London, Beijing, Tokyo, Istanbul, etc. (Calderon & Stern, 2015, p. 79). The recent pandemic has had significant global and local impacts on cities, with radical changes observed around the world. The pandemic started in Wuhan, China, and then spread to major cities in Europe such as London, Paris, Milan, Istanbul, and eventually even American cities. COVID-19 affected almost the whole world on a global basis and started to spread with increasing multiply cases very short period.

The initial measures implemented to combat the pandemic, such as social distancing and shutdowns, had significant socio-economic impacts on cities and their neighborhood and communities. These impacts were further compounded by subsequent measures, such as changes to transportation systems at both the metropolitan and neighborhood levels, alterations to work and living norms, and modifications to housing policies. These changes sparked new debates about the ways in which people live and work in cities, including discussions about transportation infrastructure, urban networks, transformational technologies, and the impact of information technologies on cities. The changes on social forms and adaptations to newly transformed urban norms and patterns have come to the forefront of city agendas, recognized as the "new normal" to describe the newly confronted circumstances. Initially, it was thought that this was a temporary occurrence, and it was not predicted that it would become a permanent part of city life. However, people in cities had to alter their lifestyles by transforming their living and working systems during the social distance and long shutdown later in the post-pandemic era.

In the post-pandemic era, housing design and planning in worldwide have developed specifically and varies by cities differently. While some American cities experienced a rapid population loss toward nearby towns during the COVID-19 period, urban-based planning and measures became more prominent in some European cities, such as France and Italy. The COVID-19 pandemic has unexpectedly set a new milestone in increasing awareness of sustainability. The disease has accelerated the implementation of sustainable practices and design in cities, leading to a reorganization of urban lifestyles and housing models to accommodate newly urged habits. Indeed, the most ambitious plan as a 15 min city of Paris was recently put into effect by the Mayor of Paris, Ann Hidalgo, and new city models and zoning regulations related to COVID-19 and Sustainability. The city was transformed from a car-centered suburban model developed by *cité-jardins* (garden-cites) around the *périphérié* (periphery) in the early twentieth century to a city-centered immediate vicinity of facilities and houses with the pandemic restrictions. The French concept of *hyper-proximité* (hyper-proximity) which is considered the world's most ambitious community plan, challenges this old orthodox old *zone plan* (zoning) model of the twentieth century that divided the city into separate areas; for residential, retail, entertainment, manufacturing, and office districts (O'Sullivan, 2020, p. 4). This frantic urban design proposal overlaps with the concept of COVID-19 and sustainability, and it was only possible to be implemented in the city of Paris.

Overall, pandemic has acted as a catalyst for many cities and made to rethink their urban planning and design approaches, and to prioritize sustainability, health, and resilience in their plans for the future. Also, various measures were taken like, Italy, Spain, Netherland, India, etc. along with COVID-19. In some Italian cities allowed developing measures by socio-economic and environmental factors such as sources of air pollution, urban ventilation, and population density difficult to measure can be proxi in social interaction (Coccia, 2020, pp. 4–13), (Bontempi et al., 2020). While constant depopulation has been observed in some American cities, European cities like Spain and the Netherlands, as well as India, experienced a significant decrease in working and social mobility due to government restrictions (Aloi et al., 2020, p. 6). The post-COVID-19 era has brought significant changes that have led people to migrate from big cities, impacting living, work, and social life, and led to specifying and recreating new office and housing models. When the effects of pandemic began to diminish, there was a radical turn in the housing market with the demand for new housing models that differ from those of last century. The pandemic emphasized the necessity of radical-change in densely populated and urbanized cities adapting to new living conditions to ensure the health and safety of residents such as New York, Singapore and Hong Kong which are prominent their cultural richness to urban environments (Badger, 2020). The COVID-19 pandemic brought about new problems, such as the need for social distancing and urban lockdowns. These measures transformed cities from social spaces to anti-socialized ones, as people were confined to their homes and urban networks, including roads, metro systems, and information technologies.

Similar to some American cities, Türkiye and Istanbul also experienced a significant population decline, and as newly implemented restrictions and regulations with remote working system became widespread, many chose to move to nearby villages and small towns. In addition, those living in high-rise buildings and apartments, fleeing the chaotic city centers that have been popular for many years, have started to prefer low-rise buildings and houses with gardens in more spacious and green-areas outside the city close to nature. With this situation change it has witnessed radical alterations, and developments that have not been seen for years in many places around the world. Indeed, the newly set regulations and conditions, which is regarded as temporary, is expected to be permanent and continue in the pandemic and post-pandemic period in Istanbul, Türkiye also in the world cities.

While COVID-19 was initially seen as a transient disease affecting only people's health, its long-term impacts on cities and residences cannot be insignificant. As Bill Gates has warned that, we may not have even seen the worst case scenario yet (Gates, 2021, p. 287). The world encountered new, urban planning dynamics with the problematic with the pandemic, focused on creating habitable cities and environments that prioritize human well-being, sustainability, and resilience. The problems faced by developing cities are still awaiting solutions, as is the case with Istanbul, which has transformed in sprawling mega-city fraught with various issues. In order to addition this issues, world cities need to adopt new planning dynamics and establish new norms and habits that prioritize healthy living conditions, especially in light of the ongoing COVID-19 pandemic.

Over the past 30–40 years, the world has been focused on issues of sustainability and resilience, the emergence of the COVID-19 pandemic in the last 2 years has abruptly shifted priorities and reminded us of the urgent need to address these issues. As a result of the pandemic, new regulations and restrictions specified as new living conditions have emerged that have required compulsory changes in our daily lives, such as social isolation measures, distance living, and remote work. These changes have led to a radical shift in the demand for housing models and typologies.

The disease not only affected people's lives but also led to a shift away from big cities to the nearest small towns and villages. This has caused a demographic shrinkage in cities and a shift in the urban population from city centers to the periphery or outskirts in search of more open and sunnier areas, among other factors. Furthermore, other effects were observed, such as an increase in housing prices, rents, and mortgage rates, even after people returned to the cities.

2 Problem Statements

This study focuses on a multidisciplinary research approach to identify and discuss the main problematic issues in the context of design and planning. Methodologically, the discussion is based on a conceptual framework that is developed through analysis and discourse.

The increasing interest in the subjects of sustainability and resiliency has led to an examination of how the recent COVID-19 pandemic has triggered new concepts of sustainability. Specifically, this includes an examination of new norms and patterns of lifestyles in cities, and how they have influenced the emergence of new city and housing models.

In addition, it is aimed to explain how these two issues discussed in this research, although they are methodologically separated and independent from each other. Indeed, these two different issues have some features analogous, and parallel to each other, specifically in terms of the effects they create. The research presents the following themes: Redefining the problematic: studies in the context of the sustainability and resiliency proposed to solve the city's current problems and intricate effects of the COVID-19 to the housing problem, housing models, and typologies.

In this context, the following subheadings have been identified as the main issues to be explored: With the Climate change, the main problematics, and threats for Istanbul city's urban eco-system.

1. Examinig the latest studies on sustainability, resiliency, and coronavirus impacts on cities, submitting an updated research.
2. To obtain outputs specifically on the main effects and consequences of COVID-19 in the last 2–3 years in the cities and city life, livingstyles, working models, houses, and housing models.
3. Examination of these two different issues and whether they have compatibility with each other or not.
4. The evaluation and comparison of the outputs and effects of sustainability and COVID-19 on the city, and on the changing housing models and people's demand recently.

3 Methodology

The research methodology is based on two main pillars. Firstly, it focused on sustainability, and secondly, on the recent COVID-19 era, with a particular emphasis on modern housing typologies in world cities and Istanbul. Therefore, two objectives are pursued. First, the study examined worldwide issues and investigated how they are being addressed in world cities, in terms of both problems and solutions. The study also examined the connection, and compatibility with parallel applications concerning sustainability and COVID-19 pandemic (Fig. 1).

To present new and updated research on sustainability, resilience, and coronavirus issues, a literature review is conducted. Firstly, recent studies on sustainability and resilience were presented to provide an updated view of these issues. The second part of the study focused on COVID-19

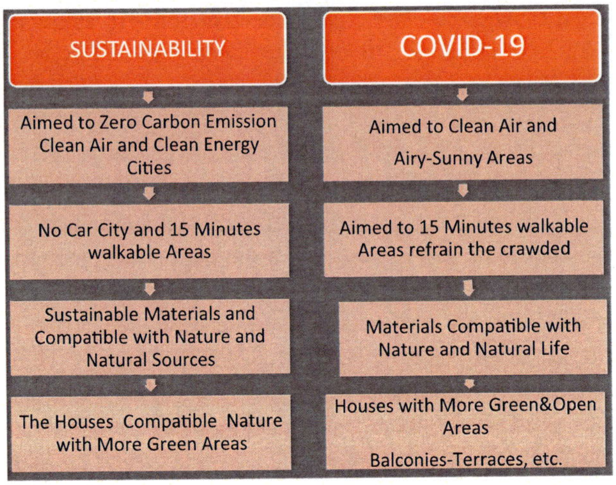

Fig. 1 The compatibility of sustainable housing design and COVID-19. Figure by Author

pandemic. The study examined the effects of the pandemic on people's lifestyles and how these changes have impacted the city and housing. Additionally, the study investigated new housing models and typologies that have emerged in response to these changes in the last few years.

In conclusion, within the scope of the findings obtained after the research, are presented new knowledge according to the people's changing housing demands, housing models, and typologies that emerged both in the context of sustainability and with COVID-19 and revealed their similarity and their compatibility.

4 Hypothecal Solutions

The research specified two main problematics: first focused on sustainability and the recent COVID-19 pandemic era, with a particular emphasis on modern housing typologies in world cities and Istanbul. Therefore, two objectives are pursued. First, the study examined worldwide issues and investigated how they are being addressed in world cities, in terms of both problems and solutions. The study also explored the connection and compatibility with parallel operations between sustainability and COVID-19.

The primary hypothesis of this study is to examine the changes and transformations in housing, housing models, and typologies in the post-pandemic era in world cities and Istanbul, focusing primarily on sustainability and then on the COVID-19 period. The study scrutinized urban planning and architectural design and proposed that housing issues be established as a main research axis, from the "urban-scale" to the "architectural-scale", in a trans-disciplinary dimension. The study used analyses to explore these issues in depth.

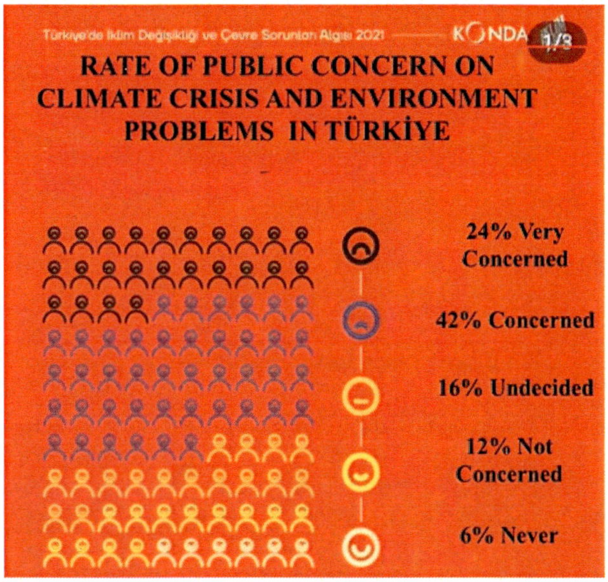

Fig. 2 According to recent research by Konda, only 24% of Turkish people are very concerned about climate change, and Environment—42% people concerned, 16% are undecided, and 12% are not concerned. Graphic, by Konda, Iklimhaber

The research examines the changes and transformations in cities in the post-pandemic period and how they have impacted evolved housing models and typologies. The study explored the effects of the disease on urban and housing planning in world cities and Istanbul. Notably, although the same disease is fought worldwide in the post-pandemic era, its reflection and impacts on cities developed differently according to the cities' dissimilar dynamics. Many studies have been carried out on sustainability and resiliency since the 1980s, with a particular focus on climate issues via climate-change conferences and declarations by Habitat, which have pioneered important steps in this regard.

Despite many climate-change conferences realized in world cities since the 1980s, their effects have been unsatisfactory, and the steps taken have been insufficient, except for certain cities and countries. Unfortunately, even though one of these conferences was held in Istanbul in 1996, it did not create any significant impact on the city's problems (Fig. 2). According to recent research published, it was declared that 66% of Turkish people do not have adequate awareness of climate-change issues (Sözcü Newspaper, 2022). However, the post-research rate appears to be high it is more likely to be much less. Even if it is accepted that there is awareness in the theoretical base, the rate of reflection in practice does not seem to be quite high as a percentage.

According to the World Intellectual Property Report and the 2022 CO Sustainability Report, the most significant factor in innovation has been determined to be the pandemic, the struggle against climate-change, and the digital revolution (Borusan Holding Reports, 2022). In this context, when examining the effects of both sustainability and the COVID-19, the recent pandemic has unexpectedly triggered vast radical effects on climate issues, increasing interest in sustainability. Furthermore, these two critical subjects, which were previously evaluated independently of each other, have similar effects. The focus of this study is on the similarities between sustainability and COVID-19 issues, particularly those related to people's lives and housing models and typologies.

In particular, the increase in CO_2 emissions in megacities such as Istanbul, where the city center is highly concentrated and overpopulated, causes air pollution. This issue is relevant to both sustainability and the impact of pandemic on cities, as there is a growing demand for sunnier and airier green-areas that can help combat air pollution. Recent research has shown a correlation between CO_2 and COVID-19, as ambient environments that are conducive to the survival of the virus may enhance transmission dynamics as a virus carrier (Zoran et al., 2020, p. 2) (Fig. 3). Evidence from countries such as China, Italy, the US, Brazil, Iran, Norway, and Turkey, which feature different climatic conditions, indicates that the results reported regarding the influence of environmental factors on COVID-19 are not uniform across different cities and regions.

The research emphasizes the significant impact of pandemic on people's lives, housing models, and typologies, and highlights the similarities between the issues of sustainability and COVID-19. These studies demonstrate that the effects of COVID-19 are more profound than previously thought and that the pandemic can be viewed as being as crucial as climate change. Although scientists have been emphasizing the importance of climate change and raising awareness about it, it was the pandemic that brought this fact to our attention once again in an outrageous way.

The COVID-19 pandemic, with its rapid spread and devastating effects on human life, was a grim reality observed daily on television worldwide. In response to this situation, people have been urged to take precautionary measures such as adhering to general lockdowns and disclosures, to prevent the disease from spreading further. To refrain the transmission of COVID-19, numerous local governments have implemented partial or complete lockdowns and mobility restrictions (Carteni et al., 2020, p. 3).

Several studies have observed that there is no significant relationship between urban density and COVID-19 infection rates in the Netherlands (Boterman, 2020). Despite initial concerns about the impact of pandemic on densely populated cities, measures such as general lockdowns were implemented immediately. However, subsequent studies have failed to establish a direct correlation between urban

Fig. 3 Istanbul. View from the Haliç Bridge to the Historical Peninsula. Air Pollution. *Photo* www.airpollutioninistanbul

density and pandemic. Nevertheless, governments have had to adopt stringent measures to mitigate the life-threatening consequences of COVID-19 in cities.

4.1 The Post-COVID-19 Era, in the World Cities; Paris, NewYork, London, Beijing, Istanbul

- **Cities Transformation with Zoning Applications as Metropolitan Model**

Since the beginning of the twentieth century, cities have been developed according to certain outdated urbanistic forms and norms. The COVID-19 pandemic has forced people to confront a new reality that conflicts with the old, rigid metropolitan model. During the pandemic, a new city model was required that divided the cities into functions and ensured that people could reach everywhere in the city in a short time. Indeed, one of the old city models as the *zone plan* (zoning planning) as the hundred-year-old idea was developed by the first generation of French-originated architect-urbanists, which involved separating the city's functions and differentiating between residential, commercial, entertainment, and industrial areas. This model, to some extent, overlapped with the city concept wanted to develop during the pandemic as well as the sustainability that has been developed until now.

As a response to the COVID-19 pandemic, measures such as general lockdowns and promoting distance living have led to the implementation of new urban planning ideas such as the 15 min city, which has been urgently put into effect in Paris under the leadership of Hidalgo et al., (2017, p. 2) (Fig. 4). The 15 min city concept known as *hyper proximité plan)* (hyper-proximity plan) aims to mix as many uses as possible within the same space, challenging the planning orthodoxy of separating residential areas from retail, entertainment, manufacturing, and office districts (based on old zoning plans of French architect-urbanists). Although this geographical division of uses may have made sense at the dawn of the industrial era, when polluting factories posed health hazards, it no longer aligns with the current need for sustainable and livable cities. Living and working in proper conditions are significant factors for overall well-being, along with access to provisions, education, and leisure (O'Sullivan, 2020, p. 4). It is based on meeting the basic needs of urban life in modern cities and the new city model aims to meet the simple needs in the shortest way. So, the municipality of Paris has replaced this old metropolitan model with an accessible within 15 min city model by a short walk or a bike ride to grocery stores, work, parks, schools' cafes, and gyms, within a quarter which made public transportation (Zacharia, 2021, p. 138) (Fig. 4).

Like France, Italy has also implemented similar practices to promote sustainability and mitigate the effects of COVID-19, including the development of integrated

Fig. 4 Istanbul, the view from Bosphorus. Photo, depositphotos. Down, the 15-min city. *Photo* ıntertraffic.com

assessment approaches to facilitate better response strategies. Also, some of the Italian cities have allowed developing measures by considering various socio-economic and environmental factors such as sources of air pollution, urban ventilation, and population density (Coccia, 2020, pp. 4–13).

Similar accessibility to other circumstances emerged in cities with the COVID-19, leading to the need for re-planning urban areas within a new urban network. These included creating arranged, walkable urban environments, providing more public spaces, parks, and children's playgrounds, as well as developing architectural and urban solutions for the future that prioritizes safety (Sennett, 2016). However, the pandemic has also made cities less social and more anti-social, with social distancing measures limiting interactions between people. In other cities where governmental restrictions were implemented, transportation has significantly decreased. Recent empirical evidence has shown a decrease in social mobility of up to 76% in Spain (Aloi et al., 2020, p. 6) and India (Saha et al., 2020). Moreover, significant reductions in other sectors and people's activities, such as retail and entertainment, supermarket and pharmacy visits, and park visits have been observed in the Netherlands (de Haas et al., 2020, p. 2).

The COVID-19 pandemic had a significant impact on the residential real estate market in the United States during the spring months. Due to health concerns, stay-at-home orders, and economic uncertainty, many metropolitan areas experienced a noticeable drop in their population, which in turn led to fewer buyers looking for homes and fewer sellers willing to list their properties or allow strangers to enter their homes during a pandemic (Gascon & Haas, 2020). This approach was also observed in daily house renting system like Airbnb further reduced their activities. Thus, the pandemic has driven people to escape from big cities and move to the less-density suburbia or vicinity of small towns (Wall Street Journal, 2021). However, the effects of this trend are not limited to geographical changes, and they have broader social and sociological implications. For instance, the sudden emptying of big cities and subsequent return of people has led to sociological complications in the post-COVID-19 period, in the housing market.

Thus, the exodus of people from big cities during the COVID-19 pandemic led to difficulties in finding new homes upon their return, resulting in various consequences in the housing market. The effects of these radical changes were reflected in various areas, including the unexpected increase in rents and rental rates in American cities such as Austin, New York, and New Jersey, etc. (Bloomberg, 2022). The housing prices pending home sales escalating mortgage rates as in New York City (New York Times, 2022). COVID-19 even effected banking system, and mortgage rates, in recently Banks in the UK have withdrawn a record number of mortgages overnight as in London, contributed to extend the other areas of this problem (Financial Times, 2022). All of this had unexpected implications for how effective and far-reaching the effects of COVID-19 could be, including cities, houses, and housing problems.

Dense urbanization and the growing human population have emerged as a threat not only after the pandemic but also due to the peak of carbon emissions in big cities, which poses a growing threat to human health. Consequently, the recent pandemic has made more visible deterioration of living and working conditions and forced people to make radical changes. The COVID-19 pandemic has accelerated the adoption of green and healthy lifestyles, which were once seen as potential solutions to climate issues in the 1990s. This trend is evident in various countries including the United States, England, Europe, and Türkiye. As people are forced to stay at home for extended periods, there is a growing realization that more comfortable, car-free, and healthy living options are needed. In addition, the pandemic necessitated changes in work patterns, with remote working becoming the norm for many. As a result, people began to question their living spaces and how these could be adapted to support remote work and more sustainable lifestyles. As a result, the effects of COVID-19 have led to a redefinition of the demands for the arrangement of workspaces and houses in cities. The pandemic has not only changed housing demands and models but also working conditions, resulting in the rise of remote and hybrid working models. Consequently, big cities and metropolises, which were previously popular for their high-rise buildings, chic cafes, and shopping areas, now seem to be losing their appeal.

Indeed, the COVID-19 pandemic has accelerated transformation to the sustainability and resiliency living norms and housing models in cities, as seen as the threat to human health and the disruption of urban life emphasized the importance of creating more livable and sustainable environments. While efforts to promote sustainability have been ongoing for many years, the pandemic has forced rapid changes and increased attention to issues such as green spaces, walkability, and alternative modes of transportation. As a result, there has been a growing body of research aimed at understanding the impacts of the pandemic on cities and identifying key lessons that can inform post-COVID urban planning efforts. Related to COVID-19 and cities, there still exist many studies to understand the impacts of the pandemic on cities and to highlight major lessons that can be learned for post-COVID urban planning (Sharifi et al., 2020).

4.2 Istanbul City, Post-Pandemic Era and the Compatibility with Sustainable Planning

As a metropolis, Istanbul's influence extends beyond its vast administrative boundaries, which are located on the hinterland of western end of Turkey in the Marmara Region. Istanbul is situated on both sides of the Bosphorus Strait, which connects the Black-Sea in the north and the Sea of Marmara in the south and separates the two continents of Asia and Europe (Fig. 4). In the twentieth century, the world cities such as Tokyo, Shanghai, Beijing, Sao Paulo, Mexico City, Mumbai, and Calcutta have developed into megacities with endless urban landscapes due to orderly planning and strategic regulation (Lehmann, 2011, p. 244). Like Istanbul, there are 33 megacities in the world characterized by populations above 10 million, high levels of knowledge-based services and trade, and incomes ranging from 2,000 to 20,000 dollars, including capital cities such as London, Beijing, and Tokyo, etc. (Calderon, Stern, 2015, p. 78). According to the Istanbul Mayor's website on sustainable transportation, the population of Istanbul is expected to reach approximately 18.8 million by 2040, up from the current population of around 16.6 million. Depending on the unit of analysis, Istanbul has evolved from a mono-centric metro-region to a polycentric megacity with a dominant core and multiple urban nuclei, like most metro regions (Perouse, 2008, p. 30).

Istanbul has undergone significant urban growth, expanding into new neighboring quartiers and regions, and developing new *banlieues* (suburbs) around the old Historical Peninsula through suburbanization (Coskun, 2021b). The city of Istanbul has developed and extended initially along the southern axis, from the European side to the Asian side, and along the coast of the Marmara Sea, finally reaching as far as Izmit Bay and city in the east and Tekirdağ to the west. As fast-growing mega-cities, Istanbul has faced severe traffic congestion since the central city was split by the strait first in the 1970s by the Bosphorus Bridge between European Side and Asian Sides transformed the city into a metropolitan (Tekeli, 2013, p. 358) According to the French planner Henri Prost Plan, the city was mainly divided into two sides: the northern part which has grown toward the Northern-forest and the green-areas, and the southern part which has grown as a linear residential area along the coast of the Marmara Sea. Indeed, Istanbul was promoted as having an anti-metropolitan density which was derived from inherently unsustainable countries that had industrialized quickly (Bozdoğan & Akcan, 2012, p. 162). Additionally, in this rapidly developing urban form, many people commute from the Asian side, which is mostly known as a residential area, to the European side, which is mostly considered as the business area (Coskun, 2017, p. 309) with the connecting of these two parts through the Bosphorus from Asian side to European side with the bridges, Bosphorus, the Fatih S. Bridge, and the Yavuz Selim Bridge) (Fig. 4).

Every day more than 380 000 vehicles cross over the Istanbul bridges and the number of motor vehicles has increased by 647%, while population has rised by 142% between 1980 and 2005 when the comparison, the number of persons in Istanbul per car is 4.2 against 4 in New York, 2.5 in Paris and 2.2 in Tokyo (Perouse, 2008, pp. 101–102).

- **Istanbul Municipality's Visionary Plan Until 2040, For Sustainability**

Similarly, to Paris Mayor Ann Hidalgo, the Istanbul Municipality has declared that aim to transform the city into a sustainable city like Paris in the future. The Istanbul Mayor has emphasized a visionary plan for the city and stated their goals within the scope of the "Climate Action Plan", which was announced last year with the title of "Green Solution". The plan aims to ensure that Istanbul's transportation system is environmentally friendly and to promote a sustainable, active, and healthy lifestyle. The plan also focuses on the "Low Carbon Transition" theme, which is based on the target of achieving carbon neutrality (surdurulebilirulasim.istanbul, 2022) (Fig. 3). The Istanbul Municipality has planned to reduce carbon emissions from transportation by 60% by 2040 and achieve carbon neutrality by 2050 through the adoption of electric vehicles, promotion of carbon-free transportation, development of walking, cycling, and micro-mobility infrastructure. Environmentalists have proposed the electrification of vehicles, bus fleets, metro buses, City Lines ships, electric taxis, electric water taxis, and even hydrogen-based buses, and have initiated the implementation of this vision. The Istanbul Municipality has an extensive sustainability plan, but they cannot slow down any plans for the post-pandemic COVID-19 era. The Istanbul Mayor has listed the 9 main objectives of the Istanbul Sustainable and Smart City Planning and Urban Design Center (SKHP) as follows:

- An accessible, affordable, integrated, and inclusive transportation system;
- An environmentally sustainable transport system;
- An economically sustainable and resilient transport system;
- A transportation system that increases safety in transportation and confidence in travel;
- A transportation system that reduces traffic congestion and automobile dependency;
- A transportation system that facilitates the transition to public transportation;
- A transportation system that promotes alternative modes such as cycling and walking;
- A transport system that supports compact and multi-center development.

- **Istanbul Municipality's Housing Planning Vision, For Sustainability and COVID-19**

According to the OECD territorial plans for Istanbul in 2008, an city wide attention was drawn to the issue of uncontrolled urban growth in residential areas, which was identified as one of the most significant problems facing Istanbul long years. The 1995 Istanbul Master Plan produced by IMP, *Istanbul Büyük Şehir Belediyesi*, (Istanbul Metropolitan Municipality) estimates to calculate that around 55% of city's settled areas were several types of informal settlements, with the potential to rise to 75% uncontrolled urban sprawl (Perouse, 2008) (Figs. 4, 5, 6, 7 and 8). Due to increasing awareness of recent sustainability problems, the spatial development of Istanbul has raised

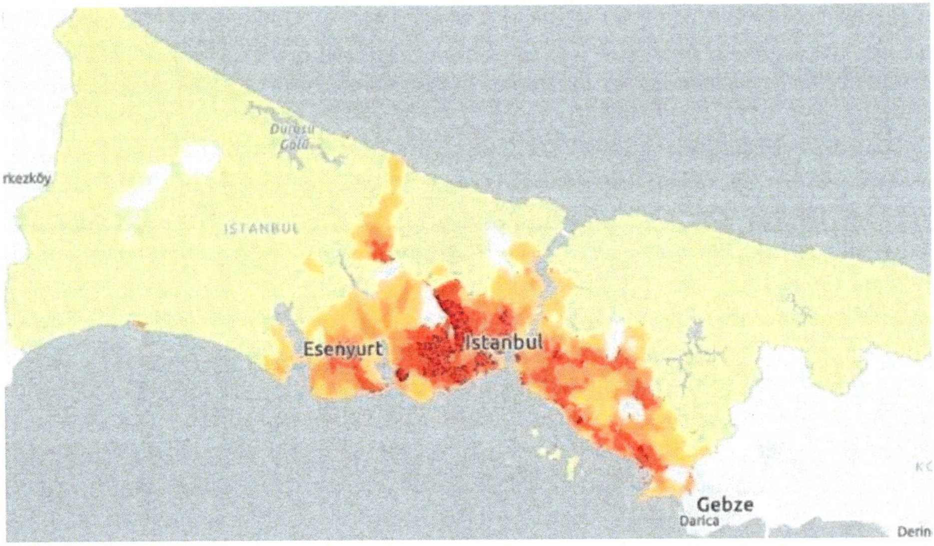

Fig. 5 Istanbul and districts, housing and urban densification during pandemic COVID-19. Map, Science Direct.com

Fig. 6 The 2020s, the post-pandemics, newly created a healthy-city concept, newly launched modern garden-cities with 90% green areas, European Side, Levent, Istanbul. *Photo* Dapyapı.com.tr website

serious concerns for the city's environment. Informal settlements have spread toward water reserves and forest areas in the northeastern periphery (Coskun, 2021a) where these new settlements have since been recognized as official sub-provinces, particularly over the past 20 years (Perouse, 2008, p. 35). The Habitat II UN Conference contributed to the acceptance and spread of Local Agenda 21 in Istanbul and its adoption as a model for local sustainability planning worldwide and this process was strengthened by strong international foundations (Arar, 2021). Some of the projects that were introduced as environmentally friendly have caused impacts and damage to Northern-forests and green-areas (Aykan, 2014) even water basins. With increasing awareness, the city's master plans have started to incorporate environmental considerations. However, the city's growing population and informal and uncontrolled settlements also raise important concerns for the future provision of basic services, including water supply. At the recent Climate-Summit held in November 2021 in Glasgow, it was predicted that Istanbul would experience severe water scarcity and drought (CNNTürk, 2021).

4.3 Istanbul City, Post-Pandemic Era the Housing Design and Planning

Due to COVID-19, depopulation was observed in American cities, as people relocated to areas outside the big cities, seeking more spacious, greener, and sunnier surroundings in nearby small towns. During the closure the population of major American cities like New York, and Washington, relatively shrunk (Vox, 2022). (Examination of the densification map of Istanbul during the COVID-19 lockdown revealed that the city's population concentrated in coastal areas along the southern axis of the Marmara Sea or in areas close to green spaces in the Northern forests (Fig. 5).

Fig. 7 Post-pandemic Era İstanbul, healthy-city concept and new housing models and typologies with garden (on the left) and Houses with Expansive-Terraces nearly 50 m^2 (On the Right). *Photos* www.Mesa.com., www.Sinpas.com

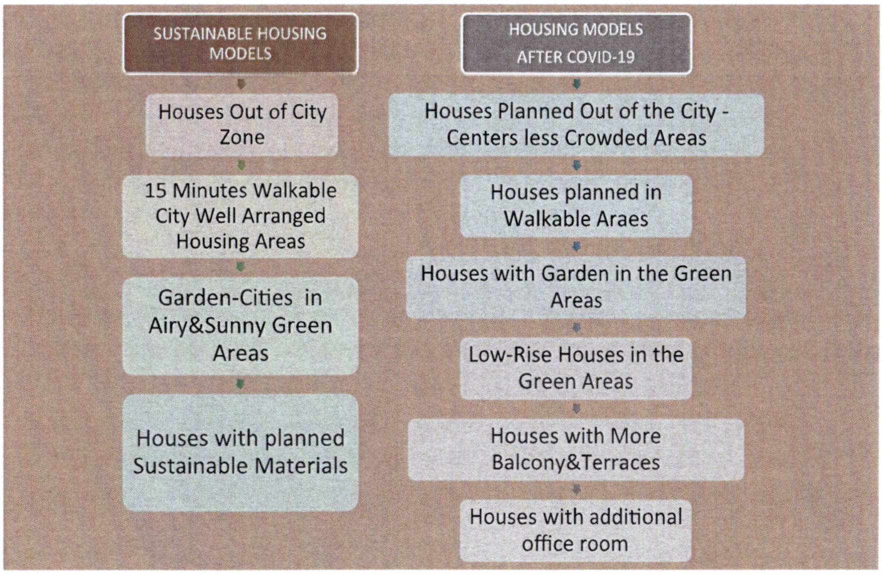

Fig. 8 Consequences and compatibility of sustainable housing design and COVID-19. Figure by Author

Although a new migration wave occurred from these big cities to nearby towns was also realized in Istanbul similarly, like Tekirdağ, Çanakkale, Kocaeli, etc., and the major change occurred in housing models. Especially, after COVID-19, people have preferred to leave the city centers and live in areas outside the city and in greener and sunnier areas (Figs. 6 and 7).

Considerable impacts of the COVID-19 as symbolized by vacant office buildings, empty shopping malls and abandoned flats in metropolitan areas, the new coronavirus disease 2019 has severely impacted real estate markets (Balemi et al., 2021). The Turkish housing market was highly affected during the pandemic period, showing a sharp drop graphically starting from 2019 when COVID-19 started as a disease, and until the end of 2021. In contrast, other countries such as the UK, France, Germany, Italy, and Spain showed a longer period of fluctuation between 2019 and 2021 during the pandemic. The COVID-19 pandemic led to a slowdown in housing construction due to measures affecting working conditions and people who worked, as well as falling or changing demands for housing purchases. The number of newly planned houses and completed houses decreased by almost half in the period from 2017 to 2021 (Dünya, 2022).

However, in the world of architecture, the COVID-19 pandemic in European cities such as Italy was seen as a possible opportunity to rethink on housing architecture. Contrary to American cities, many European cities have a numerous populations living in high-rise buildings and apartments; like Europe, nearly 88% population of Türkiye living in high-rise buildings and apartments (Kulisi, 2023). Most of these high-rise buildings, whose number is estimated to be 1.5 million, are in poor conditions. In American cities and Canada, 29.4% of the population have lived in apartment buildings (Government Reports of Canada, 2017) and spend most of their time at home, their apartments, thus during the pandemic to support resident well-being became critical, considerations such as window replacement, qualities of views, and access to nature (Peters & Halleran, 2020, p. 8). Similarly, in European cities in Italy, 28.8% of the population have lived in overcrowded apartments and faced challenges accessing basic services (D'Alessandro et al., 2020, p. 2).

In Italy, due to COVID-19, they required to rethink on housing architecture and models, such as housing with semi-private, collective, and healthy for physical distancing. Through the pandemic, due to the impossibility to reach public parks, green-areas houses are planned with private terraces, collective spaces, shared courtyards and to support daily physical activity while keeping social distances (Cappolongo et al., 2020, p. 8). Chicago, Toronto, Seattle, Paris, and other EU cities have recently carried out significant redevelopment projects at urban scale, increasing the number of green-roofs to improve the quality of life (Buffoli et al., 2018). Also the green roofs were significant as the benefits, both physically and visually, and vegetable gardens, during the pandemic as playgrounds or simple resting and reading places as example new constructions adopt similar typological solutions (Cappolongo et al., 2020, p. 8).

Likewise, Italy and Canada COVID-19 also have significantly impacted the real-estate market in Türkiye, particularly the housing models and typologies. After the COVID-19, *Türk Inşaatçılar Birliği* (Turkish Builders Association) has introduced a new construction plan that emphasizes sustainable and environment-friendly projects to meet people's changing demands and requirements (Hürriyethaber, 2021). The post-pandemic period has witnessed a shift in the housing demand from high-rise buildings to lower-rise buildings, and even houses with garden models built in green-areas outside the city, which have been implemented by private contractors (Figs. 6 and 7). However, it is worth noting that the state tends to focus on building multi-story mass-production houses, particularly for low-income groups.

In response to the changing demand in the housing market in Istanbul, private contractors have shifted their focus toward new housing ideas and policies. The newly emerging innovative concept of "healthy-houses" with more green-areas and garden reflects the shifting demands and requirements of people after the COVID-19 pandemic. The post-pandemic period has seen a surge in people's preference for residences in natural surroundings outside of the cities, with features such as gardens, expansive terraces, and large balconies and low-rise buildings. As a result, new and innovative housing concepts have emerged, including green- houses, healthy-houses, terrace-houses, lake-houses, forest-houses, and more, catering to people's new desires for green spaces, healthy living, and proximity to nature out of city.

The construction market, architects, and constructors have responded to the newly changing demands of people by developing new and innovative projects. To create a sustainable relationship between urban dwellers and urban areas in line with United Nations' requirements, United Nations (2021) they have launched more low-rise, horizontal buildings and garden-cities in green-areas outside the city instead of high-rise buildings. Future housing projects now have specified new criteria with a focus on gardens, terraces, and balconies. In Istanbul, there have been alterations in housing models, typologies, dimensions, and the number of rooms to accommodate the new remote working model, which has made it almost mandatory to have an additional study room in new housing models (Habertürk, 2022).

Overall, the COVID-19 pandemic has accelerated the trend toward more sustainable and healthy housing models in Istanbul. However, there is currently no adequate information available regarding the supply and demand of these houses in Istanbul. Nevertheless, more recently, it is observed an increasing interest in houses compatible with sustainability, and COVID-19 in the advertisements of construction companies (Figs. 6 and 7). The most popular housing models have developed as; houses in green-areas, forest, large parks, houses with gardens, houses on the ground floor and with gardens, and houses with expanded terraces or large balconies (Figs. 6, 7, 8 and Table 1). While the private contractors have been primarily focused on the rapidly changing demand for housing market due to COVID-19, the government seems more focused on sustainability. Recent reports from the Turkish government have emphasized the importance of sustainability and environmental issues in creating livable, healthy, durable, and safe housing, to meet the United Nations Climate-Change declaration.

The COVID-19 pandemic has had a profound impact on the housing market in Istanbul and throughout Türkiye, leading to a unimagined vast reconfiguration of housing models and typologies, emphasizing on sustainability and environmental issues. Today, Türkiye's vision for housing is to provide adequate, livable, healthy, durable, safe, affordable, and sustainable housing for everyone, along with access to basic services as specified in the *Onbirinci Kalkınma Planı* Eleventh Development Plan (Turkish Government Housing Reports, 2018). Given the principles of sustainability, the housing needs of people, particularly those from low-income groups, will be addressed through the development of healthy and alternative solutions while also considering supply and demand balance. The government's response to the decline in the housing market due to COVID-19 includes plans to enhance the existing housing stock to better cater to the needs and requirements of various income groups in the market (Turkish Government Housing Report, 2018).

5 Conclusion

The research focused on the interrelated issues of sustainability and COVID-19, which have gained significant interest and attention in cities worldwide. The study examined the impact of these issues on architectural design and urban planning, particularly with regards to housing design, and planning. The research found notable changes in the housing market and the development of housing models and typologies, particularly in the post-pandemic period. The evolution of housing models in Istanbul and the world was examined in the context of sustainability starting from the 1990s to the present, including the COVID-19 era. Istanbul is a mega-city facing various problems, and its location makes it significant in the region. The recent socio-spatial impacts of pandemic have compelled cities to undergo changes, transforming them into globally developed.

Although the concept of sustainability was first introduced in Istanbul during the Climate-Change Conferences in 1996, it was realized the unexpected emergence of COVID-19 in 2019 that it gained prominence on the global urban agenda. Housing models had already begun to shift toward sustainable model since the 1990s, with the holistic approach of sustainability use of renewable energy, natural resources, and sustainable materials. However, the pandemic has led to further changes in housing design and planning at a new level. Hence, in the world of architecture, the pandemic in European cities such as Italy is considered as a possible opportunity to rethink housing design and architecture. The impact of COVID-19 on world cities, including Istanbul, has prompted a reevaluation of urban and housing planning models and practices.

Since the twentieth century, in the 1990s, while the housing models have developed toward the idea of sustainability, the post-2020 COVID-19 pandemic has accelerated this change even more and unexpectedly. The measures taken in cities, such as social isolation and public closure, confined people to their homes and transformed cities from social places to anti-social spaces. This also led to a transformation of urban networks, roads, metro systems, and even information technologies with remote working systems. The COVID-19 pandemic led to radical differences in living and working patterns, styles, and system such as remote working and privatized housing models and typologies due to social distancing. The post-pandemic world brought significant changes to the way we live and work, and the spatial attributes of our homes and workplaces needed to adapt to meet these new requirements which led to the evaluation of the housing models. This led to new living and working patterns emerged, that reflect these radical changes in cities and housing typology and models.

As urbanization continues to accelerate globally, the health of cities and the model and the quality of housing have become major concerns for people. The prioritization of health issues in world cities, particularly in Istanbul, has received considerable support from multiple sectors, with architecture and construction playing pivotal roles. The World Health Organization Health Committee had already called for other professions and disciplines to join the Healthy Cities agenda recognizing the critical contribution they could make toward achieving health and well-being (WHO, 2008, p. 6). In the aftermath of the pandemic, health issues have been thrust into the spotlight to achieve better outcomes in cities. This led to innovative developments in architecture specifically just after the COVID-19, that prioritize health, sustainability, and resilience in urban environments. Thus, even before the pandemic, the concept of "healthy-city" had already begun to be generated and the pandemic made this purpose much more visible. The people have begun to question the healthiness of the cities and homes they live in, prompting new concepts such as "healthy-city" and "healthy-houses " to emerge as promising solutions to address these concerns.

The significant changes in the demands of the urban population and housing have led contractors to initiate the to search for design and construction of new "healthy-cities" and "healthy-housing" models in the post-pandemic period in accordance with the idea of World Health Organization Health Committee's declarations. (Figs. 6, 7 and Table 1) There has been a growing interest in producing new and healthy housing models with the changing housing demands in cities, especially in Istanbul, with both the state and private contractors taking notice. While private contractors have more focused on the changing housing models in response to COVID-19, the government has adopted a more sustainability-oriented approach.

Consequently, in the world cities and Istanbul in the post-COVID-19 pandemic period, the housing models have undergone a recent reconfiguration and changing supply and demand, as well as an increased focus on sustainability. This led process of evaluation of the housing models

Table 1 The evolution of the housing models and typologies from the 1990s to today

Housing models	The transformation of housing models
Before 1990s Housing Model and Typologies: (Modern Design) –Modern Design –High-Rise or Low-Rise Houses –Concrete or Steel Buildings –Green Social and Collective Areas, playgrounds, etc. –Small Windows, and Balconies –Traditional Roofs	
Sustainability Housing Model and Typologies: (Energy-Efficient) –Low-Rise Houses with Expanded Terrace and Balconies –Sustainable Materials, etc. –Sunny and Green Areas –Using Natural Resources –Renewable Energy –Green Social and Collective Areas—Green Roofs and Terraces	
Sustainability & COVID-19 Housing Model and Typologies: (Green Design) –Low-rise houses with Expanded Terrace and Balconies –Sustainable materials –Using natural resources –Sunny and green areas –Green social and collective areas, playgrounds, etc. –Green-roofs and terraces	
Post-COVID-19 Housing Model and Typologies: (Privatized Model) –Low-rise houses with Expanded terraces and balconies –Sunny and green areas –Green social and collective Areas, playgrounds, etc –Green roofs and terraces –Private gardens—Terraces –Private pPools	

Figures by Author

known in the modern era that we have witnessed since the mid-twentieth century and onward and housing models have gone through a recent process of evaluation, changed, and transformed as green, sustainable and healthy models (Table 1).

References

Aloi, A., Alonso, B., Benavente, J., Cordera, R., Echániz, E., González, F., Ladisa, C., Lezama- Romanelli, R., López-Parra, A., Mazzei, V., Perrucci, L., Prieto-Quintana, D., Rodríguez, A., & Sañudo, R. (2020). Effects of the Covid-19 lockdown on urban mobility: empirical evidence from the city of santander (Spain). *Sustainability (Switzerland), 12*(9). Retrieved January 19, 2023. [Google Scholar]. https://mdpi.altmetric.com/details/81737242

Arar, A. (2021). TC. Dış İşleri Başkanlığı, Daire Başkanı, Uluslararası Ekonomik Sorunlar Dergisi, Sayı VI, Yerel Gündem 21.

Aykan, S. (2014). *İstanbul'da Kentsel Sürdürülebilirlik*. Metis, First Publication.

Badger, E. (2020). Density is normally good for us: That will be true after Coronavirus, too. The Upshot—The New York Times (24 March 2020). Retrieved July 30, 2022. [Google Scholar]. https://www.nytimes.com/2020/03/24/upshot/coronavirus-urban-density-risks.html

Balemi, N., Füss, R., & Weigand, A. (2021). Covid-19's impact on real estate markets: review and outlook. Axel & Springer.

Bloomberg. (2022). Retrieved September 22, 2022. https://www.bloomberg.com/news/articles/2022-07-18/where-is-rent-most-expensive-austin-nyc-apartment-prices-surge?leadSource=uverify%20wall [Google Scholar]

Bontempi et al. (2020). E. Bontempi, S. Vergalli, F. Squazzoni Understanding COVID-19 Diffusion Requires an Interdisciplinary. *Multi-dimensional Approach Environmental Research, 188*, 109814. https://doi.org/10.1016/j.envres.2020.109814. Retrieved September 20, 2022.

BorusanHolding Reports. (2022). Sürdürlebilirlik Rehberi, Retrieved September 25, 2022, https://www.borusan.com/Assets/Media/PDF/BorusanSurdurulebilirlikRehberi.pdf

Boterman, W. R. (2020). Urban-rural polarisation in times of the corona outbreak? The early demographic and geographic patterns of the SARS-CoV-2 epidemic in the Netherlands. *Tijdschrift voor Economische en Sociale Geografie, 111*, 513–529. https://doi.org/10.1111/tesg.12437. Retrieved January 15, 2023 [Google Scholar]. https://scholar.google.com/scholar?hl=en&as_sdt=0,5&q=Boterman,+W.R.,+2020.+Urban-rural+polarisation+in+times+of+the+corona+outbreak%3F+The+early+demographic+and+geographic+patterns+of+the+SARS-CoV-2+epidemic+in+the+Netherlands.+Tijd.+voor+econ.+en+Soc.+Geog.+111,+513-529

Bozdoğan, S., & Akcan, E. (2012). *Turkey*. Redaktion Books, First Publication.

Buffoli, M., Rebecchi, A., Gola, M., Favotto, A., Procopio, G.P., Capolongo, S., Green. (2018). SOAP. A calculation model for improving outdoor air quality in urban contexts and evaluating the benefits to the population's health status. In: G. Mondini, E. Fattinnanzi, A. Oppio, M. Bottero, S. Stanghellini (Eds.) *Integrated evaluation for the management of contemporary cities*. Springer Green Energy and Technology, pp. 453–467. Retrieved September 15, 2022 [Google Scholar]. https://doi.org/10.1007/978-3-319-78271-3_36

Calderon, & Stern. (2015). La Nouvelle Economie Climatic, Une Meilleure Croissance Un Meilleure Climatique, pour le traduction Française, Les Petit Matins. ISBN 976-2-36383-178-1

Cappolongo, et al. (2020). COVID-19 and cities: From urban health strategies to the pandemic challenge. A Decalogue of Public Health opportunities. *ActaBiomedica* https://doi.org/10.23750/abm.v91i2.9515. Retrieved January 12, 2023 [Google Scholar]. https://www.ncbi.nlm.nih.gov/pmc/articles/PMC7569650/

Carteni, A., Di Francesco, L., & Martino, M. (2020). How mobility habits influenced the spread of the Covid-19 pandemic: Results from the Italian case study. *Science of the Total Environment, 741*, 140489. Retrieved January 23, 2023 [Google Scholar] https://www.sciencedirect.com/science/article/pii/S0048969720340110?casa_token=dwTSlnMgzJUAAAAA:yvd-SkQOZBP8Y9FJUqAxFj9jfVRRQkVM2eDvVCgX3jtKXfZlHmtWOvSxUG8jqezmjbNXAURh6jO0

Coccia, M. (2020). An index to quantify environmental risk of exposure to future epi-demics of the Covid-19 and similar viral agents: Theory and practice. *Environmental Research,* 110155. Retrieved January 5, 2023 [Google Scholar] https://www.sciencedirect.com/science/article/pii/S0013935120310525?casa_token=jo0oKYlvG7AAAAAA:E9OXnzIkryoJQFfIcsa0-zay81ChujqMxah-SOxXuRWU3NmIlFI7lJEoXZTSJIng3LJYcHZA26Sc

Coskun, H. (2017). In the beginning of the 20th century, analyzing methods of the housing problem and an Example: Henri Prost's Istanbul plannings, Ph.D. Thesis, 2017, MSGSU, Mimar Sinan Fine Arts University, The Institute of Science, Faculty of Architecture, Building Design

Coskun, H. (2021a). Re-planning of the İstanbul in the 21th century post-prost period and the future city and housing projects. In *4th international conference of contemporary affairs and urbanism ICCAUA-2021*, 20–21 May 2021.

Coskun, H. (2021b). Sustainable and resilient plannings, developed housing models for Istanbul. In *International conference, rrau, resilient and responsible architecture and urbanism*, 4th Edition, 03–05 December 2022. Ximanen University Malaysia, Southwest Jiatotong University

CNN, News, Turkish Edition. (2021). Retrieved October 23, 2021.

D'Alessandro, D., et al. (2020). COVID-19 and Living space challenge. Well-being and Public Health recommendations for a healthy, safe, and sustainable housing. *Acta Biomedica, 91*(9–S), 61–75. Published online 2020 Jul 20. https://doi.org/10.23750/abm.v91i9-S.10115. Retrieved January 3, 2023 [Google Scholar] https://www.ncbi.nlm.nih.gov/pmc/articles/PMC8023091/

de Haas, M., Faber, R., & Hamersma, M. (2020). How Covid-19 and the Dutch 'intelligent lockdown' change activities, work and travel behaviour: Evidence from longitudinal data in the Netherlands. *Transportation Research Interdisciplinary Perspectives, 6*, 100150. https://www.sciencedirect.com/science/article/pii/S2590198220300610, Retrieved January 19, 2022 [Google Scholar]

Dünya. (2022). Retrieved February 21, 2022. https://www.dunya.com/sektorler/emlak/markali-konutcular-uretim-atagina-gecti-haberi-649508

Kulisi, E. (2023). Retrieved January, 2023, https://emlakkulisi.com/turkiyede-nufusun-yuzde-81i-apartmanlarda-yasiyor/464997

Financial Times. (2022). Retrieved September 21, 2022. https://www.ft.com/content/d9a62b84-4a51-40d6-9002-a1310ce3c8dc

Forsyth, A. (2020). What role do planning and design play in a pandemic? News–Harvard's Graduate School of Design 2020. Retrieved 20 August, 2022. Reference Source, Retrieved 5 September, 2022 [Google Scholar]. https://www.gsd.harvard.edu/2020/03/what-role-do-planning-and-design-play-in-a-pandemic-ann-forsyth-reflects-on-covid-19s-impact-on-the-future-of-urban-life/

Gates, B. (2021). *Climat comment Eviter Un Desastre*. Flammarion.

Gascon, C.-J., & Haas, J. (2020). *The impact of Covid-19 on the residential real estate market*. Federal Reserve Bank, St. Louis.

Government Reports of Canada. (2017). Retrieved 20 October, 2022. https://lop.parl.ca/sites/PublicWebsite/default/en_CA/ResearchPublications/201916E

Habertürk. (2021). Retrieved October 10, 2021. https://www.haberturk.com/cevre-ve-sehircilik-bakanliginin-adi-cevre-şehircilik-ve-iklim-değişikliği-bakanlığı-oldu

Habertürk. (2020). Retrieved October 5, 2022. https://www.haberturk.com/dap-yapi-yonetim-kurulu-baskani-ziya-yilmaz-dan-korona-virus-aciklamasi-haberler-2649021-ekonomi

Hidalgo, et al. (2017). Cities have the boldest plans to deliver a sustainable future. *Finacial Times*. Retrieved April 20, 2021, [Goggle Scholar]. https://scholar.google.com/scholar?hl=en&as_sdt=0,5&q=Hidalgo,+et.al.,+(2017),+Cities+have+the+boldest+plans+to+deliver+a+sustainable+future,+Financial+Times.+January+22,+2017

Hürriyethaber. (2021). Retrieved February, 2021.https://www.hurriyet.com.tr/ekonomi/muteahhitler-surdurulebilir-sehirlere-odaklandi-41934020

Iklimhaber. (2021). Instagram photos and videos from Iklim Haber Gazetesi (@iklimhabergazetesi), (Instagram Retrieved October 25, 2021), Newly declared Laws about Newly opened green areas

Istanbul Planning Office. (2022). Reports for Housing Research, Istanbul Existing Situationa and Proposals. Istanbul Planlama Ajansı, 9.10.2022

Lehmann, S. (2011). *What is green urbanism? Holistic principles to transform cities for sustainable*

New York Times. (2022). Retrieved September 15, 2022. https://www.nytimes.com/2022/09/15/business/mortgage-rates.html

Perouse, J.-P. (2008). OECD territorial reviews. Istanbul, Turkey. www.oecd.org/publishing/corrigenda

Pérouse, J.-P. (2014). *İstanbul'da Sürdürülebilir Kalkınma: Sekteye Uğramış, Kısmi ve Fırsatçı Bir Uygulama*. Metis, First Publication.

Peters, T., & Halleran, A. (2020). How our homes impact our health: using a COVID-19 informed approach to examine urban apartment housing. *IJAR, 15*(1). Archnet-IJAR, ISSN: 2631-6862. [Google Scholar].

Rice, L. (2020). The nature and extent of healthy architecture: The current state of progress. *Archnet-IJAR: International Journal of Architectural Research, 13*(2), 244–259. https://doi.org/10.1108/ARCH-11-2018-0005. [Google Scholar]. Retrieved January 18, 2023, https://www.emerald.com/insight/content/doi/10.1108/ARCH-11-2018-0005/full/html

Saha, J., Barman, B., Chouhan, P. (2020). Lockdown for Covid-19 and its impact on community mobility in India: An analysis of the Covid-19 community mobility reports, 2020. *Children and Youth Services Review, 116*, 105160. Retrieved September 3, 2022, [Goggle Scholar]. https://www.sciencedirect.com/science/article/pii/S1201971220325029

Sharifi, K.-G. (2020). The COVID-19 pandemic: Impacts on cities and major lessons for urban planning, design, and management. Elsevier. Retrieved 10 September, 2022 [Google Scholar]. https://scholar.google.com/scholar?hl=en&as_sdt=0,5&q=Sharifi,+Kharavan-Gamesir,+(2020),+The+COVID-19+pandemic:+Impacts+on+cities+and+major+lessons+for+urban+planning,+design,+and+management+,+Elsevier

Sennett, R. (2016). *Buildings and dwellings: Ethics for the city*. Penguin Books-Allen Lane.

SözcüNewspaper. (2022). Retrieved August 2022. https://www.sozcu.com.tr/2022/dunya/carpici-anket-turkiye-iklim-degisikligiyle-ilgili-endiselerin-en-fazla-oldugu-ikinci-ulke7365748/

Surdurulebilirulasim.istanbul. (2022). Retrieved September 25, 2022, www.istanbulbelediyesi.com.tr

Tekeli. (2013). Istanbul'un Planlamasının ve Gelişmesinin Öyküsü, TTVY. Yay., Istanbul.

Turkish Goverment Housing Reports. (2018). 11th Turkish development and housing policy comission report. Retrieved October 09, 2021, https://www.sbb.gov.tr/wp-content/uploads/2022/07/On_Birinci_Kalkinma_Plani-2019-2023.pdf

O'Sullivan, F. (2020). *Paris mayor, its time for 15 minutes city*. TheBloomberg CityLab.

United Nations. (2021). Conference Sustainable Development, Rio de Janeiro, Brazil, 20–22 June 2012.

Vox. (2022). Retrieved April 14, 2022. https://www.facebook.com/Vox/posts/2131439300376994

Wall Street Journal (2021). Retrieved September 20, 2021. https://www.wsj.com/articles/pandemic-supercharged-changes-in-where-americans-live-11619536399

WHO. (2008). *Zagreb declaration for healthy cities*. WHO Regional Office for Europe.

World Economic Forum. (2020). 7 innovative projects making cities more sustainable. World Economic Forum 2020–09–22.

Zacharia, F. (2021). *Ten lessons for a post pandemic world*. Penguin Books, US-UK, ISBN 978-C-141-99562-5.

Zoran, M. A., Savastru, R. S., Savastru, D. M., & Tautan, M. N. (2020). Assessing the relationship between ground levels of ozone. Retrieved April 22, 2022, [Google Scholar], https://www.sciencedirect.com/science/article/pii/S0048969720335257. (O_3) and Nitrogen Dioxide (NO_2) with Coronavirus (COVID-19) in Milan, Italy. *Science of the Total Environment, 740*

Decarbonisations of Built Environment for Health Livable Cities and Climate Change Mitigation

Risk Assessment of Green Reuse of Abandoned Industrial Buildings: A Case of Steel Mill in Xi'an City

Qing Chai, Bao-Jie He and Simei Wu

Abstract

Green reuse of abandoned industrial buildings (GRAIBs) is one of the critical ways to promote sustainable urban renewal and reduce the urban heat island. In order to improve the effect of GRAIBs, this paper aims to consider the risk assessment of the whole process of GRAIBs. Firstly, a risk assessment index system for the GRAIBs was constructed on the basis of literature review, field research, and in-depth expert interviews. Then, the weight of each index is calculated by the analytic hierarchy process (AHP), and the risk assessment model of GRAIBs is established based on uncertainty measurement theory. Finally, taking a GRAIB project in Xi'an as an example, the rationality and feasibility of the proposed risk assessment index and risk assessment model for the GRAIBs are verified. This study helps stakeholders understand the risks of GRAIBs better, provides suggestions for projects to reduce risks, and also proposes ideas for promoting the successful implementation of GRAIBs.

Keywords

Abandoned industrial buildings · Green reuse · Urban heat island · Climate change · Risk assessment

1 Introduction

With the adjustment of the urban industrial structure and the rapid economic development, many abandoned industrial buildings (AIBs) that were originally located in the urban center are gradually idle (Martinat et al., 2018). At present, the reuse of valuable AIBs has become one of the important forms of disposal of AIBs (Yung & Chan, 2012). Reusing is the process of retaining the original structure of some buildings and performing functional replacement, which could bring economic, social, cultural, and environmental benefits (Vardopoulos, 2019; Samadzadehyazdi et al., 2018). Especially, green reuse of abandoned industrial buildings (GRAIBs) can not only improve the safety and comfort of buildings but also have great benefit for reducing building energy (Sanchez et al., 2020). The GRAIBs, rather than demolition, helps to retain the hidden energy of existing buildings, reduce carbon dioxide emissions, and mitigate the impact of urban heat island (UHI) (Chen et al., 2018). Therefore, the GRAIBs have become an important form of sustainable urban renewal and circular economic development. As the importance of energy conservation and emission reduction increases, GRAIBs receive more and more attention. However, the risk of GRAIBs project hinders the enthusiasm of stakeholders. To successfully promote the implementation of GRAIBs projects, it is necessary to focus on the risks of GRAIBs.

Many scholars have studied the risks of the RAIBs. For example, based on field research on 24 abandoned industrial buildings in Wu et al. (2014) used principal component analysis to identify six major risk factors for the RAIBs and proposed countermeasures. Li et al. (2019) introduced the rule-based reasoning (RBR) into the case-based reasoning

Q. Chai
School of Civil Engineering, Xi'an University of Architecture and Technology, Xi'an 710055, China
e-mail: cqjy@xauat.edu.cn

B.-J. He
Centre for Climate-Resilient and Low-Carbon Cities, School of Architecture and Urban Planning, Chongqing University, Chongqing 400045, China

Key Laboratory of New Technology for Construction of Cities in Mountain Area, Ministry of Education, Chongqing University, Chongqing 400045, China

S. Wu (✉)
School of Management, Xi'an University of Architecture and Technology, Xi'an 710055, Chinae-mail: simeiwu@xauat.edu.cn

B.-J. He et al. (eds.), *Resilient Horizons: Building Sustainable Environments for Climate Adaptation and Health*, Advances in Science, Technology & Innovation, https://doi.org/10.1007/978-3-031-46109-5_7

(CBR) to construct an emergency management model for the construction risk of RAIBs. Jiang and Zhang (2021) established the safety risk assessment model of the RAIBs during the construction stage based on the matter-element extension method. However, these studies mainly focus on the risks in the construction stage or the whole process of the RAIBs. There are few reports on the risk of GRAIBs. With the improvement of building performance and energy efficiency requirements in China, many scholars began to integrate green concepts into the RAIBs. Lee (2019) discussed the green design methods and specific strategies for transforming the post-industrial era sites into urban parks in Korea. Zhang et al. (2021a) proposed that green regeneration of industrial brownfield is an inevitable choice under the collision of urban industrial structure adjustment and ecological civilization construction, and further explored the impact mechanism of green regeneration of industrial brownfield under the constraint of ecological security. Sessa et al. (2022) put forward policy suggestions on green regeneration of industrial brownfield in Europe. Tian et al. (2022) established a satisfaction evaluation model for GRAIBs based on the three decision-making bodies of the government, the developer, and the resident merchants using the fuzzy comprehensive evaluation method. Wu et al. (2022) established the governance risk evaluation model of the GRAIBs project based on the SNA perspective. These researches on GRAIBs have promoted the development of the theory of GRAIBs and the increase of GRAIBs projects. However, there are few risk studies on the whole process of GRAIBs.

At present, the risk research on GRAIBs mainly comes from the studies on the existing buildings' green retrofits. For example, Remoy and Van der Voordt (2014) revealed the risk of converting offices into residential buildings. Sun et al., (2018a, 2018b) took a project as an example to investigate the green retrofit process of existing buildings and analyzed the advantages, disadvantages, opportunities, and challenges of existing buildings' green retrofits. Wang et al. (2018) used the failure mode and effects analysis (FMEA) method of fuzzy inference system to evaluate the risk of existing buildings green retrofits. Issa et al. (2021) systematically determined the risk factors of existing buildings' green retrofits from the aspects of economy, society, environment, management, sustainable operation, sustainable design, and transformation. However, different from the general existing buildings, the AIBs, as a type of historical building, have profound historical and cultural values. In particular, the historical characteristics of the buildings are required to be preserved in the reconstruction process, which increases the complexity of the GRAIBs and the risk level is higher. Therefore, the risk study of existing buildings' green retrofits may not be completely applicable to AIBs.

Based on the above analysis, this paper summarizes the risk factors affecting the whole process of GRAIBs through literature review, field research, and in-depth expert interviews with experts and constructs a risk assessment model based on the uncertainty measurement theory. This paper will contribute to realize the systematic risk evaluation of the whole process of GRAIBs and provide ideas for the risk management of GRAIBs.

2 Literature Review

2.1 Concept of GRAIBs

The research group conducted several field surveys on the RAIBs in major typical cities of China in the last 10 years and conducted in-depth interviews with relevant stakeholders such as the government, developers, designers, and constructors. The following problems in the RAIBs were found:

First, the AIBs were built at an early age, the interior and exterior of the buildings are dilapidated and the structural strength design does not meet the requirements of the current code (Chen et al., 2019; Tian et al., 2021). The use of existing building structures and materials without reasonable design will also affect the safety of users. Second, in order to preserve the historical original appearance of AIBs, the GRAIBs often do not pay attention to the thermal insulation inside the building, which eventually leads to high energy consumption of GRAIBs and uncomfortable indoor environment (Chen et al., 2022). Third, due to the original process requirements of AIBs, many buildings have leftover pollutants, and pollutants that have not been tested and treated by professional organizations will affect human health. In addition, most GRAIBs only focus on the transformation of existing buildings, roads, and pipelines, resulting in incomplete supporting facilities, affecting the user's happiness (Blagojević & Tufegdžić, 2016; Chan et al., 2015). Therefore, on the premise of ensuring the safety of the building, it is urgent to integrate the green concept and optimize the reuse method by improving the architectural appearance, environmental performance, and user comfort, so as to effectively improve the comprehensive benefits of the GRAIBs. According to the core requirements of green buildings, the GRAIBs is to make full use of the characteristics of original structure, open space, and historical and humanistic value of industrial buildings during the whole life cycle (Zhang et al., 2021b). While meeting the new functional requirements and reasonable economy, it saves resources to the maximum extent, protects the environment, and reduces pollution, and provides people with healthy, efficient, applicable, and cultural space. That is, GRAIBs is are harmony with the economy, society, and the environment.

2.2 Characteristics of GRAIBs

Compared with the reuse of general buildings, the GRAIBs have the following characteristics:

(1) Different goals. The reuse of general buildings mainly emphasizes the quality, cost, and construction period. In order to achieve the requirements of saving resources, the GRAIBs need to take into account the quality, cost, construction period, environment, resources, and humanity. The health and comfort of the building is considered the goal of green building.
(2) Different technical methods. The general building retrofit technology is mainly oriented by function, and the choice of technology is only based on the satisfaction of function. In the choice of green reuse technology, in addition to meeting the functional needs, it is also necessary to take into account the requirements of energy saving, and environmental protection. Also, active and passive energy-saving technologies should be joined in order to achieve a better user experience (Liao & Li, 2022).
(3) Different control points. The reuse of general buildings only attaches importance to economic benefits to meet functional requirements and ensure maximum profits as the control points, ignoring the huge consumption of energy and the resulting environmental problems. Green reuse emphasizes on the premise of ensuring and improving the ecological environment, achieving the coordination of multi-stakeholder interests, and taking functional applicability, energy saving, environmental protection, health, and comfort of building (Mushi et al., 2022).
(4) Different level needs. The reuse of general buildings mainly meets the basic use needs, and green reuse is to meet the functional needs on the basis of ensuring the low-carbon environmental protection, health, and comfort of building, which is a higher level of use needs (Sun et al., 2018a, 2018b).

3 Methodology

3.1 Establishment of Risk Assessment Model

The uncertainty measurement theory proposed by Guangyuan Wang is an uncertainty information theory (Wang, 1990). It can quantitatively analyze uncertainty factors effectively and reliably. Since the risk of GRAIBs projects has the characteristic of uncertainty, therefore, this study establishes a risk assessment model of GRAIBs based on uncertainty measurement theory.

(1) **Determine single-index unascertained measure matrix**

Let the evaluation object be $T = \{T_1, T_2, \ldots, T_n\}$, each evaluation object has an m evaluation index, and the evaluation index is $V = \{V_1, V_2, \ldots, V_m\}$. If t_{ij} denotes the measured value of the ith assessment object T_i with respect to the jth assessment index V_j, then T_i can be expressed as an m-dimensional vector $T_{ij} = \{t_{i1}, t_{i2}, \ldots, t_{im}\}$. Suppose that the assessment grade space $G = \{G_1, G_2, \ldots G_P\}$, where G_k is the rating value of T_{ij}. Supposing that the k-th grade is higher than the $k+1$th grade in the sustainability assessment process, i.e., $G_k > G_{k+1}$. If $G_1 > G_2 > G_3 \cdots > G_P$ or $G_1 < G_2 < G_3 \cdots < G_P$ is satisfied, then $G = \{G_1, G_2, \ldots G_P\}$ is an ordered segmentation class of assessment space.

In order to better evaluate things, according to the needs of the evaluation object, it can be divided into each part as a relatively independent whole. The unascertained single measurement index is taken as the smallest unit factor in the overall evaluation for independent analysis. Let $\mu_{ijk} = \mu\left(t_{ij} \in G_k\right)$ denote the degree to which the measurement value t_{ij} belongs to the kth evaluation grade G_k, and the following equation.

$$0 \le \mu\left(t_{ij} \in G_k\right) \le 1(i = 1, 2, \ldots n; j = 1, 2, \ldots m; k = 1, 2, \ldots p) \tag{1}$$

$$\mu\left(t_{ij} \in G_k\right) = 1(i = 1, 2, \ldots n; j = 1, 2, \ldots m; k = 1, 2, \ldots p) \tag{2}$$

$$\mu\left|t_{ij} \in \sum_{k=1}^{k} G_k\right| = \sum_{k=1}^{k} \mu(t_{ij} \in G_k)(i = 1, 2, \ldots n; j = 1, 2, \ldots m; k = 1, 2, \ldots p) \tag{3}$$

According to the definition of unascertained measure, the single-index measure function $\mu\left(t_{ij} \in G_k\right)$ is determined first, and then each assessment index measure value μ_{ijk} of a factor T_i is obtained. μ_{ijk} indicates the degree to which the measured value T_{ij} belongs to the eth assessment grade, and satisfies formulas (1)–(3). The assessment matrix composed of each single-index measurement value is as follows:

$$\left(\mu_{ijk}\right)_{m\times p} = \begin{pmatrix} \mu_{i11} & \mu_{i12} & \cdots & \mu_{i1p} \\ \mu_{i21} & \mu_{i22} & \cdots & \mu_{i2p} \\ \vdots & \vdots & \vdots & \vdots \\ \mu_{im1} & \mu_{im2} & \cdots & \mu_{imp} \end{pmatrix} \tag{4}$$

(2) **Confirm the weight of indexes**

Analytic hierarchy process (AHP) is based on the decision-making objectives of the system, many factors are hierarchized and organized, so as to form a comprehensive assessment index system of hierarchical levels (Saaty, 1980). The weight of each assessment index is determined by the relative importance of factors between the upper and lower levels. In this study, AHP approach is used to determine the weight of the risk assessment index of the GRAIBs. The specific steps of AHP to determine the index weights are as follows.

(1) Build the framework of risk hierarchy analysis
(2) Construct judgment matrix A
The nine-scale method is used to construct the judgment matrix, that is, the elements of the criterion layer and the index layer are compared and scored to form a judgment matrix. The form of the judgment matrix is shown in Formula (5). The definition of the nine-scale method is shown in Table 1.

$$A = \begin{bmatrix} a_{11} & a_{12} & \cdots & a_{1n} \\ a_{21} & a_{22} & \cdots & a_{2n} \\ \vdots & \vdots & \vdots & \vdots \\ a_{n1} & a_{n2} & \cdots & a_{nn} \end{bmatrix} \tag{5}$$

where a_{ij} takes the values of 1–9, 1/2–1/9.

Table 1 Definition of–nine-scale method

Intensity of importance	Definition
1	Equal importance
3	Somewhat more importance
5	Much more importance
7	Very much more importance
9	Absolutely more importance
2, 4, 6, 8	Intermediate values between the two adjacent judgements

(3) Calculate the weight of indexes
① The matrices were normalized by columns,

$$a'_{ij} = \frac{a_{ij}}{\sum_{i=1}^{n} a_{ij}} (i = 1, 2, \ldots, n; j = 1, 2, \ldots, m) \tag{6}$$

② Add the normalized matrix by rows, then

$$w' = \sum_{j=1}^{n} a'_{ij} (i = 1, 2, \ldots, n) \tag{7}$$

③ Normalize the vectors $w' = \left(w'_1, w'_2, \ldots w'_n\right)^T$,

$$w_i = \frac{w'_i}{\sum_{j=1}^{n} w'_j} (i = 1, 2, \ldots, n) \tag{8}$$

Then the obtained eigenvector is $w = (w_1, w_2, \ldots w_n)^T$.
④ Calculate the maximum eigenvalue of the judgment matrix λ_{max},

$$\lambda_{\max} = \frac{\sum_{i=1}^{n} (Aw)_i}{nw_i} (i = 1, 2, \ldots, n) \tag{9}$$

where is the ith element of the vector obtained by multiplying matrix A with weight vector w.
⑤ Consistency test of judgment matrix **A**
The judgment matrix is mostly constructed by the knowledge and experience of the experts, therefore, there will be some deviation. In order to make the judgment result consistent with the actual situation, it is necessary to check the consistency of the judgment result. The consistency test of the judgment matrix is as follows:

$$C.I. = \frac{\lambda_{max} - n}{n - 1} \tag{10}$$

$$C.R. = \frac{C.I.}{R.I.} \tag{11}$$

where n is the order of the judgment matrix, $R.I.$ is the average random consistency index and its values are shown in Table 2. When $C.R. < 0.1$, the consistency of matrix A can be accepted, otherwise the factors of the matrix need to be adjusted until the consistency test is satisfied.

Table 2 The average random consistency index

Order	2	3	4	5	6	7	8	9	10	11	12	13	14
R.I	0.00	0.52	0.86	1.10	1.26	1.34	1.40	1.43	1.49	1.51	1.54	1.56	1.58

(3) **Multi-index comprehensive unascertained measure matrix**

If μ_{ik} satisfies $0 \leq \mu_{ik} \leq 1$, $\mu_{ik} = \sum_{j=1}^{m} w_j \mu_{ijk} (k = 1,2,\ldots,p)$, where w_j for the weight of index, $(\mu_{ik})_{n\times p}$ is called the multiple-index unascertained measure matrix, then

$$(\mu_{ik})_{n\times p} = \begin{bmatrix} \mu_{11} & \mu_{12} & \cdots & \mu_{1p} \\ \mu_{21} & \mu_{22} & \cdots & \mu_{2p} \\ \vdots & \vdots & \vdots & \vdots \\ \mu_{n1} & \mu_{n2} & \cdots & \mu_{np} \end{bmatrix} \quad (12)$$

(4) **Determination of risk level**

In order to accurately and objectively obtain the final risk level of the evaluation object, the confidence identification criterion is used to analyze it. $\{G_1, G_2, \ldots G_P\}$ is an ordered segmentation class of the evaluation rating space. Let λ be confidence ($\lambda \geq 0.5$, usually $\lambda = 0.6$ or 0.7). When $G_1 > G_2 > G_3 \cdots > G_P$,

$$P_0 = min\left|p : \sum\nolimits_{k=1}^{p} \mu_{ik} > \lambda, i = 1,2,\ldots,n\right| \quad (13)$$

then the risk level of the assessment object is considered to be G_{P0}.

(5) **Risk ranking of first-level indexes**

Let $G_1 > G_2 > G_3 \cdots > G_P$ and the assignment of G_k be Z_k, then $Z_k > Z_{k+1}$ and

$$Q_{T_i} = \sum_{k=1}^{p} Z_k \mu_{ik} \quad (14)$$

where Q_{T_i} is the unascertained importance of the assessment object T_i, and $Q_T = \left(Q_{T_1}, Q_{T_2}, \ldots Q_{T_n}\right)$ is the unascertained importance vector, which is sorted according to the magnitude of Q_{T_i}.

3.2 Risk Assessment Indexes and Grading Standards

3.2.1 Establishment of Risk Assessment Indexes

It's critical for risk assessment to establish scientific and comprehensive risk assessment indexes for GRAIBs, which affects the accuracy and feasibility of the assessment results. The GRAIBs are characterized by complicated approval procedures, numerous participants, limited operation sites, high costs, complex construction technologies, and large risks. Therefore, it is urgent and necessary for GRAIBs to construct risk assessment indexes using a comprehensive method. In this study, based on literature review, extensive field research, and in-depth interviews with experts, the risk assessment index system for GRAIBs is determined. The risk assessment index system of GRAIBs is divided into five categories: institutional aspect, economic aspect, management aspect, technical aspect, and environmental aspect. The specific risk assessment index system is shown in Fig. 1.

3.2.2 Indexes Grading Standards

This paper determines the indexes grading standards through literature review, project investigation, and expert consultation, they are all divided into five grades, i.e., Grade I (very low risk, controllable), Grade II (low risk, acceptable and targeted treatment is required), Grade III (general risk, acceptable and treatment is required), Grade IV (high risk, preventable and treatment is required), and Grade V (very high risk, unacceptable and immediate action is required). The assessment grading space is $G = \{G_1, G_2, G_3, G_4, G_5\}$. The specific classification standards are shown in Table 3.

4 Case Study

To verify the effectiveness and feasibility of the risk assessment model, a GRAIBs project in Xi'an is selected. The experts are required to evaluate each risk index to determine the risk grade of the project. This case was located in a steel mill built in 1950, which was originally used as a steel rolling shop, and then green-reused as a two-story school building (Figs. 2, 3, 4 and 5). It covers an area of more than 0.6km^2 and has a floor area of about 0.2 km^2. The plant was renovated without changing the original structural system, and the space was sorted out and new functions were defined. In this case, two workshop buildings were renovated, and the length of the two school buildings was over 300 m. In the green reuse of the workshops, the main body of the original building structure was preserved as much as possible, and the original architectural features were not destroyed as much as possible. A part of the buildings with recent age, blurred characteristics, and low spatial elasticity were demolished to release the base space. The extra space of the demolished buildings was partly used as outdoor space for environment arrangement and partly used to build additional auxiliary and

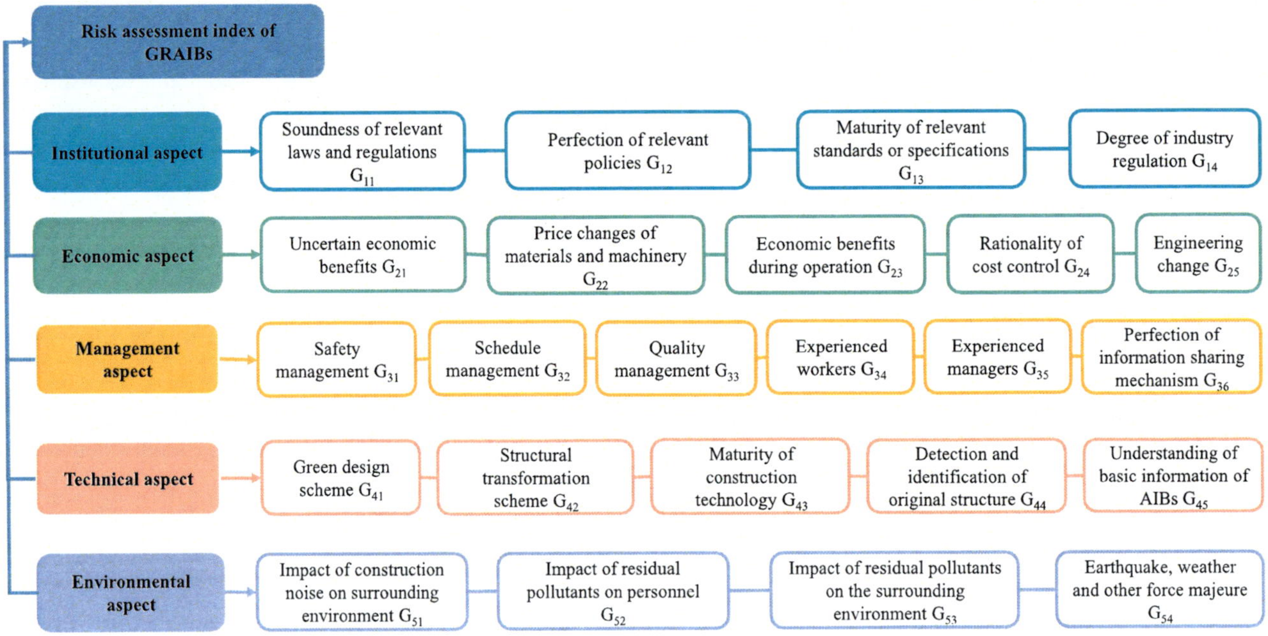

Fig.1 Risk assessment indexes of GRAIBs

Table 3 Specific classification standards of indexes grading for GRAIBs

Grade	Situation	Standard
Grade I	Very low risk	(9.0, 10]
Grade II	Low risk	(8.0, 9.0]
Grade III	General risk	(7.0, 8.0]
Grade IV	High risk	(6.0, 7.0]
Grade V	Very high risk	(0, 6.0]

Fig. 2 Building exterior after green reuse 1

Fig. 3 Building exterior after green reuse 2

equipment rooms. After green reuse, the teaching building retains the reinforced concrete frame structure of the main body of the original workshop. The bull-leg columns, crane beams, roof frames, channel roof panels, and other components with plant characteristics of the original workshop were also cleaned and retained intact. To meet the needs of the applying functions in construction, the teaching building is increased by adding a frame structure inside the space to increase the building space. The upper and

Fig. 4 Building interior after green reuse 1

Fig. 5 Building interior after green reuse 2

lower floors are connected by stairs. The exterior façade is decorated with lightweight wall materials and orange-red bright-framed curtain walls.

4.1 Data Sources

In this paper, nine experts in the field of GRAIBs were invited to evaluate the risk assessment indexes of the project. Finally, the average value of the scores was taken as the final assessment value of indexes after removing the highest and lowest scores. The basic information of 9 experts is shown in Table 4. The final values of each risk assessment index are shown in Table 5.

Table 4 Basic information of experts

No. expert	Job field	Working year	Education level
1	Professor	10	Doctor
2	Professor	8	Master
3	Professor	11	Doctor
4	Owner	6	Undergraduate
5	Owner	5	Master
6	Owner	10	Undergraduate
7	Contractor	8	Undergraduate
8	Contractor	9	Undergraduate
9	Contractor	11	Undergraduate

4.2 Calculation Process

4.2.1 Constructing the Single-Index Unascertained Measure Function

According to the definition of the single-index measure function and the evaluation standard of each index in Table 3, the single-index unascertained measure function can be constructed. Currently, the main construction forms of measure functions include linear, parabolic, sinusoidal, and exponential types. This study uses linear unascertained measure functions for analysis, then the single-index unascertained measure function is established as shown in Eqs. (15)–(19), and the single-index unascertained measure function is expressed graphically, as is shown in Fig. 6.

$$\mu_{ij1} = \begin{cases} 0 & x_{ij} \le 8.5 \\ \frac{x_{ij}-8.5}{1.0} & 8.5 < x_{ij} \le 9.5 \\ 1 & x_{ij} > 9.5 \end{cases} \tag{15}$$

$$\mu_{ij2} = \begin{cases} 0 & x_{ij} \le 7.5 \,\text{or}\, x_{ij} > 9.5 \\ \frac{x_{ij}-7.5}{1.0} & 7.5 < x_{ij} \le 8.5 \\ \frac{9.5-x_{ij}}{1.0} & 8.5 < x_{ij} \le 9.5 \end{cases} \tag{16}$$

$$\mu_{ij3} = \begin{cases} 0 & x_{ij} \le 6.5 \,\text{or}\, x_{ij} > 8.5 \\ \frac{x_{ij}-6.5}{1.0} & 6.5 < x_{ij} \le 7.5 \\ \frac{8.5-x_{ij}}{1.0} & 7.5 < x_{ij} \le 8.5 \end{cases} \tag{17}$$

$$\mu_{ij4} = \begin{cases} 0 & x_{ij} \le 3.0 \,\text{or}\, x_{ij} > 7.5 \\ \frac{x_{ij}-3.0}{1.0} & 3.0 < x_{ij} \le 6.5 \\ \frac{7.5-x_{ij}}{1.0} & 6.5 < x_{ij} \le 7.5 \end{cases} \tag{18}$$

Table 5 The final value of the risk assessment index

Index	Value	Index	Value
Soundness of relevant laws and regulations G_{11}	8.52	Experienced workers G_{34}	8.87
Perfection of relevant policies G_{12}	8.34	Experienced managers G_{35}	8.79
Maturity of relevant standards or specification G_{13}	8.11	Perfection of information sharing information mechanism G_{36}	8.76
Degree of industry regulation G_{14}	8.23	Green design scheme G_{41}	8.43
Uncertain economic benefits G_{21}	8.36	Structural transformation scheme G_{42}	8.56
Price changes of materials and machinery G_{22}	7.98	Maturity of construction technology G_{43}	8.43
Economic benefits during operation G_{23}	7.88	Detection and identification of original structure G_{44}	8.86
Rationality of cost control G_{24}	8.21	Understanding of basic information of AIBs G_{45}	8.98
Engineering change G_{25}	8.45	Impact of construction noise on surrounding environment G_{51}	8.65
Safety management G_{31}	8.64	Impact of residual pollutants on personnel G_{52}	8.87
Schedule management G_{32}	8.35	Impact of residual pollutants on environment G_{53}	8.79
Quality management G_{33}	8.34	Earthquake, weather, and other force majeure G_{54}	8.53

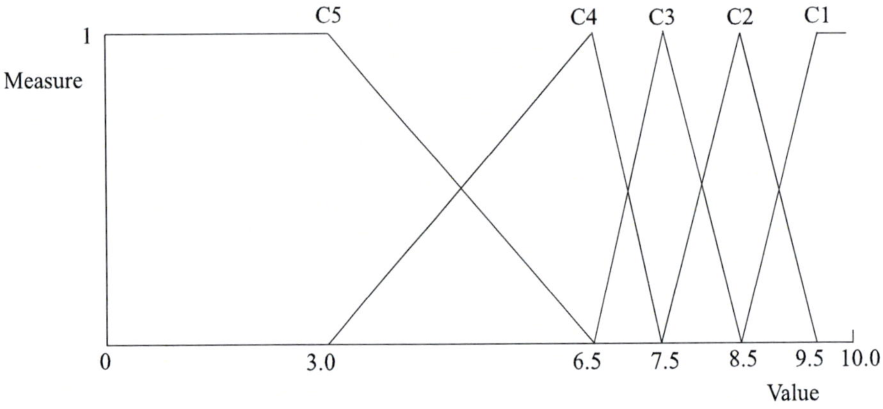

Fig. 6 The single-index unascertained measure function

$$\mu_{ij5} = \begin{cases} 1 & x_{ij} \leq 3.0 \\ \frac{x_{ij}-3.0}{1.0} & 3.0 < x_{ij} \leq 6.5 \\ 0 & x_{ij} > 6.5 \end{cases} \tag{19}$$

4.2.2 Calculating the Single-Index Unascertained Measure Matrix

According to the single-index unascertained measure function constructed in Fig. 6 and the final risk assessment value of each index in Table 5, the second-level index unascertained measure matrix can be obtained, as shown in Eqs. (20)–(24) respectively.

$$\mu_1 = \begin{bmatrix} 0.02 & 0.98 & 0 & 0 & 0 \\ 0 & 0.84 & 0.16 & 0 & 0 \\ 0 & 0.61 & 0.39 & 0 & 0 \\ 0 & 0.73 & 0.27 & 0 & 0 \end{bmatrix} \tag{20}$$

$$\mu_2 = \begin{bmatrix} 0 & 0.84 & 0.14 & 0 & 0 \\ 0 & 0.48 & 0.52 & 0 & 0 \\ 0 & 0.38 & 0.62 & 0 & 0 \\ 0 & 0.71 & 0.29 & 0 & 0 \\ 0 & 0.95 & 0.05 & 0 & 0 \end{bmatrix} \tag{21}$$

$$\mu_3 = \begin{bmatrix} 0.14 & 0.86 & 0 & 0 & 0 \\ 0 & 0.85 & 0.15 & 0 & 0 \\ 0 & 0.84 & 0.16 & 0 & 0 \\ 0.37 & 0.63 & 0 & 0 & 0 \\ 0.29 & 0.71 & 0 & 0 & 0 \\ 0.26 & 0.74 & 0 & 0 & 0 \end{bmatrix} \tag{22}$$

$$\mu_4 = \begin{bmatrix} 0 & 0.93 & 0.07 & 0 & 0 \\ 0.06 & 0.94 & 0 & 0 & 0 \\ 0 & 0.93 & 0.07 & 0 & 0 \\ 0.36 & 0.64 & 0 & 0 & 0 \\ 0.48 & 0.52 & 0 & 0 & 0 \end{bmatrix} \tag{23}$$

Table 6 Consistency check of judgement matrix

Matrix	G_{11}–G_{14}	G_{21}–G_{25}	G_{31}–G_{36}	G_{41}–G_{45}	G_{51}–G_{54}
C.R	0.098	0.111	0.053	0.031	0.059

$$\mu_5 = \begin{bmatrix} 0.15 & 0.85 & 0 & 0 & 0 \\ 0.37 & 0.63 & 0 & 0 & 0 \\ 0.29 & 0.71 & 0 & 0 & 0 \\ 0.03 & 0.97 & 0 & 0 & 0 \end{bmatrix} \tag{24}$$

4.2.3 Calculate the Weight of Each Index

The nine-scale method is used to establish a judgment matrix for first-level indexes and second-level indexes in turn, and then the weight of each index is calculated. As is shown in Table 6, the consistency test meets the requirements. The general weights of the first-level indexes and second-level indexes are shown in Table 7.

4.2.4 Calculating the First-Level Index Comprehensive Measure Matrix

According to the second-level index unascertained measure matrix and the weight of the second-level index, the first-level index comprehensive measure matrix can be obtained as follows:

$$\mu = \begin{bmatrix} 0.006 & 0.845 & 0.149 & 0.000 & 0.000 \\ 0.000 & 0.670 & 0.330 & 0.000 & 0.000 \\ 0.157 & 0.793 & 0.050 & 0.000 & 0.000 \\ 0.203 & 0.781 & 0.016 & 0.000 & 0.000 \\ 0.258 & 0.742 & 0.000 & 0.000 & 0.000 \end{bmatrix} \tag{25}$$

4.2.5 Calculating the Total Target Comprehensive Measure Matrix

According to the first-level index unascertained measure matrix and the weight of the first-level index, the total target comprehensive measure matrix can be calculated, as shown in Eq. (26).

$$\mu_{total} = \begin{pmatrix} 0.068 & 0.773 & 0.159 & 0 & 0 \end{pmatrix} \tag{26}$$

Table 7 Weight of risk assessment index for GRAIBs

First-level indexes	Weight	Second-level indexes	Weight	General weight
Institutional aspect	0.3815	Soundness of relevant laws and regulations G_{11}	0.2939	0.112
		Perfection of relevant policies G_{12}	0.4790	0.183
		Maturity of relevant standards or specifications G_{13}	0.0937	0.036
		Degree of industry regulation G_{14}	0.1334	0.051
Economic aspect	0.2865	Ease of investment and financing G_{21}	0.3522	0.101
		Price changes of materials and machinery G_{22}	0.0668	0.019
		Economic benefits during operation G_{23}	0.3048	0.087
		Rationality of cost control G_{24}	0.1810	0.052
		Engineering change G_{25}	0.0952	0.027
Management aspect	0.1066	Safety management G_{31}	0.3210	0.034
		Schedule managementG_{32}	0.0512	0.005
		Quality management G_{33}	0.2621	0.028
		Experienced workers G_{34}	0.1106	0.012
		Experienced managers G_{35}	0.1667	0.018
		Perfection of information sharing mechanism G_{36}	0.0884	0.009
Technical aspect	0.1684	Green design scheme G_{41}	0.0748	0.013
		Structural transformation scheme G_{42}	0.3322	0.056
		Maturity of construction technology G_{43}	0.1489	0.025
		Detection and identification of original structure G_{44}	0.2507	0.042
		Understanding of basic information of abandoned industrial buildings G_{45}	0.1934	0.033
Environmental aspect	0.0569	Impact of construction noise on surrounding environment G_{51}	0.1636	0.009
		Impact of residual pollutants on personnel G_{52}	0.3479	0.020
		Impact of residual pollutants on the surrounding environment G_{53}	0.3479	0.020
		Earthquake, weather and other force majeure G_{54}	0.1406	0.008

4.2.6 Decision of Risk Assessment Level

According to the confidence recognition criterion, take the confidence $\lambda = 0.7$. According to Eq. (13), when $P_0 = 2$, from small to large, $0.068 + 0.773 = 0.840 > 0.7$, that is, the risk grade is rated as Grade II. From large to small, $0.773 + 0.159 = 0.932 > 0.7$, that is, the risk grade is rated as Grade II. Combined with the two results, it is judged that the risk assessment grade of GRAIBs is Grade II. The assessment results are in good agreement with the actual situation. Accordingly, the assessment results are reliable.

4.2.7 First-Level Index Risk Ranking

If five first-level indexes are taken as evaluation objects for horizontal comparison, since $G_1 > G_2 > \cdots > G_P$, $G_1 = 5$, $G_2 = 4$, $G_3 = 3$, $G_4 = 2$, $G_4 = 1$. According to Eq. (14), the unascertained importance of the risk of these five first-level indexes can be obtained. Among them, the unascertained importance of institutional risk is shown in Eq. (27).

$$Q_{TV1} = 0.06 \times 5 + 0.845 \times 4 + 0.149 \times 3 + 0 \times 2 + 0 \times 1 = 3.857 \tag{27}$$

Similarly, the unascertained importance of other first-level indexes is calculated as follows:

$$Q = (Q_{TV1}, Q_{TV2}, Q_{TV3}, Q_{TV4}, Q_{TV5}) = (3.857, 3.670, 4.108, 4.187, 4.258) \tag{28}$$

5 Result and Discussion

According to the above calculation, the risk assessment level of this project is Grade II. That is, the risk is low, which indicates that it is within the acceptable range and needs to take targeted treatment. In addition, by calculating the risk ranking of the first-level indexes, the degree of risk from small to large is: environmental risk, technical risk, management risk, institutional risk, and economic risk. Through field research of the project, we found that the risk assessment results of the project were consistent with the actual situation.

In the first-level indexes, the environmental risk is the lowest. Before the construction of the project, the pollutants of the AIBs are treated carefully until the indoor pollutant concentration of the buildings meets the current relevant standards, especially contaminated soil. Therefore, the impact of pollutants on the human and the surrounding environment is low. The second lowest risk is technical risk. The project was investigated in detail before construction. For example, due to the lack of original architectural design drawings, the developer organized professionals to carry out structural inspection and appraisal of the AIBs and recorded the basic information of the relevant equipment and facilities in the building. In addition, during the planning process of the project, strict bidding requirements were put forward, so that contractors experienced in GRAIBs projects were selected. The biggest risk is economic risk. The relevant laws and regulations on GRAIBs are not perfect. Also, the GRAIBs are lack of supporting funds from government. As a result, stakeholders are skeptical and hesitant to participate in GRAIBs.

Based on the assessment results and the actual situation analysis, some targeted measures are proposed from the following aspects: First, the project management organization should set up a special external liaison department to fully understand the preferential policies and government subsidies related to the GRAIBs, expand the publicity of the green GRAIBs, and attract the attention of investors and the public; Second, considering the uncertainty of the economic benefits of the new functions, the developer should conduct a detailed investigation of the functional positioning of the surrounding environment at the project planning stage to determine the new functions of the project. This new function should not only keep the historical connotation of the old industrial buildings from being destroyed but also adapt to the surrounding environment. Third, GRAIBs project should select experienced managers and operators to carry out project management.

Accordingly, the risk assessment model based on uncertainty measurement theory in this paper is feasible. The model can solve the uncertain problems existing in the GRAIBs, and the calculation of the model is relatively simple. Therefore, the model is very suitable for the risk assessment of the GRAIBs.

6 Conclusions

Based on the perspective of the whole risk process of GRAIBs, this paper establishes the GRAIBs risk assessment model based on the uncertainty measurement theory. This model provides a way of thinking for the quantitative risk assessment of GRAIBs project and provides a reference for the sustainable implementation of GRAIBs project. The conclusions of this study are as follows:

(1) Through literature review, extensive field research, and in-depth interviews with experts, 23 risk assessment indexes of GRAIBs are constructed from five aspects, including institutional aspect, economic aspect, management aspect, technical aspect, and environmental aspect.
(2) Considering the uncertainty and fuzziness of risk of GRAIBs, a risk assessment model of GRAIBs is established based on the uncertainty measurement theory.

(3) The established risk assessment model is applied to a GRAIBs project in Xi'an, which the team participated in personally. The assessment grade is Grade II, and the risk is low. In addition, the environmental risk is the lowest and the biggest risk is economic risk. The results of risk assessment are consistent with the actual situation of the project, which verifies the feasibility and rationality of the model. This shows that this method can accurately identify the risk level of GRAIBs and provide a reference for scientific risk management of GRAIBs.

References

Blagojević, M. R., & Tufegdžić, A. (2016). The new technology era requirements and sustainable approach to industrial heritage renewal. *Energy and Buildings, 115*, 148–153.

Chan, A., Cheung, E., & Wong, I. (2015). Recommended measures on the revitalizing industrial buildings scheme in Hong Kong. *Sustainable Cities and Society, 17*, 46–55.

Chen, C. S., Chiu, Y. H., & Tsai, L. (2018). Evaluating the adaptive reuse of historic buildings through multicriteria decision-making. *Habitat International, 81*, 12–23.

Chen, W. Y., Li, Y. X., Cao, D., & Bi, X. (2022). Research on indoor daylighting optimization strategy for renovated old industrial buildings with large spans. *Industrial Construction, 52*(10), 139–145.

Chen, X., Guo, X., Tian, W. (2019). Game analysis of green transformation evolution of old industrial building recycling project. *Journal of Civil Engineering and Management, 36*(4), 69–74+95.

Issa, U., Sharaky, I., Alwetaishi, M., Balabel, A., Shamseldin, A., Abdelhafiz, A., Al-Surf, M., Al-Harthi, M., & Osman, M. M. A. (2021). Developing and applying a model for evaluating risks affecting greening existing buildings. *Sustainability, 13*(11).

Jiang, H. Y., & Zhang, F. (2021). Risk assessment of construction safety of recycling of old industrial buildings based on entropy weight matter-element method. *Journal of Safety and Environment, 21*(6), 2378–2385.

Lee, M. J. (2019). Transforming post-industrial landscapes into urban parks: Design strategies and theory in Seoul, 1998–present. *Habitat International, 91*, 102023.

Li, R. P., Wang, H. W., & Chen, X. (2019). Research on emergency management of construction risk for recycling of old industrial buildings. *Journal of Safety Science and Technology, 15*(2), 151–156.

Liao, B., & Li, L. (2022). How can green building development promote carbon emission reduction efficiency of the construction industry?–Based on the dual perspective of industry and space. *Environmental Science and Pollution Research, 29*(7), 9852–9866.

Martinat, S., Navratil, J., Hollander, J. B., Trojan, J., Klapka, P., Klusacek, P., & Kalok, D. (2018). Re-reuse of regenerated brownfields: Lessons from an Eastern European post-industrial city. *Journal of Cleaner Production, 188*, 536–545.

Mushi, F. V., Nguluma, H., & Kihila, J. (2022). A critical review of African green building research. *Building Research and Information, 50*(6), 610–627.

Remøy, H., & van der Voordt, T. (2014). Adaptive reuse of office buildings into housing: Opportunities and risks. *Building Research and Information, 42*(3), 381–390.

Saaty, T. L. (1980). *The analytic hierarchy process: Planning, priority setting, resources allocation*. McG raw-Hill Company.

Samadzadehyazdi, S., Ansari, M., Mahdavinejad, M., & Bemaninan, M. (2018). Significance of authenticity: Learning from best practice of adaptive reuse in the industrial heritage of Iran. *International Journal of Architectural Heritage, 14*(3), 329–344.

Sanchez, B., Rausch, C., Haas, C., & Saari, R. (2020). A selective disassembly multi-objective optimization approach for adaptive reuse of building components. *Resources, Conservation and Recycling, 154*, 104605.

Sessa, M. R., Russo, A., & Sica, F. (2022). Opinion paper on green deal for the urban regeneration of industrial brownfield land in Europe. *Land Use Policy, 119*, 106198.

Sun, X., Gou, Z., Lu, Y., & Tao, Y. (2018a). Strengths and weaknesses of existing building green retrofits: Case study of a LEED EBOM gold project. *Energies, 11*(8), 1936.

Sun, X., Gou, Z., Lu, Y., & Tao, Y. (2018b). Strengths and weaknesses of existing building green retrofits: Case study of a LEED EBOM gold project. *Energies, 11*(8).

Tian, W., Wang, A. D., Guo, X., Guo, P., & Chen, X. (2022). Study on satisfaction evaluation for green regeneration decision of old industrial buildings. *Journal of Safety and Environment, 22*(1), 371–278.

Tian, W., Zhong, X., Zhang, G., & Goh, Y. M. (2021). Sustainability analysis of reused industrial buildings in China: An assessment method. *Journal of Civil Engineering and Management, 27*(1), 60–75.

Vardopoulos, I. (2019). Critical sustainable development factors in the adaptive reuse of urban industrial buildings. A fuzzy DEMATEL approach. *Sustainable Cities and Society, 50*.

Wang, G. Y. (1990). Uncertainty Information and its mathematical treatment. *Journal of Harbin Architecture and Engineering Institute, 23*(4), 1–8.

Wang, Y., Dou, L., & Liu, S. T. (2018). Risk assessment studies on the green retrofit projects based on improved FMEA. *Ecological Economy, 34*(1), 139–145.

Wu, Q., Sun, M., Huang, T. X., Geng, J., & Xiang, R. (2022). Research on project governance risk of green regeneration of old industrial plant based on SNA. *Journal of Safety and Environment, 22*(3), 1121–1131.

Wu, Q., Wang, C., Zheng, D. Z., & Ma, H. F. (2014). Research on risk factors of reuse of old industrial building in Xi'an based on principal component analysis. *Industrial Construction, 44*(10), 61–63+90.

Yung, E. H. K., & Chan, E. H. W. (2012). Implementation challenges to the adaptive reuse of heritage buildings: Towards the goals of sustainable, low carbon cities. *Habitat International, 36*(3), 352–361.

Zhang, Y., Ye, F., & Zhang, G. (2021a). Determinants of green regeneration of industrial brownfield under ecological security constraints: Model and empirical analysis based on grounded theory. *Sustainability, 13*(20), 11396.

Zhang, Y., Zhang, G., & Guo, P. (2021b). Regeneration path of abandoned industrial buildings: The moderating role of the goodness of regeneration mode. *Journal of Cleaner Production* 297, 126668.

Urban Energy Consumption in the City of Naples (Italy): A Geographically Weighted Regression Approach

Gerardo Carpentieri, Floriana Zucaro and Carmen Guida

Abstract

The energy transition has been advocated at a worldwide level as a policy priority in promoting sustainability and delivering the Green Deal. The nexus between raw materials supply and related GHG emissions of the production process makes the energy issue relevant for all decision-makers. If there is now a widespread acceptance that action at the local level is essential and decisive to realize low-carbon cities that can efficiently save energy, a lack of integration between energy-saving solutions and urban planning continues to affect the work of local decision-makers, technicians and practitioners. According to this, the work investigates the relationships between urban characteristics and residential energy consumption data, through the integration of statistics and spatial statistics techniques. The principal component analysis and the geographically weighted regression were applied on an urban scale to support the policy-makers in increasing the energy sustainability due to climate challenges and new emerging geopolitical scenarios regarding energy supply. All spatial analyses were based on a set of variables related to the main urban characteristics: socio-economic, physical, functional and environmental ones. The test area refers to the City of Naples (Italy) where the results show the GWR prediction model estimating the higher residential energy consumption values related to the urban input variables. The outcomes of the analysis can be useful for local decision-makers in defining strategies and interventions oriented to reduce energy consumption.

Keywords

Urban energy planning · Residential energy consumption · Geographically weighted regression

List of Abbreviations

AIC	Akaike information criterion
GIS	Geographical information system
GWR	Geographically weighted regression
IEA	International Energy Agency
IPCC	International Panel on Climate Change
OLS	Ordinary least squares regression
PCA	Principal component analysis
VIF	Variance inflation factor

1 Introduction and State of Art

The sustainable development goal related to "Sustainable cities and communities" (n. 11) strengths the role of urban systems in the improvement of the quality of life by increasing its resilience. Urban policy and decision-makers are required to use the resources more efficiently and to reduce energy consumption. The attention to the energy issue is drawn as energy production and consumption trigger about two-thirds of GHG emissions, despite technological progress mainly related to the supply of renewable energy sources and the distribution to the final users. According to IEA and IPCC (2022), the world energy demand will increase by 50% to 2050 such as the GHG emissions without policy actions. These data stress

G. Carpentieri (✉) · F. Zucaro · C. Guida
Laboratory Territory, Mobility and Environment, Department of Civil, Architectural and Environmental Engineering, University of Naples Federico II, Naples, Italy
e-mail: gerardo.carpentieri@unina.it

F. Zucaro
e-mail: floriana.zucaro@unina.it

C. Guida
e-mail: carmen.guida@unina.it

B.-J. He et al. (eds.), *Resilient Horizons: Building Sustainable Environments for Climate Adaptation and Health*, Advances in Science, Technology & Innovation, https://doi.org/10.1007/978-3-031-46109-5_8

the urgency of saving energy by improving lifestyles and improving the organization and functioning of the urban systems (Asarpota & Nadin, 2020; Gargiulo & Lombardi, 2016; Papa et al., 2016a; Tira, 2020). Nevertheless, the study of energy consumption distribution related to the different kinds of activities and areas within a city makes the energy issue a challenging task for academics and institutional bodies.

Several scholars conducted different reviews of the influences of building and urban features on energy consumption to provide systematic and comprehensive comparisons of strengths, limitations and applications of the most spread approaches and models. For instance, Quan and Li (2021) identified two temporal periods characterized by different approaches to the study of the urban energy consumption issue: the range included between the first half of the 1980s and the first years of the 2000s when the attention was dedicated to one component of the urban system (e.g. land uses) and its influence on energy consumption levels of single or clusters of buildings; the next period until 2020 when the study area was extended from a few buildings to district level by considering the more complex relationships between the building, geometry and morphology features of urban fabrics and heating/cooling energy variations.

Currently, two main categories of models investigating, simulating and predicting energy consumption at the urban level can be identified (Fig. 1): top-down and bottom-up models (Jebaraj & Iniyan, 2006; Liu et al., 2022; Prina et al., 2020; Swan & Augural, 2009). Top-down models can be defined as aggregated models as they consider cities at the same regional or national scale and predict energy use and consumption variations by correlating energy data to economic (e.g. GDP, fuel prices), climate and main construction materials (e.g. envelopes) historical data series (Barker et al., 2016; Hunt & Ryan, 2015; Lund et al., 2017). For instance, Bianco et al. (2009) and Lin and Liu (2015) investigate the influence of socio-economic and demographic characteristics on yearly electricity use, respectively, in Italy and China. This kind of models mainly assess the effects of energy price control policies on sectors such as the economic-financial and public welfare ones (Andersen et al., 2019; Böhringer & Rutherford, 2008; Shaaban et al., 2022). They are market-oriented approaches and lack detail in the description of technology advances that strongly affect energy consumption trends.

If top-down models include system dynamic and agent-based ones according to the calculation technique (input–output analysis and optimizing decision rules), the bottom-up model categorization refers to the kind of data used. Data-driven models rely on detailed characteristics of buildings (e.g. building age, building envelope properties, building volume) and their energy-use data and are based on statistical analysis or machine learning techniques (Ferrando & Causone, 2019; Liu et al., 2022). These data-driven models identify predictors as they relate energy consumption values (outputs) to the depending variables (inputs). Recently this category of bottom-up models has been incorporating GIS analysis to assess the energy saving obtained by the retrofit intervention (Mastrucci et al., 2014; Moghadam et al., 2019; Walter & Sohn, 2016). Data-driven models are often used when limited thermal and physic building data can be gathered and this is their

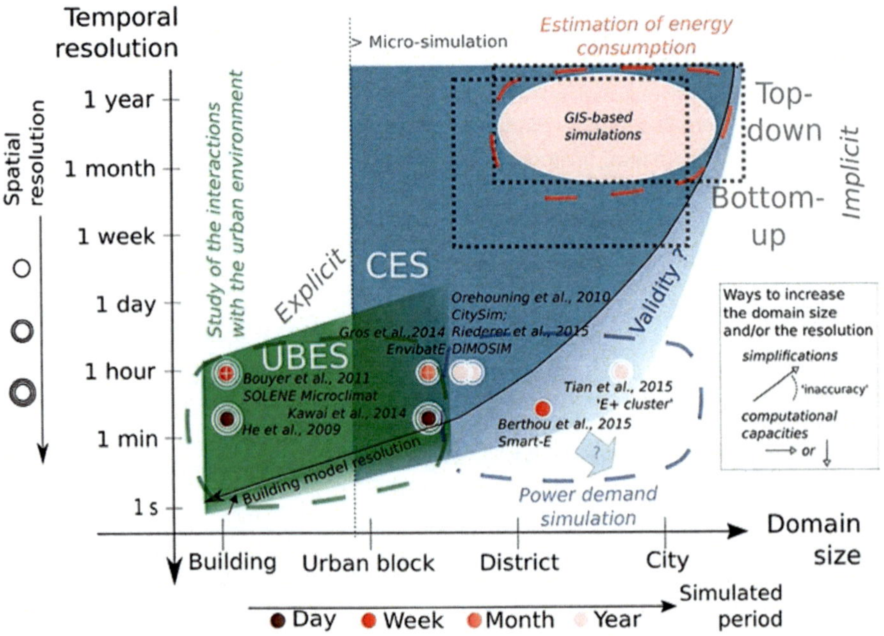

Fig. 1 The summary framework of the main energy simulation models. Retrieved from Frayssinet et al. (2018)

main advantage compared to physics-based models. These last ones require a huge amount of data collection and processing mainly related to climate variables and hourly cooling and heating load of buildings and occupants' characteristic parameters. The relevance of the impacts of the heat exchange among built elements and the urban asset has represented a major trend research topic of the physics-based models, such as heat island impacts and outdoor thermal comfort (Ferrando & Causone, 2019). For instance, many works and reports assert that the UHI intensity is responsible for the related cooling demand increase between 10 and 120% (Li et al., 2019) according to thermal variations.

Such as data-driven models, also the physics-based ones integrate GIS analysis (remote sensing) to facilitate the collection of detailed building shape data and their distribution and localization (Chen et al., 2021a; Deb & Schlueter, 2021).

The flexibility and interpretation capability of GIS allow to merge efficiently the statistical and mathematical aspects of bottom-up models with the different components of the urban areas, paving the way for GIS-based tools oriented to forecast and simulate the energy behaviour of the whole city. Innovative tools such as Cityish (Walter & Kämpf, 2015) SEMANCO (Corrado et al., 2015) and Urban SEM (Gargiulo & Zucaro, 2015; Papa et al., 2016b) are among the urban energy models developed to overcome the sectorial approach and the scales of detail like that in construction that has characterized the scientific debate on energy issue for a long time. In this regard, it is worth referring to the work of Allegrini et al. (2015) who identify and classify the three main clusters of features to study energy consumption at the district level: district energy systems (such as heat distribution infrastructure), renewable energy production and local microclimate which affects building energy demands. These scholars, such as the authors of these work, emphasize the fact that the neighbourhood level may facilitate the difficult interpretation of the numerous and complex relations between the urban system and energy consumption (Batty, 2008; Gargiulo & Lombardi, 2016; Papa et al., 2016a; Ferrando and Causone, 2019).

Within the energy issue studies at the city level, the ones based on a systemic approach that relate the urban characteristics and energy consumption have started to be more spread only in recent years, but they are still rare in the Mediterranean area. In this perspective, this work describes the case study of Naples Municipality where statistical and spatial statistical analysis techniques were applied for investigating the relationship between residential energy consumption in this densely stratified city with the physical, functional and socio-economic characteristics. The authors are oriented to provide local decision-makers with the identification of priority areas of intervention, in terms of reducing energy consumption. The paper is organized as follows. Section 2 describes the methodology; Sect. 3 illustrates the results obtained for the study area of the City of Naples; Sect. 4 provides the main conclusions of the work.

2 Methodology

This section describes the methodology aimed to identify the highest energy-consuming portions of a city by defining and measuring the intensity of interactions between urban characteristics and residential energy consumption. To pursue this goal, the methodology is based on using multivariate statistical and spatial statistics techniques (Fig. 2). These last ones allow for overcoming the limitation of traditional statistics which do not always interpret into account and interpret the physical and social aspects of the territory where the studied phenomenon occurs (Chen et al., 2020; Kashi et al., 2021).

The first step concerned the definition of the urban characteristics and the corresponding explanatory variables, which, based on the scientific literature and the authors' previous works, appear to be those that have the greatest influence on energy consumption (Carpentieri & Favo, 2017; Gargiulo & Russo, 2017; Nguyen et al., 2021; Papa & Zucaro, 2014; Papa et al., 2016a, 2016b; Soltani et al., 2012; Steemers & Yun, 2009; Wang et al., 2022). The selected 23 variables refer to the urban system in its entirety, to consider the different features of which it consists of and the related interactions among them and also the energy consumption (Papa et al., 2016a). The list of variables and the urban characteristics that they allow taking

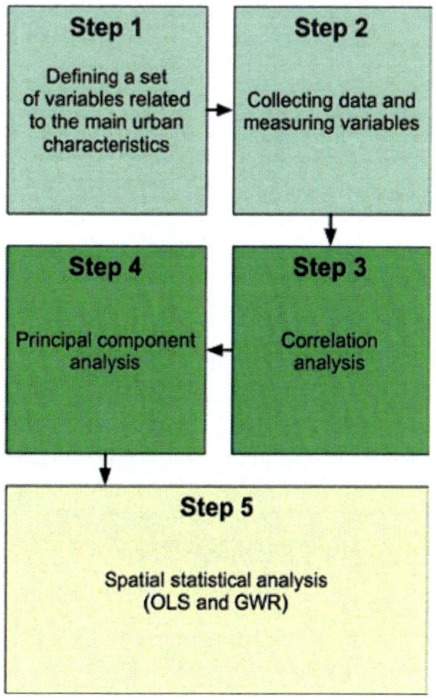

Fig. 2 Flowchart of the proposed methodology

Table 1 The list of selected variables for statistical and spatial statistical analysis

ID	Category	Short name	Description	Data source
A01	Energy	Energy consumption	kWh consumed annually by urban residents	Italian revenue agency
A02	Physical characteristics	Average age of buildings	The average building period	ISTAT
A03	Physical characteristics	Built-up area	The total surface of built areas for census tract	ISTAT
A04	Socio-economic characteristics	Total population	No. of total inhabitants for census tract	ISTAT
A05	Socio-economic characteristics	Population under 15	No. of inhabitants under the age of 15 years old for census tract	ISTAT
A06	Socio-economic characteristics	Population between 15 and 65	No. of inhabitants aged between 15 and 65 years old for census tract	ISTAT
A07	Socio-economic characteristics	Population over 65	No. of inhabitants over the age of 65 for census tract	ISTAT
A08	Physical characteristics	Census section surface	The total surface of each census tract	ISTAT
A09	Socio-economic characteristics	Average age of the population	The average age of the population for census tract	ISTAT
A10	Socio-economic characteristics	Average number of members per household	The average number of members per household for census tract	ISTAT
A11	Physical characteristics	Average number of floors	The average number of floors for census tract	ISTAT
A12	Physical characteristics	State of conservation	The state of conservation of buildings for census tract	ISTAT
A13	Physical characteristics	Number of residential buildings	The total number of residential buildings for census tract	ISTAT
A14	Physical characteristics	Number of buildings in masonry	The total number of masonry buildings for census tract	ISTAT
A15	Physical characteristics	Number of buildings in concrete	The total number of concrete buildings for census tract	ISTAT
A16	Physical characteristics	Number of dwellings occupied	The total number of dwellings occupied for census tract	ISTAT
A17	Physical characteristics	Number of empty dwellings	The total number of empty dwellings for census tract	ISTAT
A18	Physical characteristics	Surface of occupied dwellings	The total surface of occupied dwellings for census tract	ISTAT
A19	Physical characteristics	Number of buildings	The total number of buildings for census tract	ISTAT
A20	Physical characteristics	Number of buildings used	The total number of used buildings for census tract	ISTAT
A21	Functional characteristics	Number of employees	The total number of employees for census tract	ISTAT
A22	Environmental characteristics	Green areas	The total surface of green areas for census tract	Urban Atlas
A23	Functional characteristics	Temperature	The mean radiant temperature for census tract	City of Naples

into account are reported in Table 1. In the second step, the variables have been measured by accessing different freely accessible databases. The open access web portal of Urban Atlas, ISTAT and Naples Municipality provided the data to measure the variables. The energy consumption data have been retrieved from the Italian Revenue Agency collecting the data of the tax registry for any individual or company. The energy consumption data concern the year 2011 and refer to sales (€) and kWh used each year by urban residents, identified with cadastral data. All the data were then associated with census tracts that are the minimum territorial reference area of this study.

The $3{,}403 \times 23$ matrix was processed in the third step to normalize the whole dataset. The min–max scaling technique allowed redefining the values of the variables that are characterized by a non-Gaussian distribution on a fixed interval from 0 to 1. In other words, the normalization step transforms the values of the numeric columns of the dataset into a common scale, without deforming the differences in the ranges of values.

To define the strength (magnitude of closeness) and the kind (positive or negative) of a relationship between the variables, a correlation analysis was made in the fourth step. The measure of the correlation degree among the variables is a preparatory for the Principal Component Analysis (PCA), as this multivariate statistics technique is based on the variance in common between the variables. It means that only variables that are at least moderately correlated with the others are to be considered useful for the creation of the model. In fact, in the absence of correlation, "forcing" a variable to "merge" with others would result in too high a loss of information, thus compromising the pursuit of the initial goal. The Pearson correlation has been elaborated, as it allows to examine variables expressed by different measurement scales. This measures the degree of the relationship between linearly correlated variables by calculating a correlation coefficient between -1 and 1, where the values of -1 and 1 correspond to an increase or decrease of a fixed proportion between the two variables, while a value equal to 0 indicates no correlation. The Pearson correlation coefficient is calculated by the ratio between the covariance and the product of the standard deviations of the two variables considered.

Then, in the fifth step, the statistical analysis aimed at reducing the complexity of the dataset to obtain better interpretable results has been developed. The Principal Component Analysis (PCA) is a multivariate statistic which decreases the initial dataset size and provides a reduced number of linear combinations of variables explicating most of the variance in the correlation matrix. The PCA has been developed by Pearson (1901) but the description of extracting components for a data matrix is due to Hotelling (1933). The principal components are the most relevant correlations among the n initial variables Y_1, Y_2, Y_3, …, N_y: the analysis return vectors Z_1, Z_2, Z_3, …, Z_n, describing a different underlying variation or trend found in the input dataset. The Z vectors are ordered by their importance: the component Z_1 is the most prevalent trend identified all over the data and accounts for more variation than Z_2 and so on. This aim is pursued by projecting the input variables into a Cartesian system (linear transformation), equal to or smaller than the original one, in which the variables are classified in decreasing order of variance. These new variables are the principal components that synthesize the input variables and that are linear combinations of them through weights. From a geometric perspective, the goal is to identify a space of reduced size on which to project the cloud of original points in such a way as to deform the distances between the PCA points as little as possible.

Following the PCA, the spatial regression analysis was developed to identify the urban characteristics that have the greatest relationship with energy consumption (sixth step). This kind of analysis allows to define and interpret «the spatial dynamics of resource consumption as a comprehensive representation of urban physical and social processes by highlighting the autocorrelation (spatial dependency), heterogeneity (spatial nonstationary) and weighting (spatial correction) of explanatory variables» consumption (Chen et al., 2021b). In other words, the Geographically Weighted Regression (autocorrelation GWR) aims at defining «whether the underlying data generating process exhibits spatial heterogeneities (the distribution of energy consumption on an urban scale) or local deviations from a global regression model» (Wheeler & Tiefelsdorf, 2005). Furthermore, the GWR is also useful for identifying hotspots for energy consumption according to the distribution of the PCA output values in the test area. The GWR overcomes the limits of traditional OLS regression, as it provides a local model of the dependent variable y by estimating a specific regression equation for each territorial unit of the dataset. Each unit constitutes an observation point in which an equation is estimated. To build these equations, the GWR uses the values assumed by the explanatory variables of the territorial units that fall in the proximity to each of them. These units are defined through a weighted distance matrix that is elaborated by the kernel function, which uses a fixed or variable distance criterion. The matrix attributes a weight and a different significance level to the contiguous spatial units depending on the distance from the observation point, unlike the OLS model which implicitly assumes all equal weights.

The main advantages of GWR model compared to other regression models such as OLS refer to: (i) GWR moves from the global level of analysis to the local one; (ii) GWR allows the relationships to vary over space and (iii) estimated errors in GWR are lesser than the OLS and the R^2 value (local coefficient of determination) increase (Chen et al., 2021b; Fotheringham & Ohsan, 2016).

The GWR model for each regression point is as follows:

$$y_{i=\beta_0}\left(\mu_j v_j\right) + \sum\nolimits_{i=1}^{p} \beta_i\left(\mu_j v_j\right) x_{ij} + \varepsilon_j \quad (1)$$

where y_d is the dependent variable; u_p and V_t are the coordinates for location j; β_0 $(u_p\ V_t)$ represents the intercept for location j; β_I $(u_p\ v_j)$ is the local estimated coefficient for the independent variable X_{in} and j_{ob} refers to the random spatial error coefficient (Comber et al., 2020; Gomes et al., 2017; Kitchin & Thrift, 2009).

The GWR outputs are obtained from the single regression results: every spatial dataset has its regression equation with an R^2 value, representing the variance for a dependent variable, and the adjusted R^2 value defines the better fitting results by checking Akaike information criterion (AIC). The AIC compares statistical models by evaluating the amount of information lost when a given model is used to describe a real phenomenon. The best model is the one with the lowest AIC value.

AIC together with Global Moran's I p-value is the performed tests to assess the fit of the OLS and GWR. In fact, before elaborating on the GWR the very best OLS model needs to be identified to draw inferences about the first (global)- and second (local)-order process (Anselin & Getis, 2010). Moran's I measures the spatial non-stationarity among predictors and the independent variables (Ivanic et al., 2014). It estimates the intensity of this correlation as a function of the distance separating the observations (correlograms) through the ratio between the covariance and the sample variance. Its values vary from +1 (strong positive spatial autocorrelation) to 0 (random pattern) to −1 (strong negative spatial autocorrelation). It is worth noting Moran's I tends «to average local variations in the strength of spatial autocorrelation» (Ivajnsic et al., 2014).

In summary, the sixth step of the overall methodology is based on the following stages (Fig. 1).

- OLS to select independent variables and eliminate multicollinearity;
- GWR to identify relationships between the dependent and explanatory variables.

The methodology was tested in the City of Naples (southwest of Italy), according to the high complexity of social, economic and environmental characteristics that characterize this urban area that is the fifth largest Italian city in terms of population density.

3 Results

Naples is the third Italian city in terms of the number of inhabitants and the fifth in population density. The inhabitants that live in the City of Naples are 959,188 (ISTAT, 2020), on a surface equal to 119.02 km^2; therefore, the density is 8,059 inh/km^2 (ISTAT, 2020). The city is divided into 12 neighbourhoods and 10 borough authorities. Within Naples' municipality it is possible to recognize different urban textures on the base of physical and functional characteristics (Fig. 3). The historical area is located in the central districts (Monte Calvario, San Giuseppe, Porto and San Ferdinando), which have a high population and budling density. The peripheral areas (the external districts) were mainly built between the nineteenth and twentieth centuries with an ununiform urban structure and low-quality buildings. The social and economic characteristics are significantly different between the different districts as Posillipo,

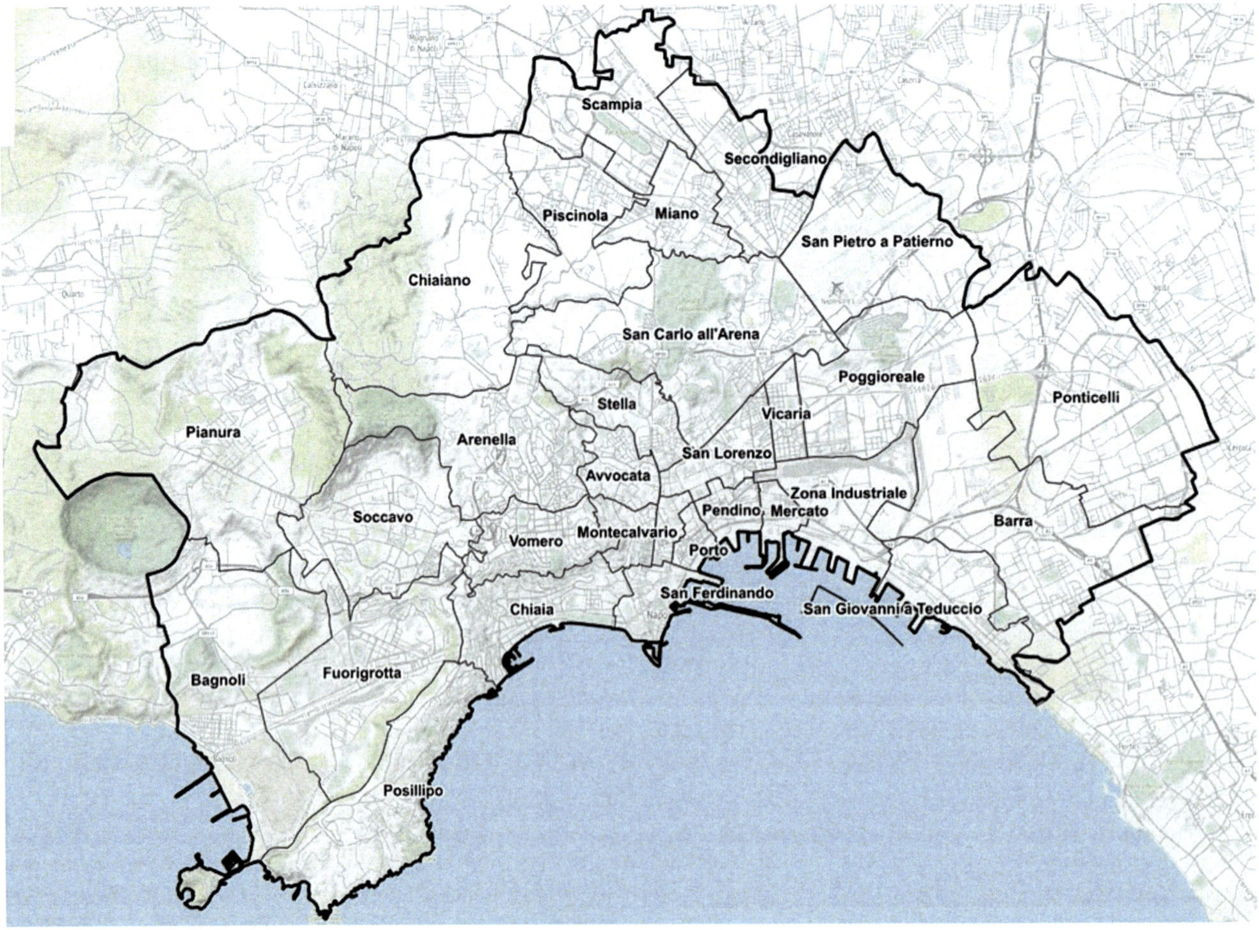

Fig. 3 The municipal boundaries and districts of the City of Naples

Vomero and Chiaia are considered the richest areas and Pinera, Scampia and Ponticelli are poor and disadvantaged. These different conditions significantly influence the citizen quality of life.

Table 2 lists the descriptive statistics for 23 explanatory factors. The reliability of the dataset is guaranteed by basic statistical analysis that represents the starting point for developing the following spatial regression analysis, as they show data distributions and patterns in the study area. The average building age seems to show that the building complex of the city is quite recent (the maximum value is equal to 45) and it can be explained by the urban expansion that occurred intensely during the 60–80s period. Furthermore, the masonry and concrete building maximum presence do not differ much. The average number of floors varies from 0 to 5. The average values of green surfaces and temperatures seem to suggest that the increase of unbuilt spaces could be an efficient strategy to decrease thermal stress and the consequent energy consumption. Finally, the demographic distribution shows that the presence of the elderly is significant, and this data should be considered to define efficient actions oriented to promote more sustainable energy behaviours.

Table 2 Main explanatory statistics of the selected variables

ID	Name	Mean	Standard deviation	Min	Max
A01	Energy consumption	179,544	240,782	34.00	2,914,395
A02	Average age of buildings	1.59	2.19	0.00	45.07
A03	Built-up area	5,259.89	6,051	0.00	100,581
A04	Total population	247.42	281.78	0.00	3,052
A05	Population under 15	38.54	46.33	0.00	588.00
A06	Population between 15 and 65	163.90	187,083	0.00	2,060
A07	Population over 65	44.98	55.46	0.00	725.00
A08	Census section surface	27,289	83,423	59.11	2,183,276
A09	Average age of the population	41.07	12,661	0.00	90.00
A10	Average number of members per household	2.48	0.88	0.00	5.00
A11	Average number of floors	3.25	1,327	0.00	5.00
A12	State of conservation	2.47	0.88	0.00	4.00
A13	Number of residential buildings	10.36	14,191	0.00	286.00
A14	Number of buildings in masonry	5.49	9,856	0.00	177.00
A15	Number of buildings in concrete	4.33	8,000	0.00	108.00
A16	Number of dwellings occupied	90.33	100,203	0.00	1,207
A17	Number of empty dwellings	3.82	8,566	0.00	126.00
A18	Surface of occupied dwellings	7,836	9,224,453	0.00	117,111
A19	Number of buildings	12.67	16,377	0.00	286.00
A20	Number of buildings used	12.41	16,059	0.00	286.00
A21	Number of employees	84.81	503,833	0.00	18,822
A22	Green areas	3,633	39,067	0.00	1,380,794
A23	Temperature	51,449	58,126	33.00	63.00

A correlation analysis was conducted on the 23 variables to evaluate the level of correlation between the variables and to ensure, therefore, the results accuracy of the subsequent multivariate statistical analysis. The collinear variables don't provide additional evidence and it is difficult to identify the weight that each of them has on the dependent variable.

Based on the outcomes of correlation analysis, five variables have been deleted: Average age of buildings (A02), Population under 15 (A05), Census section surface (A08), Number of buildings in masonry (A14) and Number of empty dwellings (A17).

Step 5 of the proposed methodology includes the PCA. For the implementation of the Principal Components Analysis (PCA), data related to the 18 variables were used as inputs to the relations between the urban characteristics and energy consumption in the City of Naples.

The eigenvalues useful to selecting the optimal number of components are reported in Fig. 4. The presence of an "elbow", corresponding with component number five, recommends stopping the research of significant number of components at this value.

The values in Table 3 show the percentages of variance of eigenvalues compared to the rotated matrix of the total explained variance. The distribution of the variables on the components evidences a uniformity as they are grouped around values that are not mostly different from each other. Therefore, it was necessary to consider sorting the variables in the rotated factor matrix, which affords a more significant sorting of the variables. The outputs consist of the eigenvalue vector that permits the percentages of variance explained by each component of the total to be obtained. In this application, the system is structured by five components explaining over 70% of the whole system variance. The first component explains more than 31%, and the second component over 15%, with a cumulative variance of more than 46%. The remaining three components explain over 25%, according to Table 3.

Table 4 shows the positive and the negative weights for each variable included in the five components. In relation to the first component, 17 are positively correlated and 1 is inversely correlated (temperature). Overall, there are two variables close to zero (number of employees and average number of floors). The variables that most characterize the first component are number of buildings used, number of buildings, population between 15 and 65 and total

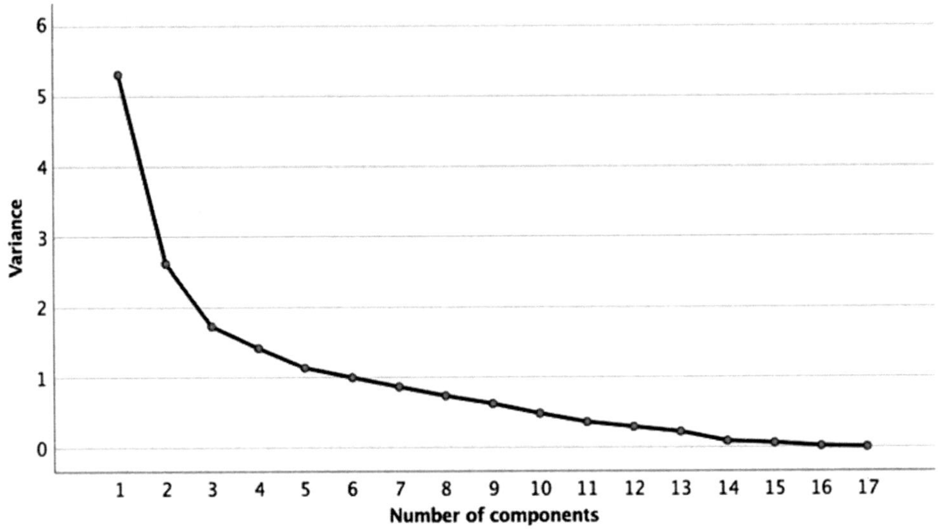

Fig. 4 Diagram of the variance trend of the components

Table 3 Variance of PCA eigenvalues

Factor	Rotation sums of squared loadings		
	Total	% of variance	% Cumulated
1	5,315	31,263	31,263
2	2,620	15,414	46,677
3	1,735	10,206	56,884
4	1,424	8,378	65,262
5	1,140	6,707	71,969

population. This first component is considered representative of socio- and settlement urban characteristics based on weight values.

For the second component, 8 variables result to be positively correlated and 10 are inversely correlated. The weight values for the temperature, number of employees, number of dwellings occupied, total population and population between 15 and 65 variables are close to zero. The variables that most characterize the second component are average number of floors, average age of the population and state of conservation. This component is considered representative of the residential buildings with high quality.

The weights obtained for the third components are representative of elderly population and green area. For this component, the PCA analysis shows that six variables are close to zero (average number of floors, number of residential buildings, number of buildings in concrete, number of buildings, number of buildings used and temperature).

Table 4 The structure of each principal component

Variables	Components				
	1	2	3	4	5
Built-up area	0.607	−0.272	0.183	−0.126	0.244
Total population	0.838	0.055	-0.295	−0.305	0.084
Population between 15 and 65	0.840	0.096	-0.302	−0.362	0.096
Population over 65	0.318	−0.367	0.755	−0.175	0.138
Average age of the population	0.303	0.771	0.306	0.102	−0.021
Average number of members per household	0.428	0.574	0.246	0.233	−0.133
Average number of floors	0.094	0.808	0.044	0.053	0.259
State of conservation	0.359	0.690	0.304	0.059	−0.168
Number of residential buildings	0.628	−0.223	−0.022	0.465	0.099
Number of buildings n concrete	0.658	−0.164	−0.026	0.035	−0.406
Number of dwellings occupied	0.311	0.050	−0.171	0.173	0.653
Surface of occupied dwellings	0.791	0.131	−0.284	−0.428	0.093
Number of buildings	0.848	−0.294	0.028	0.366	−0.167
Number of buildings used	0.850	−0.295	0.028	0.363	−0.165
Number of employees	0.042	−0.046	0.148	−0.112	0.266
Green areas	0.110	−0.261	0.751	−0.261	0.152
Temperature	−0.146	−0.081	−-0.074	0.553	0.415

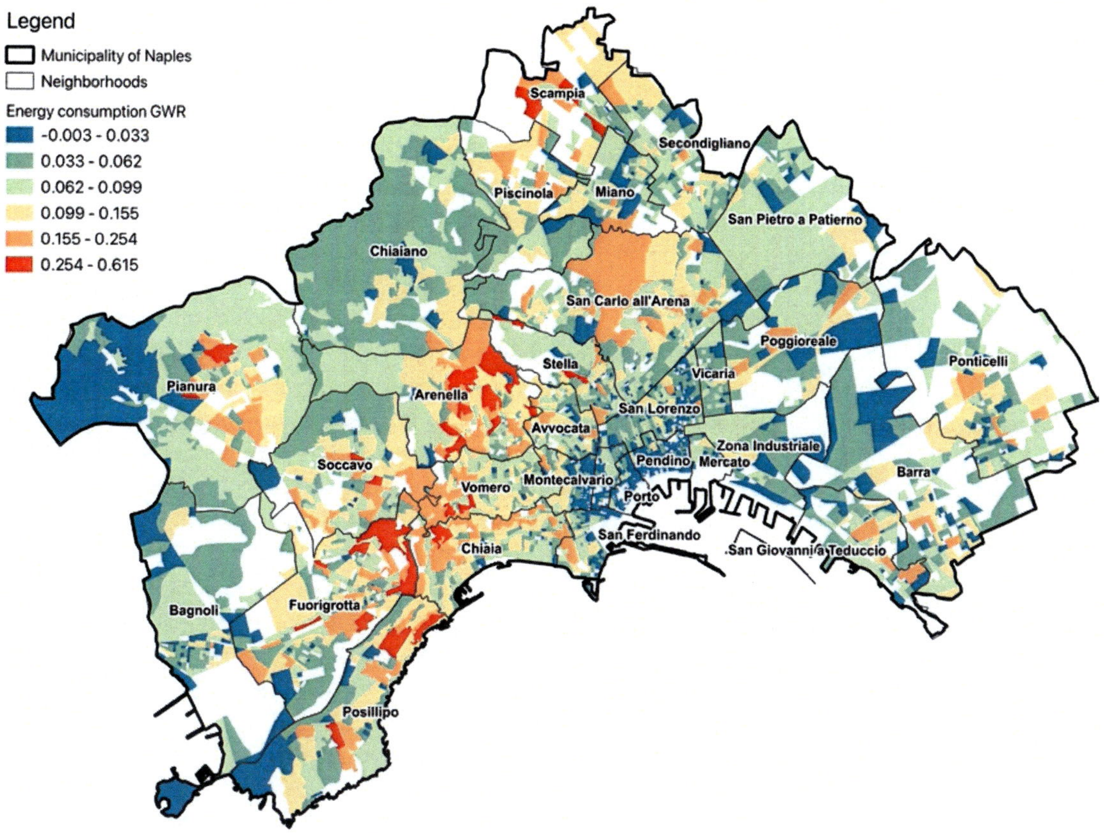

Fig. 5 The energy consumption prediction using GWR

For the last two components, the values of the weights indicate a common high value of the temperature variable, but they present different values of weight for the other variables. On the base of the weight values, these components represent high-density residential buildings and high-density settlements in urban areas.

The values of rotated variance explained for each component were used for step six of the methodology. This step includes ordinary least squares analysis, global Moran's I p-value analysis and geographically weighted regression analysis.

Adjusted R-squared, Akaike information criterion (AIC) and global Moran's I of the OLS and GWR regression results by ArcGIS Pro are reported in Table 5. AIC is the relative regression quality estimator. The residual from GWR and OLS regression is used to evaluate the global Moran's I. The result suggests that GWR performs better than the OLS method as the adjusted R^2 value from GWR is higher than OLS, and the AIC of the GWR method is lower than the OLS ones. Furthermore, Global Moran's I value from the OLS is higher than the GWR, showing that the GWR method has less spatial positive correlation than OLS. In conclusion, all Table 5 values guarantee that GWR is a better spatial model than the OLS.

Table 5 Model fit summary of OLS and GWR

	OLS	GWR
Multiple R-squared (R^2)	0.3262	0,3818
Akaike Information Criterion (Amick)	−8,644.28	−8,832.83
Global Moran's I p-value	0.020765	−0.006402

The GWR returned 3,403 different regression equations, one for each census section of Naples Municipality, in which the dependent variable is represented by normalized residential energy consumption, while the independent variables are the five principal components obtained by PCA. Figure 5 evidences the spatial distribution of energy consumption obtained by the GWR prediction model. The results evidence the city census tracks where the GWR prediction model estimates the higher energy consumption value linked to the variables. The neighbourhoods with the high values of energy consumption prediction are Arenella, Fuorigrotta, Posillipo, Vomero, Chiaia, Scampia, Soccavo and Pianura. It can be stated that the central-western part of the city between Avvocata and Fuorigrotta and Posillipo Districts is characterized by the highest value classes of energy consumption and this spatial concentration underlines the strict interrelation among the energy issue and the different characteristics of the urban system considered in this study. The northern part of the city stretching from Scampia District to Stella District is characterized by medium–low values of energy consumption due to the higher presence of unbuilt areas and the consequent less presence of anthropic activities, such as in Chiaiano District. Compared to these two extended parts of the city just described the eastern part (between Pendino to Barra Districts) is characterized by a more disaggregated distribution of energy consumption and this can be due to the high mix of activities (especially the industrial ones).

4 Conclusions and Policy Implications

The current energy crisis in the EU, caused by the war and the increase in the price of raw materials, enhances the urgency of a change in the production sources and the ways of using and saving energy. In addition, the increasing frequency and intensity of extreme climate events (e.g. floods, storms, heat waves) threaten energy supply and the related services that are essential for the organization and operation of urban systems (Groundstroem & Juhola, 2019). At the same time, energy peak demand is expected to growth in numerous OECD countries due to the temperature changes (e.g. more heat waves in USA) (IPCC, 2022).

As stated by numerous institutions and government heads, it is essential to empower the local authorities to strengthen their capacity to increase the resilience of urban areas to limit the EU's dependency on fossil fuels and reduce energy consumption.

In line with this last goal, this work provided to develop a statistical-spatial methodology to underpin local policy-makers in the identification of factors that influence the higher energy consumption in the different areas of the city. Sustainable urban planning policies, strategies and interventions can strongly reach energy-saving goals of cities if the relations between urban characteristics and energy consumption are investigated with an integrated approach and are found to be significant. In this perspective, the systemic approach allows to define and interpret the nexus between the main urban characteristics (physical, functional, geographical and socio-economic) affecting energy consumption and these last ones and the results confirm «the complexity and multidimensionality of the relationship between the urban system and energy consumption» (Carpentieri et al., 2019; Guida, 2022; Papa et al., 2016b).

An interesting future development of the work could be to apply the methodology to different cities and compare the spatial regression model (OLS and GWR) outputs to better support policy-makers in the definition of energy-saving interventions. Furthermore, the use of more recent energy consumption data also referring to other urban activities allows to provide a more detailed and efficient energy levels distribution.

To compare the performance of GWR and OLS models, the criteria of R^2, AIC and Moran's I were used (Gomes et al., 2017; Kashki et al., 2021). The outcomes indicated that the GWR model has a higher coefficient of determination and its AIC is lower than the OLS model, so the GWR can predict in a more efficient way.

In summary, this work provides outcomes of interest for urban and territorial planners, technicians and decision-makers to define specific strategies and policies for optimizing and reducing energy consumption in urban areas.

The application of the methodology in a densely built urban system such as the City of Naples was helpful to validate the methodology and evidence some critical aspects of this territorial contest useful for the local authorities in developing the general municipal master plan and the specifical energy plans at the city level and for the single districts. The outcomes of the application are also useful for policy-makers, economic operators and citizens in decision-making processes.

The energy consumption values make it possible to identify the presumably more energy-intensive areas due to the urban characteristics present. In this way, the whole urban area can be articulated according to the different levels of priority interventions to reduce energy consumption. For example, in areas such as Arenella and Scampia, where GWR outputs are the highest values (and so the highest energy consumption), solutions geared towards refurbishing low-quality buildings, raising awareness and increasing the presence of green areas can help make these neighbourhoods more energy sustainable. These possible actions and interventions can be defined based on the urban key characteristics such as low education and the significant presence of impervious surfaces. The next steps of this research will be to identify, from the outcomes of the analysis, the different practical interventions compatible with the specific characteristics and critical issues of each urban area.

References

Allegrini, J., Orehounig, K., Mavromatidis, G., Ruesch, F., Dorer, V., & Evins, R. (2015). A review of modelling approaches and tools for the simulation of district-scale energy systems. *Renewable and Sustainable Energy Reviews, 52*, 1391–1404. https://doi.org/10.1016/j.rser.2015.07.123

Andersen, K. S., Termansen, L. B., Gargiulo, M., & Gallachóirc, B. P. Ó. (2019). Bridging the gap using energy services: Demonstrating a novel framework for soft linking top-down and bottom-up models. *Energy, 169*, 277–293. https://doi.org/10.1016/j.energy.2018.11.153

Anselin, L., & Getis, A. (2010). Spatial statistical analysis and geographic information systems. In L. Anselin, & S. Rey (Eds.), *Perspectives on Spatial Data Analysis. Advances in Spatial Science* (pp. 35–47). Springer. https://doi.org/10.1007/978-3-642-01976-0_3

Asarpota, K., & Nadin, V. (2020). Energy strategies, the urban dimension, and spatial planning. *Energies, 13*(14), 3642. https://doi.org/10.3390/en13143642

Barker, T., Alexandri, E., Mercure, J.-F., Ogawa, Y., & Pollitt, H. (2016). GDP and employment effects of policies to close the 2020 emissions gap. *Climate Policy, 16*, 393–414. https://doi.org/10.1080/14693062.2014.1003774

Batty, M. (2008). *Cities as complex systems: scaling, interactions, networks, dynamics and urban morphologies* (UCL Working paper 131). Centre for Advanced Spatial Analysis, University College London.

Bianco, V., Manca, O., & Nardini, S. (2009). Electricity consumption forecasting in Italy using linear regression models. *Energy, 34*(9), 1413–1421. https://doi.org/10.1016/j.energy.2009.06.034

Böhringer, C., & Rutherford, T. F. (2008). Combining bottom-up and top-down. *Energy Economics, 30*(2), 574–596. https://doi.org/10.1016/j.eneco.2007.03.004

Carpentieri, G., Zucaro, F., Guida, C., & Granata, L. (2019). GIS-based spatial analysis for the integrated transport-land use-energy planning: An application to the great London area. *Journal of Civil Engineering and Architecture, 13*, 469–481. https://doi.org/10.17265/1934-7359/2019.11.001

Carpentieri, G., & Favo, F. (2017). The end-use electric energy consumption in urban areas: a gis-based methodology. An application in the city of Naples. *Tema. Journal of Land Use, Mobility and Environment, 10*(2), 139–156. https://doi.org/10.6092/1970-9870/5173

Chen, H. C., Han, Q., & de Vries, B. (2020). Urban morphology indicator analyzes for urban energy modeling. *Sustainable Cities and Society, 52*, 101863. https://doi.org/10.1016/j.scs.2019.101863

Chen, I. C., Cheng, K. L., Ma, H. W., & Hung, C. C. (2021b). Identifying spatial driving factors of energy and water consumption in the context of urban transformation. *Sustainability, 13*(19), 10503. https://doi.org/10.3390/su131910503

Chen, S., Zhang, G., Xia, X., Chen, Y., Setunge, S., & Shi, L. (2021a). The impacts of occupant behavior on building energy consumption: A review. *Sustainable Energy Technologies and Assessments, 45*, 101212. https://doi.org/10.1016/j.seta.2021.101212

Comber, A., Brunsdon, C., Charlton, M., Dong, G., Harris, R., & Lu, B., et al. (2020). The GWR route map: a guide to the informed application of geographically weighted regression. 2004.06070. https://doi.org/10.48550/arXiv.2004.06070

Corrado, V., Ballarini, I., Madrazo, L., & Nemirovskij, G. (2015). Data structuring for the ontological modelling of urban energy systems: The experience of the SEMANCO project. *Sustainable Cities and Society, 14*, 223–235. https://doi.org/10.1016/j.scs.2014.09.006

Deb, C., & Schlueter, A. (2021). Review of data-driven energy modelling techniques for building retrofit. *Renewable and Sustainable Energy Reviews, 144*, 110990. https://doi.org/10.1016/j.rser.2021.110990

Ferrando, M., & Causone, F. (2019). An overview of urban building energy modelling (UBEM) tools. *Building Simulation, 16*, 3452–3459. https://doi.org/10.26868/25222708.2019.210632

Fotheringham, A. S., & Oshan, T. M. (2016). Geographically weighted regression and multicollinearity: Dispelling the myth. *Journal of Geogapohical Systems, 18*, 303–329. https://doi.org/10.1007/s10109-016-0239-5

Frayssinet, L., Merlier, L., Kuznik, F., Hubert, J. L., Milliez, M., & Roux, J. J. (2018). Modeling the heating and cooling energy demand of urban buildings at city scale. *Renewable and Sustainable Energy Reviews, 81*, 2318–2327. https://doi.org/10.1016/j.rser.2017.06.040

Gargiulo, C., & Lombardi, C. (2016). Urban retrofit and resilience: The challenge of energy efficiency and vulnerability. *TeMA—Journal of Land Use, Mobility and Environment, 9*(2), 137–162. https://doi.org/10.6092/1970-9870/3922

Gargiulo, C., & Russo, L. (2017). Cities and energy consumption: A critical review. *TeMA Journal of Land Use, Mobility and Environment, 10*(3), 259–278. https://doi.org/10.6092/1970-9870/5182

Gargiulo, C., & Zucaro, F. (2015). Between the smart and the resilient city. The Urban Saving Energy Model. *TeMA Journal of Land Use Mobility and Environment*, 81–102. https://doi.org/10.6092/1970-9870/3661

Gomes, M. J. T. L., Cunto, F., & da Silva, A. R. (2017). Geographically weighted negative binomial regression applied to zonal level safety performance models. *Accident Analysis and Prevention, 106*, 254–261. https://doi.org/10.1016/j.aap.2017.06.011

Groundstroem, F., & Juhola, S. (2019). A framework for identifying cross-border impacts of climate change on the energy sector. *Environment Systems and Decisions, 39*. https://doi.org/10.1007/s10669-018-9697-2

Guida, C. (2022). Energy saving and efficiency in urban environments: integration strategies and best practices. Tema. Journal of Land Use, Mobility and Environment, 15(3), 517–531. https://doi.org/10.6092/1970-9870/9549

Hotelling, H. (1933). Analysis of a complex of statistical variables into principal components. *Journal of Educational Psychology, 24*(6 & 7), 417–441, 498–520.

Hunt, L. C., & Ryan, D. L. (2015). Economic modelling of energy services: Rectifying misspecified energy demand functions. *Energy Economics, 50*, 273–285. https://doi.org/10.1016/j.eneco.2015.05.006

IPCC–International Panel of Climate Change. (2022). *Climate change 2022. Mitigation of climate change.* Contribution of Working Group III to the Sixth Assessment Report of the Intergovernmental Panel on Climate Change. Cambridge University Press. https://doi.org/10.1017/9781009157926

ISTAT–Istituto Nazionale di Statistica. (2020). Il censimento permanente della popolazione in Campania. Available at: https://www.istat.it/it/archivio/. Last retrieved February 22, 2023

Ivajnsic, D., Kaligaric, M., & Ziberna, I. (2014). Geographically weighted regression of the urban heat island of a small city. *Applied Geography, 53*(1), 341–353. https://doi.org/10.1016/j.apgeog.2014.07.001

Jebaraj, S., & Iniyan, S. (2006). A review of energy models. *Renewable and Sustainable Energy Reviews, 10*(4), 281–311. https://doi.org/10.1016/J.RSER.2004.09.004

Kashki, A., Karami, M., Zandi, R., & Roki, Z. (2021). Evaluation of the effect of geographical parameters on the formation of the land surface temperature by applying OLS and GWR, A case study Shiraz City, Iran. *Urban Climate, 37*, 100832. https://doi.org/10.1016/j.uclim.2021.100832

Kitchin, R., & Thrift, N. (2009). *International encyclopedia of human geography*. Elsevier.

Li, X., Yao, R., Yu, W., Meng, X., Liu, M., Short, A., & Li, B. (2019). Low carbon heating and cooling of residential buildings in cities in the hot summer and cold winter zone-A bottom-up engineering stock modeling approach. *Journal of Cleaner Production, 220*, 271–288. https://doi.org/10.1016/j.jclepro.2019.02.023

Lin, B., & Liu, H. (2015). China's building energy efficiency and urbanization. *Energy and Buildings, 86*, 356–365. https://doi.org/10.1016/j.enbuild.2014.09.069

Liu, S., Zou, Y., Ji, W., Zhang, Q., Ahmed, A., & Han, X., et al. (2022). Energy-saving potential prediction models for large-scale building: A state-of-the-art review. *Renewable and Sustainable Energy Reviews, 156*, 111992. https://doi.org/10.1016/j.rser.2021.111992

Lund, H., Arler, F., Østergaard, P., Hvelplund, F., Connolly, D., Mathiesen, B., et al. (2017). Simulation versus optimisation: Theoretical positions in energy system modelling. *Energies, 10*, 840. https://doi.org/10.3390/en10070840

Mastrucci, A., Baume, O., Stazi, F., & Leopold, U. (2014). Estimating energy savings for the residential building stock of an entire city: A GIS-based statistical downscaling approach applied to Rotterdam. *Energy and Buildings, 75*, 358–367. https://doi.org/10.1016/j.enbuild.2014.02.032

Moghadam, S. T., Coccolo, S., Mutani, G., Lombardi, P., Scartezzini, J. L., & Mauree, D. (2019). A new clustering and visualization method to evaluate urban heat energy planning scenarios. *Cities, 88*, 19–36. https://doi.org/10.1016/j.cities.2018.12.007

Nguyen, C. T., Nguyen, D. T. H., & Phan, D. K. (2021). Factors affecting urban electricity consumption: A case study in the Bangkok metropolitan area using an integrated approach of earth observation data and data analysis. *Environment Science and Pollution Research, 28*, 12056–12066. https://doi.org/10.1007/s11356-020-09157-6

Papa, R., Gargiulo, C., & Zucaro, F. (2016a). Towards the definition of the urban saving energy model (UrbanSEM). In R. Papa, & R. Fistola (Eds.), *Smart energy in the smart city* (pp. 151–175). Springer International Publishing. https://doi.org/10.1007/978-3-319-31157-9_9

Papa, C. G., & Zucaro, F. (2014). Urban systems and energy consumptions: A critical approach. *TeMA Journal of Land Use, Mobility and Environment,* 783–792. https://doi.org/10.6092/1970-9870/2552

Papa, R., Gargiulo, C., Zucaro, F., Cristiano, M., Angiello, G., & Carpentieri, G. (2016b). Energy and climate change polices in Europe: Overview and selected examples from a spatial planning perspective. *Smart Energy in the Smart City*, 237–274. https://doi.org/10.1007/978-3-319-31157-9_13

Pearson, K. (1901). On lines and planes of closest fit to systems of points in space. *Philosophical Magazine, 6, 2*(11), 559–572.

Prina, M. G., Manzolini, G., Moser, D., Nastasi, B., & Sparber, W. (2020). Classification and challenges of bottom-up energy system models-A review. *Renewable and Sustainable Energy Reviews, 129*, 109917. https://doi.org/10.1016/j.rser.2020.109917

Quan, S. J., & Li, C. (2021). Urban form and building energy use: A systematic review of measures, mechanisms, and methodologies. *Renewable and Sustainable Energy Reviews, 139*, 110662. https://doi.org/10.1016/j.rser.2020.110662

Shaaban, M., Scheffran, J., Elsobki, M. S., & Azadi, H. (2022). A comprehensive evaluation of electricity planning models in egypt: optimization versus agent-based approaches. *Sustainability, 14*(3), 1563.

Soltani, A., Mehraein, M., & Sharifi, E. (2012). Urban features and energy consumption at local level. *Journal of Urban and Environmental Engineering, 6*(2), 43–47. https://doi.org/10.4090/juee.2012.v6n2.043047

Steemers, K., & Yun, G. Y. (2009). Household energy consumption: A study of the role of occupants. *Building Research and Information, 37*(5–6), 625–637. https://doi.org/10.1080/09613210903186661

Swan, L. G., & Ugursal, V. I. (2009). Modeling of end-use energy consumption in the residential sector: A review of modeling techniques. *Renewable and Sustainable Energy Reviews, 13*(8), 1819–1835. ISSN 13640321. https://doi.org/10.1016/j.rser.2008.09.033

Tira, M. (2020). About the sustainability of urban settlements. *TeMA—Journal of Land Use, Mobility and Environment*, 361–371. https://doi.org/10.6092/1970-9870/6984

Walter, T., & Sohn, M. D. (2016). A regression-based approach to estimating retrofit savings using the building performance database. *Applied Energy, 179*, 996–1005. https://doi.org/10.1016/j.apenergy.2016.07.087

Walter, E., & Kämpf, J. H. (2015). A verification of CitySim results using the BESTEST and monitored consumption values. In *Proceedings of the 2nd Building Simulation Applications Conference* (pp. 215–222). Bozen-Bolzano University Press.

Wang, M., Wang, W., & Wu, L. (2022). Application of a new grey multivariate forecasting model in the forecasting of energy consumption in 7 regions of China. *Energy, 243*, 123024. https://doi.org/10.1016/j.energy.2021.123024

Wheeler, D., & Tiefelsdorf, M. (2005). Multicollinearity and correlation among local regression coefficients in geographically weighted regression. *Journal of Geographical Systems, 7*(2), 161–187. https://doi.org/10.1007/s10109-005-0155-6

Mechanical Properties and Impact Resistance of Steel Fiber Reinforced Fly Ash and Nano-Silica Concrete

Wong King Yiin, Roszilah Hamid and Saleha Md Salleh

Abstract

Steel Fiber Reinforced Concrete (SFRC) with an additional 1.0, 1.5, and 2.0% of steel fibers was introduced onto concrete mixtures to improve the mechanical properties and ultimate energy absorption of High-Volume Fly Ash Concrete (HVFAC). The fly ash was added with a high volume of 45.0% by weight of cement. To overcome the loss of strength caused by high-volume fly ash and resistance to fire, 7.5% nano-silica and 0.2% polypropylene fibers were also added. Compressive, splitting tensile, flexural, and drop weight tests were carried out at 7, 28, 125, and 129 days. The results show that adding 2.0% steel fiber increases the mechanical strength and energy absorption to the maximum, with 27.4, 34.2%, and very significant 73.1% increments in compressive, split tensile, and flexural strengths, respectively, at 7 days compared to the control samples where it comprises 0% of steel fiber. At the later age of 125 days, the rate of increment had slowed down to 23.4, 22.0, and 76.4% for the respective strengths, showing that nano-silica addition is very effective in the development of high early strength of HVFAC as well as the impact resistance. The impact resistance of HVFAC is greatly enhanced, with the ultimate energy absorption improved by 250.0 and 200.0% at 7 and 125 days, respectively.

W. K. Yiin · R. Hamid · S. Md Salleh (✉)
Department of Civil Engineering, Faculty of Engineering & Built Environment, Universiti Kebangsaan Malaysia, UKM, 43600 Bangi, Selangor, Malaysia
e-mail: saleha.mds@uitm.edu.my

W. K. Yiin
e-mail: p112078@siswa.ukm.edu.my

R. Hamid
e-mail: roszilah@ukm.edu.my

S. Md Salleh
Centre of Foundation Studies, Universiti Teknologi MARA, Cawangan Selangor, Kampus Dengkil, 43800 Dengkil, Selangor, Malaysia

Keywords

Steel fiber reinforced concrete · High-volume fly ash · Nano-silica · Mechanical properties · Impact resistance

1 Introduction

The compressive strength of concrete is significantly higher than its tensile strength, and at higher strengths, it tends to become brittle (Abbass et al., 2018; Shende et al., 2012; Verma et al., 2015). In present construction, Steel Fiber Reinforced Concrete (SFRC) is frequently utilized in the construction of in situ or precast concrete structures to minimize the need of conventional steel bar reinforcement, especially in tunnel lining, thin-walled structures, and foundation applications (Marcos-Meson et al., 2019; Su et al., 2016). Incorporating the steel fibers (SF) as reinforcement in concrete can reduce the use of rebar and the labor cost to bend and install the rebar during construction (Andriichuk et al., 2018; Marcos-Meson et al., 2018; Sorelli & Toutlemonde, 2005). The addition of steel fibers is an effective method for enhancing its durability to resist crack formation and propagation, along with increasing the flexural strength and impact resistance (Gehlot et al., 2017; Lee, 2007; Ng et al., 2013; Zhou et al., 2018; Srikar & Kalyan, 2018). The introduction of steel fibers into the concrete mixture reduces the workability of the concrete (Afroughsabet & Ozbakkaloglu, 2015; Choi et al., 2019) and only has a small positive effect on the concrete's compressive strength (Gehlot et al., 2017; Nayan & Kisku, 2017; Niu et al., 2013; Yet et al., 2012) which have shown that the addition of steel fibers can augment its compressive, tensile, and flexural strengths. Additionally, by incorporating the steel fibers is also significantly to enhance the concrete's energy absorption capacity (Banthia et al., 1996; Barros & Cruz, 1999; Yu et al., 2016).

The incorporation of fly ash (FA) in concrete is beneficial because its pozzolanic properties can increase the

B.-J. He et al. (eds.), *Resilient Horizons: Building Sustainable Environments for Climate Adaptation and Health*, Advances in Science, Technology & Innovation, https://doi.org/10.1007/978-3-031-46109-5_9

durability of concrete. To replace cement as a binding agent in concrete, References (Begum et al., 2018; Doljikov et al., 1986; Gorai, 2018) recommended the incorporation of a high volume of fly ash (35.0% to 50.0%) as a binder replacement for cement. The use of fly ash as cement replacement in high amounts may lead to decreased compressive strength in the early stage of concrete, but adding nano-silica to the mixture can activate a pozzolanic reaction (Quercia & Brouwers, 2010; Senff et al., 2010; Singh et al., 2013; Varghese et al., 2017; Yuvaraj & Anish, 2019).

The addition of polypropylene (PP) fibers in concrete can prevent the occurrence of micro-cracks (Dharan & Lal, 2016; Saadun et al., 2016). The fire resistance of concrete can be improved by adding nano-silica and polypropylene to the concrete mixture (Ibrahim et al., 2012, Serrano et al., 2016). To prevent explosive spalling failures that can happen when concrete is subjected to heat using polypropylene fibers is a widely recognized and effective approach (Suhaendi & Horiguchi, 2005; Yazici et al., 2015; Zeiml et al., 2006).

At 28 days, the study Afroughsabet and Ozbakkaloglu (2015) found that the addition of 1.0% steel fibers to concrete mixes significantly improved the compressive, tensile, and flexural strengths by 19.0, 55.0, and 61.0%, respectively, compared to the control sample. Research by Abbass et al. (2018) demonstrated that concrete mixture with 1.5% steel fibers exhibits the most substantial strength enhancement after 28 days. In particular, the addition of steel fibers resulted in a compressive strength increase of 10.0%, tensile strength increase of 32.0%, and flexural strength increase of 149.0%, relative to the control sample. The study by Bazgir and Fu (2016) discovered that concrete mixtures containing 2.0% steel fibers had the most significant improvement in compressive, tensile, and flexural strength have the highest increase of 32.0, 35.0, and 44.0% in compressive, tensile, and flexural strengths, respectively, relative to the control sample. The concrete mixes containing 1.5% of steel fibers have higher ultimate energy absorption compared to those containing 0.5 and 1.0% steel fibers.

In this study, four HVFAC samples containing 0, 1.0, 1.5, and 2.0% steel fibers are prepared to determine their mechanical properties (compressive, splitting tensile, and flexural strengths). A drop weight test was carried out to determine the impact resistance of the HVFAC. For each mix, cement was replaced with 45.0% fly ash and 7.5% nano-silica by weight of cement, and 0.2% polypropylene fibers by volume of concrete. The objectives of this study are to determine the effect of adding 1.0, 1.5, and 2.0% steel fibers on the compressive, split tensile, and flexural strengths and the energy absorption capacity of the concrete containing a high volume of fly ash and nano-silica and PP fibers.

2 Experimental Method

2.1 Materials

The cement used in this study is Ordinary Portland Cement (OPC) type 1 with a specific gravity of 3.15 and meets the requirements stated in ASTM-C150 (2016). The fine aggregates or sand follow the specification in ASTM-C33 (2016), where 200 sieve analysis of sand is comprised with 0.0% residual material which follows the specified limit. This study uses coarse aggregates with a maximum size of 20 mm. The fly ash used in this study is class F fly ash in the ASTM-C618 (2012). Class F fly ash has pozzolanic properties. Table 1 presents the chemical components of class F fly ash.

The nano-silica used in this study is in colloidal form and has a density of 1050 kg/m^3 and a pH of 10. It contains 50% amorphous nano-silica with an average diameter of 50 nm. The polypropylene fibers are white, have a specific gravity of 0.9, and have a melting point between 160 and 170 °C. The PP fibers have a tensile strength of 0.36 kN/mm^2 and a density of 920 kg/m^3. The dose added to the concrete mix is 0.2% by volume of concrete. The hooked-end steel fiber used in this study is 35 mm long and 55 μm in diameter. It has a density of 7850 kg/m^3 and tensile strength of 1300 MPa. The water used for mixing and curing is clean, treated potable water.

2.2 Methods

2.2.1 Mixed Design

The mixed design proportion follows the British Standard (DOE). The target grade of concrete is M60 with a water–cement ratio of 0.38, but the target is difficult to achieve without the inclusion of a superplasticizer (SP). The concrete design ratio for cement: fine aggregate: coarse aggregate is 1: 1.2: 2.1. Fresh concrete is poured into six 150 × 150 × 150 mm^3 cube molds, twelve 200 mm × 100 mm Ø cylinder molds, and six 100 × 100 × 500 mm^3 beam molds. All samples are then de-molded and cured in water for 7, 28, 125, and 129 days

Table 1 The chemical component of class F fly ash (ASTM-C618, 2012)

Chemical component	Percentage (%)
Silicon dioxide + aluminum oxide + iron oxide, minimum, (%)	70.0
Sulfur trioxide, maximum, (%)	5.0
Moisture content, maximum, (%)	3.0

Table 2 Mix design proportions of concrete

Components	Sample			
	All measurements in unit kg			
	SK	K1	K1.5	K2
Cement	244.15	244.15	244.15	244.15
Water	196.6	196.6	196.6	196.6
Fine aggregate	600	600	600	600
Coarse aggregate	1066	1066	1066	1066
Fly ash	231.3	231.3	231.3	231.3
Nano-silica	38.55	38.55	38.55	38.55
PP fiber	1.84	1.84	1.84	1.84
Steel fiber	0	78.50	117.75	157.00

Table 3 Total number of samples

Age (days)	Sample			
	SK	K1	K1.5	K2
	Cube: Compressive test			
7	3	3	3	3
28/125/129	3	3	3	3
	Cylinder: Splitting tensile test			
7	3	3	3	3
28/125/129	3	3	3	3
	Cylinder: Impact test			
7	3	3	3	3
28/125/129	3	3	3	3
	Beam: Flexural test			
7	3	3	3	3
28/125/129	3	3	3	3
Total	96			

before testing. Tables 2 and 3 show the mix design proportions of the concrete and the total number of the sample.

2.2.2 Testing

In this study, the sample testing was conducted under slump, compressive, splitting tensile, flexural, and impact resistance tests. All the testing is performed in accordance with standards and specifications for concrete.

Once the sample mixture is ready, the slump test is performed following the ASTM-C143 (2015) to measure the consistency of wet concrete before it sets. The sample will be molded as specified size and cured for about 7, 28, 125, and 129 days. All cured cubes will then be de-molded to perform the compressive test. The cube test is prepared following the BS1881: Part 116: 1983. Each sample is subjected to a uniform static compression load at a constant rate of 0.2 kN/s until the cube sample fails.

The cured cylinder sample was examined under the splitting tensile test following the British Standard BS1881: Part 117: 1983. The cylinder sample is subjected to a constant compression load on its long axis until the cylinder sample is split into two halves. Two-point load flexural test follows the British Standard BS1881: Part 118: 1983.

The impact resistance of concrete is determined by a drop weight test following the ASTM-D7136 to determine the ultimate energy absorption of concrete (ASTM-D7136, 2012). This test is performed using a cylinder sample with a diameter of 100 mm and a height of 200 m. A 1.8 kg mass is applied to the cylinder sample and released from a height of 2.4 m.

3 Results and Discussion

3.1 Workability

Based on the concrete design procedure adopted in this study, the value of the selected slump of concrete is between 30 and 60 mm. Table 4 and Fig. 1 show that the slump value decreased when steel fibers were added to the concrete mixture compared to the slump value of the control sample (SK). This decrease is because the steel fibers incorporated into the concrete mix hinder the movement of coarse particles.

The highest slump value of 36 mm for SK is because of the absence of steel fibers in the concrete mixture. Sample K2 has the highest percentage decrease in slump value of 16.67%, and sample K1 has the lowest difference percentage of 5.56%. The large proportion of steel fibers in sample

Table 4 Slump of fresh mixes and the difference compared with control

Sample	Slump (mm)	Percent difference (%)
SK	36	
K1	34	−5.56
K1.5	31	−13.89
K2	30	−16.67

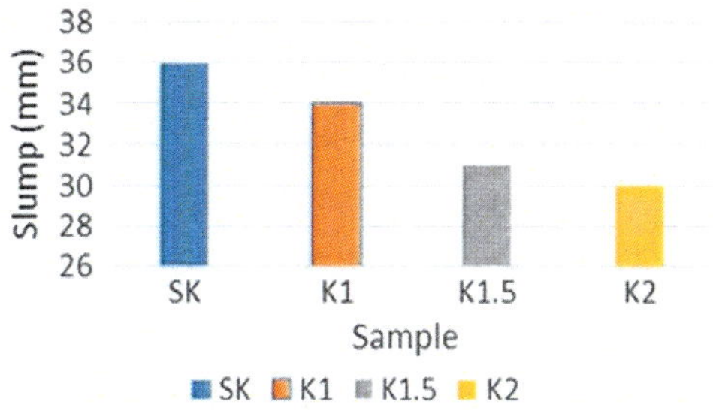

Fig. 1 Slump of fresh mixes

Table 5 Compressive, splitting tensile, and flexural strength at 7, 28, 125, and 129 days

Sample	Compressive strength, f_c				Splitting tensile strength, f_{st}				Flexural strength, MOR			
	7 days		28/125/129 days		7 days		28/125/129 days		7 days		28/125/129 days	
	f_c	% Increase	f_c	% Increase	f_{st}	% Increase	f_{st}	% Increase	MOR	% Increase	MOR	% Increase
SK	33.4	–	40.4	–	1.9	–	2.9	–	4.9	–	6.2	–
K1	37.4	12.0	43.0	6.4	2.1	11.2	3.1	8.4	5.5	12.8	6.8	9.5
K1.5	39.9	19.3	46.4	15.0	2.4	26.2	3.3	13.6	6.2	27.1	8.6	39.3
K2	42.6	27.4	49.8	23.4	2.5	34.2	3.5	22.0	8.5	73.1	10.9	76.4

SK and K1 tests were done at 28 days, K1.5 tests were done at 129 days, K2 tests were done at 125 days
SK is the control sample, K1, K1.5, and K2 are the concrete samples containing 1.0 1.5, and 2.0% steel fibers, respectively

K2 increases concrete resistance to flow. Although steel fibers do not have any absorption capacity, they can prevent the flow of concrete by overlapping over one another to form a "canopy frame".

These results are consistent with those of previous studies, such as the study by Ismail and Hassan (2017), which found that the slump value of concrete mixture without steel fiber is 715 mm. The slump value decreased to 710 mm after the addition of 0.35% steel fibers, which indicates that the addition of steel fibers reduced the workability of concrete. Iqbal et al. (2015) stated that high steel fiber content of more than 0.75% reduces workability.

3.2 Compressive Strength

Table 5 and Fig. 2 show that the control sample, SK, has the lowest compressive strength of 33.4 MPa (56.0%) of the 7 day target strength (60 MPa), and 40.36 MPa (67.0%) of the target strength (60 MPa). The sample with 2.0% SF, K2, has the highest compressive strength of 42.58 MPa (71.0%) of the 7-day target strength (60 MPa), and 49.80 MPa (83.0%) of the target strength (60 MPa) at 125 days. All samples did not reach the target strength because the mixtures were too dry during the mixing process, and, as a result, the cement was not able to effectively bind the mixture's internal components such as aggregates and fibers. Therefore, 0.5 kg of water was added during the mixing process, and the increase in water-to-cement ratio from 0.37 to 0.38 resulted in a lower target strength. At 125 days, the rate of increment in strength has slowed down with an increase of only 12.0% from the strength at day 7, which indicates that the addition of nano-silica can develop a high early strength of HVFAC.

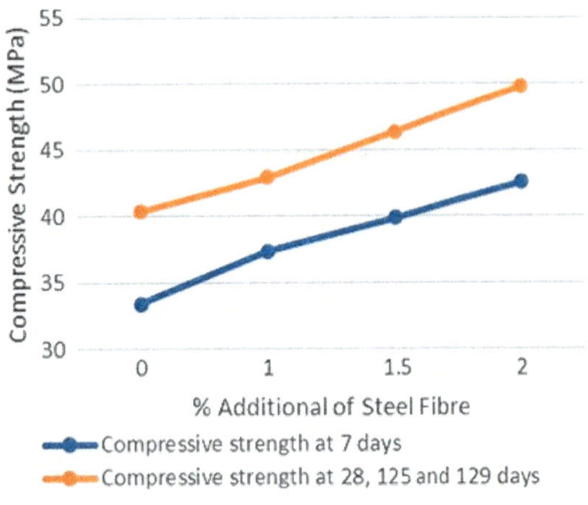

Fig. 2 Compressive strength of the concrete at 7, 28, 125, and 129 days

The addition of steel fibers also resulted in a higher compressive strength (Fig. 3). The addition of 2.0% of steel fibers in sample K2 resulted in the highest increase of 27.4 and 23.4% in the 7 and 125 day compressive strength, respectively, relative to the control sample, SK. The addition of 1.0% of steel fiber into sample K1 increased the 7 and 28 day compressive strength by 12.0 and 6.4%, respectively. The addition of 1.5% steel fiber into sample K1.5 increased the 7 and 129 day compressive strength by 19.3 and 15.0%, respectively. The increase in compressive strength is due to the random distribution of steel fibers in the concrete mixture, which gives the concrete the ability to resist the formation and growth of cracks.

These results are consistent with those of previous studies. For instance, Mohammadi et al. (2009) stated that the increase in compressive strength was shown to be directly proportional to the increase in quantity of steel fiber added to the concrete mixture. The 28 day compressive strength for the control sample is 57.82 MPa. The addition of 1.0,

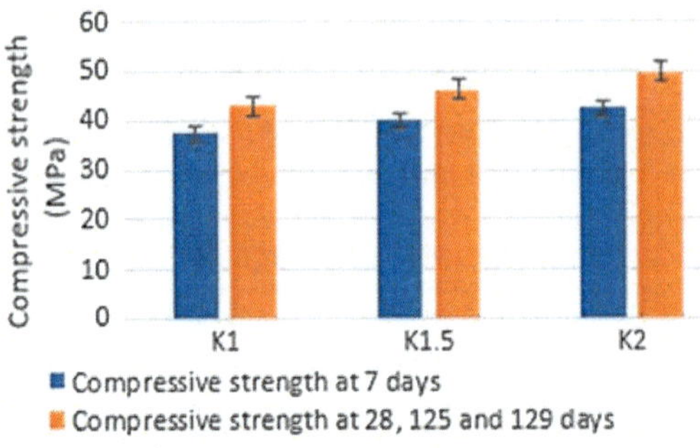

Fig. 3 Compressive strength of the steel fiber reinforced concrete

1.5, and 2.0% steel fibers resulted in 69.83, 72.13, and 72.82 MPa compressive strength, respectively, which is equivalent to an increase of 20.8, 24.7, and 25.9%. These results show that the addition of 2.0% steel fibers resulted in a maximum increase in compressive strength compared to the addition of 1.0% and 1.5% steel fibers. The addition of steel fibers resulted in a 30.0% maximum increase in compressive strength relative to the control sample.

3.3 Split Tensile Strength

Table 5 and Figs. 4, 5 show that the addition of steel fibers increased the splitting tensile strength. The addition of 2.0% steel fibers to the K2 sample resulted in a 34.2 and 22.0% increase in the 7 and 125 day splitting tensile strength relative to the control sample, SK. The addition of 1.0% of steel fibers resulted in an 11.2% and 8.4% increase in the 7 and 28 day HVFAC strength, respectively. The addition of 1.5% of steel fibers resulted in a 26.2 and 13.6% increase in the 7 and 129 day concrete split tensile strength of the K1.5 sample, respectively. The increase is the result of the steel fibers and concrete matrix acting as composite materials that improve the bonding between materials. The random distribution of steel fibers in concrete improves the concrete's resistance to the expansion and spread of cracks. When cracks begin to form in the concrete structure, the steel fibers in the concrete matrix prevent them from becoming larger.

In Nayan and Kisku (2017), claimed that the incorporation of steel fibers can raise the splitting tensile strength. The research conducted showed that at 7 and 28 days, the control sample had a splitting tensile strength of 2.22 and 3.18 MPa, respectively. The addition of 1.0 and 2.0% of steel fibers increased the 7 day splitting tensile strength to 2.71 and 2.92 MPa, respectively, while the 28 day splitting tensile strength increased to 3.40 and 3.75 MPa, respectively. Therefore, the addition of 2.0% of steel fibers resulted in a higher increase in the splitting tensile strength relative to the addition of 1.0% of steel fibers, where the increment is 31.5% at 7 days and 17.9% at 28 days.

3.4 Flexural Strength

Table 5 and Fig. 6 show that the increase in flexural strength is higher than the increase for other types of strengths as the percentage of steel fiber increases. Figure 7 shows that the addition of 2.0% of steel fibers in sample K2 resulted in the highest increment of 73.1 and 76.4% in the 7 and 125 day flexural strength, respectively, relative to the control sample, SK. The addition of 1.0% of steel fibers into sample K1 resulted in a 12.8 and 9.5% increase in the

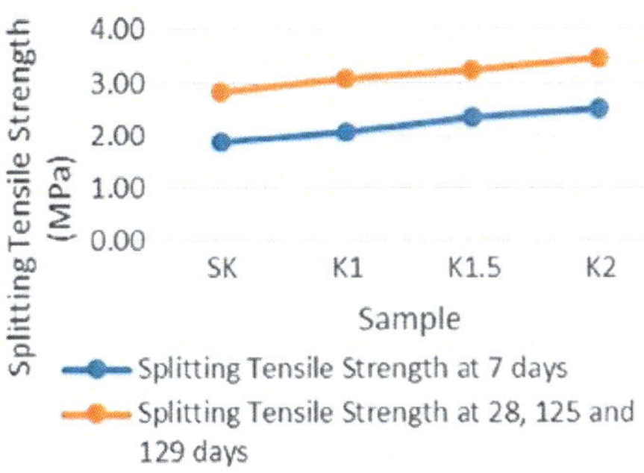

Fig. 4 Splitting tensile strength of the concrete at 7, 28, 125, and 129 days

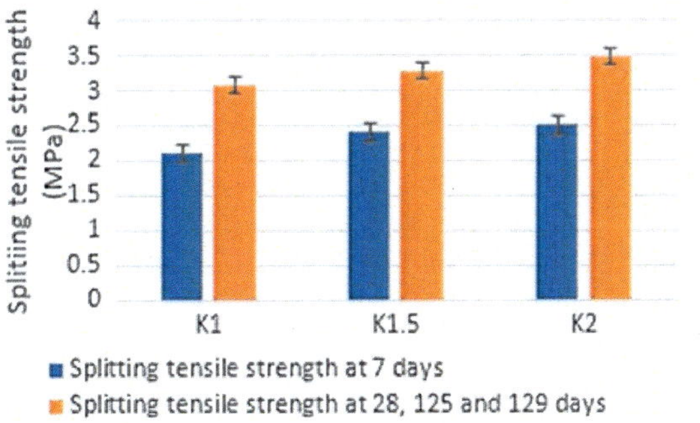

Fig. 5 Splitting tensile strength of steel fiber reinforced concrete

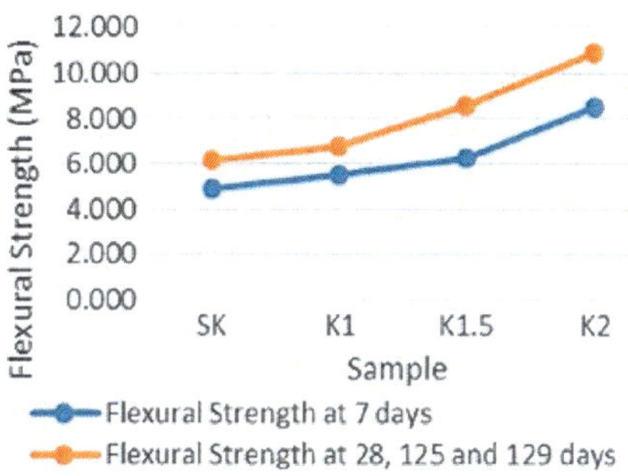

Fig. 6 Flexural strength of the concrete at 7, 28, 125, and 129 days

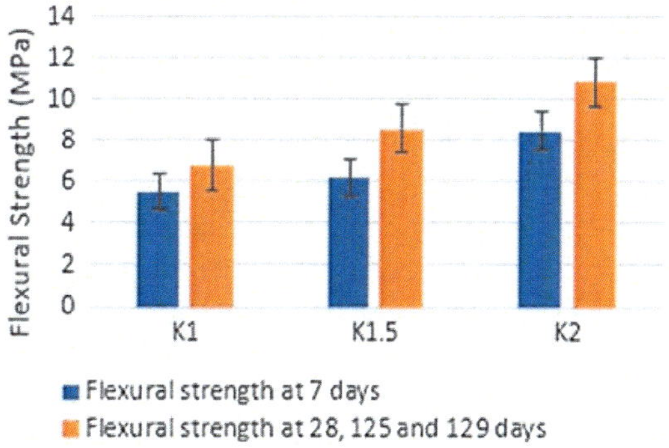

Fig. 7 Flexural strength of steel fiber reinforced concrete

7 and 28 day flexural strength, respectively. The addition of 1.5% of steel fibers into sample K1.5 resulted in a 27.1 and 39.3% increase in the 7 and 129 day flexural strength, respectively. The increase in flexural strength is due to the randomly distributed steel fibers that can hold the matrix together like a bridge across the cracks. Steel fibers are very effective in reinforcing concrete matrix, and this effect is known as the crack-bridging effect.

According to Bazgir and Fu (2016), the control sample had a flexural strength of 3.60 MPa. The addition of 0.5, 1.0, 1.5, and 2.0% steel fibers resulted in a flexural strength of 3.84, 4.51, 4.66, and 5.19 MPa, respectively, which is the equivalent of an increase of 6.7, 25.3, 29.4, and 44.2%.

The study of Solanke et al (2022) demonstrated the control sample had a flexural strength of 3.38 MPa at 28 days. The flexural strength shows the increment once the steel fiber adding with 0.5, 1.0, and 1.5% where 3.54, 3.70, and 3.96 MPa respectively. The equivalent increment by percentage is 4.7, 9.5, and 17.2%. By comparing the result of the current study to those of Solanke et al (2022), it is evident that the optimal steel fiber content for enhancing the flexural strength is 1.5%. However, the additional 2.0% of steel fiber from Bazgir and Fu (2016) shows a further increment of flexural strength when increasing the percentage of steel fiber.

3.5 Comparison of the Results of the Mechanical Properties

The results of the compressive, splitting tensile, and flexural strengths tests have proven that the addition of steel fibers into concrete mixtures can increase the strength of concrete. The most effective percentage for the incorporation of steel fibers is 2.0% (K2), followed by 1.5% (K1.5) and 1.0% (K1). Table 6 shows that the addition of steel fibers into concrete is particularly effective in increasing flexural strength compared to compressive and splitting tensile strengths. The increase in the flexural strength of concrete is due to the crack-control mechanism provided by steel fibers that can effectively close the cracks in concrete.

Table 6 The strength of K2 at 7 and 125 days

Sample	% Increase in compressive strength	% Increase in splitting tensile strength	% Increase in flexural strength
K2 (7 days)	27.5	34.6	73.1
K2 (125 days)	23.4	22.2	76.5

3.6 Impact Resistance

Figures 8 and 9 show that the increase in energy absorption occurred only after the incorporation of steel fibers into the concrete mix. Of the four concrete samples, K2 showed the highest increment in the ultimate energy absorption. Tables 7 and 8 and Fig. 10 show that, compared to the control sample, the 296.66 Nm (250.0%) increase in the energy absorption of K2 occurred after s7 days of curing, while the increase after 125 days of curing is 381.42 or 200.0%. The impact value on failure for K2 is seven times in 7 days and nine times in 125 days. For the control sample, the 7 and 28 day ultimate energy absorptions are 84.76 and 127.14 Nm, respectively. The percentages of increase in the 7 and 28 day ultimate energy absorption for K1 are 50.0 and 33.3%, respectively, while for K1.5 it is 150.0% at 7 days and 133.3% at 129 days compared with the control

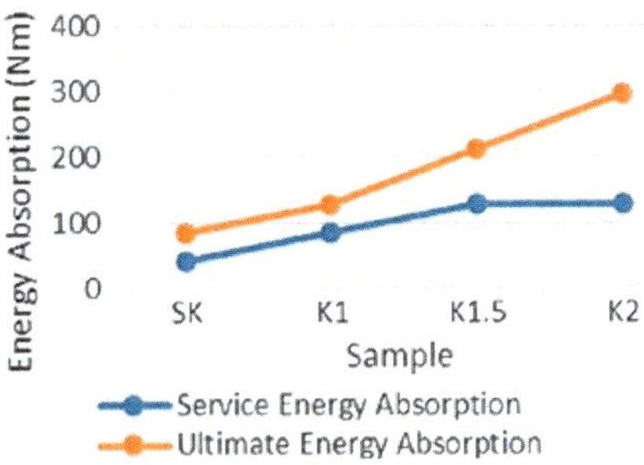

Fig. 8 Energy absorption of the concrete at 7 days

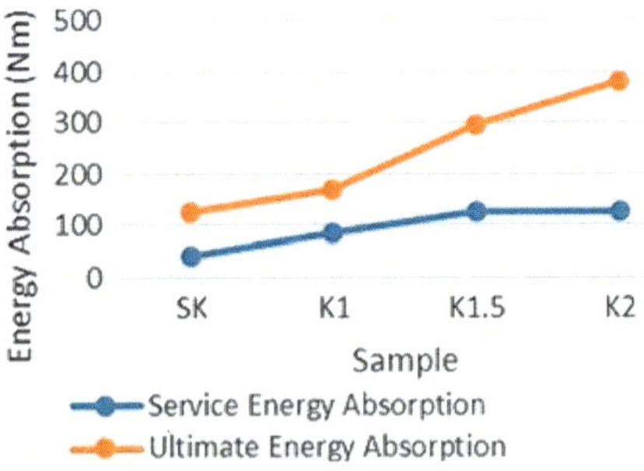

Fig. 9 Energy absorption of the concrete at 28, 125, and 129 days

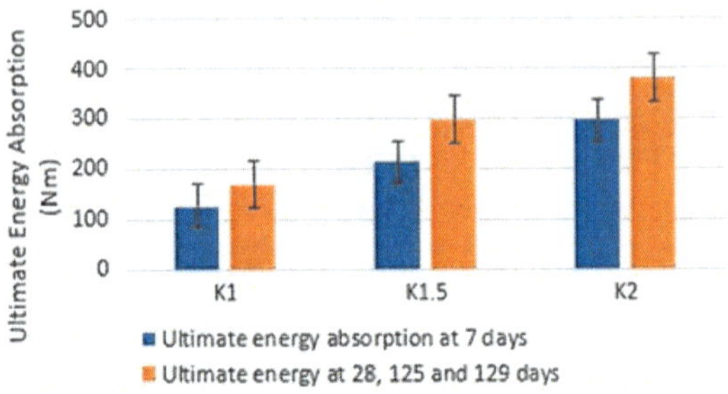

Fig. 10 Ultimate energy absorption of steel fiber reinforced concrete

Table 7 Energy absorption at 7 days

Sample	Service energy absorption (Nm)	Ultimate energy absorption (Nm)	The percentage increase for ultimate energy absorption (%)
SK	42.38	84.76	–
K1	84.76	127.14	50.0
K1.5	127.14	211.90	150.0
K2	127.14	296.66	250.0

Table 8 Energy absorption at 28, 125, and 129 days

Sample	Service energy absorption (Nm)	Ultimate energy absorption (Nm)	The percentage increase for ultimate energy absorption (%)
SK	42.38	127.14	–
K1	84.76	169.52	33.3
K1.5	127.14	296.66	133.3
K2	127.14	381.42	200.0

sample. This increase shows that the incorporation of 2.0% of steel fibers is the optimum percentage to increase energy absorption in concrete. Steel fiber reinforced concrete can arrest cracks. Thus, the steel fiber composite can increase its energy absorption capacity and its ability to withstand repetitive impacts such as shock or impact loads. The initial appearance of a crack, followed by a series of others, and the crack-bridging fibers link over this fracture, serves not only as means of absorbing energy but also as a mechanism for transferring loads.

The impact resistance of hybrid kenaf and coir fiber reinforced concrete was analyzed by Abdul Aziz and Hamid (2018). The concrete with hybrid kenaf to coir fiber ratio of 40:60 has the highest ultimate energy absorption relative to the control sample. In the present study, sample K2 has an ultimate energy absorption of 339.03 Nm and an ultimate energy absorption value of 381.42 Nm. A comparison of the two studies showed that the addition of steel fibers is more effective in increasing the energy absorption capacity by 200.0%.

4 Conclusions

This paper has investigated how steel fibers impact the mechanical properties and resistance of high-volume fly ash concrete containing nano-silica. Based on the results obtained and discussed in this article, the following conclusions are drawn:

1. The result of the slump test showed that steel fibers reduced the workability of concrete. The lower workability is because the steel fibers hinder the movement of coarse particles and make the compaction process more difficult. Sample K2 has the highest percentage difference in the slump value of 16.67%, while sample K1 has the lowest percentage difference of 5.56%.
2. Concerning the mechanical properties of the concrete, the addition of 2.0% of steel fibers resulted in a higher maximum compressive, splitting tensile and flexural strengths, compared to the 1.0% and 1.5% SF at 27.4%, 34.2%, and a very significant 73.1%, respectively, at 7 days compared to the control samples (0% steel fiber). The increments at 125 days are 23.4%, 22.0%, and 76.4% for the compressive, splitting tensile and flexural strengths, respectively.
3. At later age (from 7 days to 125 and 129 days), the rate of increase in the compressive strength of HVFAC has slowed down. The increment of about 12.0% in compressive strength from 0 to 7 days showed that nano-silica could produce a high early strength of HVFAC.
4. The addition of steel fibers into concrete is the most effective in increasing flexural strength relative to compressive strength and splitting tensile strength, where the highest increment in flexural strength is 76.4%.
5. Sample K2 has the highest ultimate energy absorption of 200.0% after 125 days of curing compared to the control sample.

Acknowledgements The authors would like to express their gratitude for the financial support provided by Universiti Kebangsaan Malaysia through grant DIP-2019-002.

References

Abbass, W., Khan, M. I., & Mourad, S. (2018). Evaluation of mechanical properties of steel fiber reinforced concrete with different strengths of concrete. *Construction and Building Materials, 168*, 556–569.

Abdul Aziz, A. N., & Hamid, R. (2018). Mechanical properties and impact resistance of hybrid kenaf and coir fiber reinforced concrete. *Journal of Advanced Research in Applied Mechanics, 47*(1), 1–10.

Afroughsabet, V., & Ozbakkaloglu, T. (2015). Mechanical and durability properties of high-strength concrete containing steel and polypropylene fibers. *Construction and Building Materials, 94*, 73–82.

Andriichuk, O., Babich, V., Yasyuk, I., & Uzhehov, S. (2018). The impact of the reinforcement percentage on the stress-strain state of the bending steel fiber reinforced concrete elements. *MATEC Web of Conferences* (Vol. 230).

ASTM-C143. (2015). Standard test method for slump of hydraulic-cement concrete. *ASTM International,* 1–4.
ASTM-C150. (2016). Standard specification for Portland cement. *ASTM International,* 1–8.
ASTM-C33. (2016). Standard specification for concrete aggregates. *ASTM International,* 1–8.
ASTM-C618. (2012). Standard specification for coal fly ash and raw or calcined natural pozzolan for use. *ASTM International,* 1–5.
ASTM-D7136. (2012). Standard test method for measuring the damage resistance of a fiber-reinforced polymer matrix composite to a drop-weight impact event. *ASTM International,* 1–16.
Banthia, N., Mindess, S., & Trottier, J. F. (1996). Impact resistance of steel fiber reinforced concrete. *ACI Materials Journal, 93*(5), 472–479.
Barros, J. A. O., & Cruz, J. S. (1999). Energy absorption capacity of steel fiber reinforced concrete. *ISISE—Comunicações a Conferências Internacionais, 1,* 1–17.
Bazgir, A., & Fu, F. (2016). The behavior of steel fiber reinforced concrete material and its effect on the impact resistance of slabs. *City University London School of Mathematics, Computer Science and Engineering,* 1–101.
Begum, S. J., Anjaneyulu, P. J. D., & Ratnam, M. K. M. V. (2018). A study on the effect of steel fiber in fly ash-based self-compacting concrete. *International Journal for Innovative Research in Science & Technology, 5*(1), 95–99.
BS1881: Part 116. (1983). Method for determination of compressive strength of concrete cubes. *British Standards,* 1–12.
Choi, W. C., Jung, K. Y., Jang, S. J., & Yun, H. D. (2019). The influence of steel fiber tensile strengths and aspect ratios on the fracture properties of high-strength concrete. *Materials, 12,* 1–11.
Dharan, D. S., & Lal, A. (2016). Study the effect of polypropylene in concrete. *International Research Journal of Engineering and Technology, 3*(6), 616–619.
Doljikov, Y. S., Letokhov, V. S., Makarov, A. A., Malinovsky, A. L., & Ryabov, E. A. (1986). Raman probing of overtone and combination bands to study the vibrational energy distribution produced by multiple-photon excitation. *Chemical Physics Letters, 124*(4), 304–308.
Gehlot, T., Seervi, D., & Chowdhary, P. (2017). Study of the flexure and split tensile strength behavior of steel fiber reinforced concrete using various percentages of steel fiber. *International Journal of Advanced Research, 6*(8), 41–47.
Gorai, S. (2018). Utilization of fly ash for sustainable environmental management. *Journal of Materials and Environmental Sciences, 9*(2), 385–393.
Ibrahim, R. K., Hamid, R., & Taha, M. R. (2012). Fire resistance of high-volume fly ash mortars with nano silica addition. *Construction and Building Materials, 36,* 779–786.
Iqbal, S., Ali, A., Holschemacher, K., & Bier, T. A. (2015). Mechanical properties of steel fiber reinforced high strength lightweight self-compacting concrete (SHLSCC). *Construction and Building Materials, 98,* 325–333.
Ismail, M. K., & Hassan, A. A. A. (2017). Impact resistance and mechanical properties of self-consolidating rubberized concrete reinforced with steel fibers. *Journal of Materials in Civil Engineering,* 1–14.
Lee, H. H. (2007). Shear strength and behavior of steel fiber reinforced concrete columns under seismic loading. *Engineering Structures, 29,* 1253–1262.
Marcos-Meson, V., Michel, A., Solgaard, A., Fischer, G., Edvardsen, C., & Skovhus, T. L. (2018). Corrosion resistance of steel fiber reinforced concrete—a literature review. *Cement and Concrete Research, 103,* 1–20.
Marcos-Meson, V., Fischer, G., Edvardsen, C., Skovhus, T. L., & Michel, A. (2019). The durability of steel fiber reinforced concrete (SFRC) exposed to acid attack–a literature review. *Construction and Building Materials, 200,* 490–501.
BS1881: Part 118. (1983). Method for determination of flexural strength. *British Standards,* 1–12.
BS1881: Part 117. (1983). Method for determination of tensile splitting strength. *British Standards,* 1–9.
Mohammadi, Y., Carkon-Azad, R., Singh, S. P., & Kaushik, S. K. (2009). Impact resistance of steel fibrous concrete containing fibers of mixed aspect ratio. *Construction and Building Materials, 23,* 183–189.
Nayan, R., & Kisku, P. N. (2017). Comparative study of different types of steel fiber over conventional concrete. *International Journal for Research in Applied Science and Engineering Technology, 5*(X), 1285–1299.
Ng, T. S., Amin, A., & Foster, S. J. (2013). The behavior of steel-fiber reinforced geopolymer concrete beams in shear. *Magazine of Concrete Research, 65*(5), 308–318.
Niu, D., Jiang, L., Bai, M., & Miao, Y. (2013). Study of the performance of steel fiber reinforced concrete to water and salt freezing conditions. *Materials and Design, 44,* 267–273.
Quercia, G., & Brouwers, H. J. (2010). Application of nano-silica (ns) in concrete mixtures. In *8th fib Ph.D. symposium* (pp 1–6).
Serrano, R., Cobo, A., Prieto, M. I., & González, M. N. (2016). Analysis of fire resistance of concrete with polypropylene or steel fibers. *Construction and Building Materials, 122,* 302–309
Saadun, A., Mutalib, A. A., Hamid, R., & Mussa, M. H. (2016). Behavior of polypropylene fiber reinforced concrete under dynamic impact load. *Journal of Engineering Science and Technology, 11*(5), 684–693.
Senff, L., Hotza, D., Repette, W. L., Ferreira, V. M., & Labrincha, J. A. (2010). Effect of nano-silica and micro silica on microstructure and hardened properties of cement pastes and mortars. *Advances in Applied Ceramics, 109*(2), 104–110.
Shende, A. M., Pande, A. M., & Pathan, M. G. (2012). Experimental study on steel fiber reinforced concrete for M-40 grade. *International Refereed Journal of Engineering and Science, 1*(1), 43–48.
Singh, L. P., Karade, S. R., Bhattacharyya, S. K., Yousuf, M. M., & Ahalawat, S. (2013). Beneficial role of nano-silica in cement-based materials—a review. *Construction and Building Materials, 47,* 1069–1077.
Solanke, S. S., Pawade, P. Y., & Ali Khan, H. (2022). An experimental study on tensile as well as flexural strength of concrete by using sugarcane baggase ash and steel fiber. *Materials Today: Proceedings, 60,* 627–637.
Sorelli, L., & Toutlemonde, F. (2005). On the design of steel, fiber-reinforced concrete tunnel lining segments. In *11th International Conference on Fracture* (pp. 1–6).
Srikar, V. V. M., & Kalyan, G. (2018). Performance of concrete with adding of steel fibers. *International Journal of Engineering Sciences & Research Technology, 7*(3), 290–308.
Su, Y., Li, J., Wu, C., Wu, P., & Li, Z. X. (2016). Effects of steel fibers on dynamic strength of UHPC. *Construction and Building Materials, 114,* 708–718.
Suhaendi, S. L., & Horiguchi, T. (2005). Fiber-reinforced high-strength concrete under elevated temperature-effect of fibers on residual properties. *Fire Safety Science,* 271–278.
Varghese, L., Kanta Rao, V. V. L., & Parameswaran, L. (2017). Nano silica-added concrete: strength and its correlation with time-dependent properties. *Proceedings of Institution of Civil Engineers: Construction Materials,* 1–32.
Verma, S. K., Dhakla, M., & Garg, A. (2015). Experimental investigation of properties of polypropylene fibrous concrete. *International Journal of Engineering and Innovative Technology (IJEIT), 4*(10), 90–94.

Yazici, Ş., Mardani-Aghabaglou, A., Tuyan, M., & Üte, A. A. (2015). Mechanical properties and impact resistance of roller-compacted concrete containing polypropylene fiber. *Magazine of Concrete Research,* 1–9.

Yet, T. C., Hamid, R., & Kasmuri, M. (2012). Dynamic stress-strain behavior of steel fiber reinforced high-performance concrete with fly ash. *Advances in Civil Engineering,* 1–6.

Yu, R., Van Beers, L., Spiesz, P., & Brouwers, H. J. H. (2016). Impact resistance of a sustainable ultra-high-performance fiber-reinforced concrete (UHPFRC) under pendulum impact loadings. *Construction and Building Materials, 107*, 203–215.

Yuvaraj, S., & Anish, V. (2019). Strength properties of nano-silica recycled sustainable concrete. *International Journal of Innovative Technology and Exploring Engineering, 8*(10), 2245–2248.

Zeiml, M., Leithner, D., Lackner, R., & Mang, H. A. (2006). How do polypropylene fibers improve the spalling behavior of in-situ concrete? *Cement and Concrete Research, 36*, 929–942.

Zhou, Y., Xiao, Y., Gu, A., & Lu, Z. (2018). Dispersion, workability, and mechanical properties of different steel-microfiber-reinforced concretes with low fiber content. *Sustainability (switzerland), 10*, 1–17.

Sensitivity Analysis Using Standardized Regression Coefficients of Roof Design Variables for Energy Performance in Residential Buildings

Hala Alaa, Mohammed Ayoub and Maye Yehia

Abstract

Buildings consume significant amounts of energy, especially in harsh climate regions. Great energy-saving potential lies in the appropriate design of the key building parameters. The roof building element contributes with about 50–70% of the total heat gain in a low-rise building due to the high amounts of received solar radiation. In order to demonstrate how a parameter affects building energy performance, changing an input and assessing the output of a simulation is a common approach. Standardized Regression Coefficients (SRCs) is a global sensitivity analysis method and one of the most reliable tools to measure sensitivity. The study aims to identify the impact of roof geometry design variables on both energy use intensity (EUI) and thermal discomfort hours (TDH) in residential buildings. The method is applied using a parametric approach where roof geometrical configurations (ranging from flat, pitched, to curved roof forms) were generated, and 24,637 energy simulations were conducted for a house prototype in the climate of Alexandria, Egypt. Global sensitivity analysis using SRCs was used to measure the sensitivity of the simulations' input variables towards EUI and TDH. Furthermore, Principal Component Analysis (PCA), which is an unsupervised machine learning algorithm, is applied to explain the underlying structure of the input variables with each other. The results implied that the tested geometrical transformations varied with a difference of 17.3 kWh.m^2 in EUI and 11% in indoor thermal comfort. The research concluded that forms in which corners tend to slope downwards and midpoints slope upwards, such as gable roofs, domes, and vaults, achieve better energy and thermal comfort performance. Findings also identified the most and least significant parameters in roof geometry design with positive and negative impact, as well as the level of each parameter's contribution to the outputs' variability, assisting energy modellers with limited time and computational effort in early design stages.

Keywords

Roof geometry · Standardized regression coefficient · Principal component analysis · Energy use intensity · Thermal comfort

1 Introduction

The building sector is estimated to be responsible for almost one-third of the global energy consumption (IEA, 2021). In Egypt, approximately 42% of the total electricity consumption is by the residential sector, while the industrial sector is responsible for 28%. Over the past few years, residential consumption has consistently increased. The Ministry of Electricity and Energy explains that this is due to the expansion of residential development, with the excessive energy consumption by domestic appliances, especially air conditioners, during the hot-summer months (MOEE, 2013). Instead of relying entirely on mechanical- and electrical-dependent means, building designers should devote time and effort to research and explore passive strategies for achieving the highest balance between low energy and natural climate control (Visser & Yeretzian, 2013). Research showed that more than 40% of the building energy-saving potential comes from determining the key design parameters in the early design phase (Lin et al., 2021).

H. Alaa · M. Ayoub (✉) · M. Yehia
Architectural Engineering and Environmental Design Department, Arab Academy for Science, Technology and Maritime Transport, Alexandria, Egypt
e-mail: dr.ayoub@aast.edu

M. Yehia
e-mail: maye.yehia@aast.edu

B.-J. He et al. (eds.), *Resilient Horizons: Building Sustainable Environments for Climate Adaptation and Health*, Advances in Science, Technology & Innovation, https://doi.org/10.1007/978-3-031-46109-5_10

Over the past decades, numerous shifts in extreme weather and climatic occurrences have been documented, such as the rise in surface air temperature, more frequent extreme events like droughts and floods, and change in precipitation patterns. The most evident change in global climate is the increase in surface air temperature (Sachindra et al., 2016). Elevated summer temperatures, along with more frequent and intense heatwaves, contribute to issues related to thermal comfort, health and even a rise in human mortality (Basu, 2009). Research indicates a direct correlation between high indoor temperatures and health hazards (Head et al., 2018). As a result, the implications of climate change highlight the importance of exploring solutions and approaches to enhance indoor thermal comfort in residential buildings.

Various parts of the building, including walls and roofs, are highly exposed to solar radiation. Therefore, many regions around the world, especially hot-arid climates, experience summer overheating and thermal discomfort (Elseragy & Elnokaly, 2007). Approximately 50–70% of the total internal heat gain in residential and low-rise buildings is accounted for by building roofs. Hence, the reduction of energy input caused by the roof's received solar radiation is necessary for achieving indoor thermal comfort (Vijaykumar et al., 2007). Consequently, the roof design, geometrical configurations, construction materials and components should differ in different regions and climatic conditions (Elseragy, 2003).

Flat roofs are popular for their ability to provide a possibility for vertical extension, but they are not ideal for areas with heavy precipitation or high winds, as they are not very good at shedding water or snow. Additionally, they can be prone to leaks and require regular maintenance to ensure they remain watertight. Pitched roof forms have gained popularity around the world due their high precipitation drainage qualities (Awwad et al., 2018). Also, they are generally more durable than flat roofs. However, they are typically more expensive to install and maintain, and they can be less suitable for areas with low precipitation or high winds. On the other hand, curved roof buildings have been the most important element of architecture in many ancient civilizations, especially for large state or religious spaces mainly cathedrals, churches, mosques and also for buildings such as single-family dwellings, and Arab houses (Hadavand & Yaghoubi, 2008). Fathy (1986) proposed that for curved roof structures, the solar radiation's intensity is distributed across a wider surface, which in turn lowers the average temperature rise of the roof and decreases the heat transfer to the interior. However, they are typically more complex, hence more costly, to construct and result in a more limited interior space.

Over the last few years, several passive building cooling energy-efficient practices have been proposed, which are summarized in Table 1. A few reports have discussed the thermal properties of vaulted-roof forms. Elnokaly and Elseragy (2006) conducted one of the early analysis studies of vaulted-roof forms in the hot desert climate of Aswan, Egypt, using a solar radiation simulation model. Findings showed that, on an average daily basis, vaulted roofs had received solar radiation, with a value of 75.4% of the received flat roofs. Hadavand and Yaghoubi (2008) developed a mathematical model using the finite volume method to compare the amount of heat transfer from the roof into the building for flat and vaulted-roof forms. The results conveyed that the impact of wind flow on the vaulted roof led to much lower convective heat transfer. An in-depth study investigated the relationship between the geometrical configuration of a vaulted roof and the received solar radiation upon its surface (Elnokaly et al., 2019). 2310 different cases derived from 6 vaulted-roof configurations were simulated. Results showed that higher cross-sectional ratio roofs have a larger self-shaded area, less internal volume and, consequently, less energy consumption.

Furthermore, Ayoub and Elseragy (2018) explored various domed-roof configurations in the hot desert climate of Aswan, Egypt. The study carried out an insolation analysis on the domed-roof form in order to minimize the received solar radiation above the roof. Findings stated that domed-roof forms receive significantly less amount of ADSI compared to flat roofs during summer and winter. Another study considering traditional curved roof forms (domed and vaulted roofs) was performed by Moustafa et al. (2018) in Cairo, Riyadh and Istanbul Cities. Fourteen roof designs for types of vaulted, domed and double-domed roofs were thermally simulated. Results conveyed that the pointed double-domed roof was the most effective for Cairo and Riyadh, while for the Istanbul case, half-circular/pointed domed and vaulted roofs showed better results. Other researchers explored pitched roof form's thermal properties. Roof inclination angles were compared by Ghaedi et al. (2012) in the hot desert climate of Bandar Abbas, south of Iran, to maximize photovoltaic system heat gain during the day and minimize energy loss during winter nights. A numerical model was developed to compare three roof types; horizontal slab roof or Roof 0°, gable pitched roof 30–60° and gable pitched roof 45–45°. Results presented the second as the most energy-efficient roof. Mazzeo and Kontoleon (2020) proposed a mathematical dynamic thermal model to evaluate the impact of roof inclination and orientation angles on thermal comfort during summer in Italy and Greece. The study achieved reductions of 12.5 °C in the roof's outer surface temperature and 4.7 °C in the roof's inner surface temperature, using an accurate design.

On the other hand, Dabaieh et al. (2015) investigated different forms and materials of cool roof designs to minimize the heat transfer from the rooftop in the hot desert climate

Table 1 Summary of the reviewed studies

Authors	Climatic zone	Focus	Analysis type	Studied space	Methods
Elnokaly and Elseragy (2006)	Aswan, Egypt; hot desert climate (Köppen BWh)	Vaulted roof forms	Simulation	N/D	Solar Radiation Simulation Model algorithm and calculating the hourly total clear sky irradiance
Hadavand and Yaghoubi (2008)	Hot regions in Iran; hot desert climate (Köppen BWh)	Vaulted roof forms	Mathematical, and Simulation	N/D	Numerical model using finite volume method for solar radiation, and a two-dimensional numerical simulation for the flow field
Ghaedi et al. (2012)	Bandar Abbas, Iran; hot desert climate (Köppen BWh)	Gable pitched roof forms	Mathematical	Flat roof, gable roof 30°, 60°, and 45°, 45° houses	Numerical model to calculate required coefficient of heat loss and total amount of energy gain by photovoltaics on three roof types
Dabaieh et al. (2015)	Cairo, Egypt; hot desert climate (Köppen BWh)	Roof forms	Simulation	The living room in a typical Cairene house	DesignBuilder simulations in the summer months for 37 roof design possibilities
Sirimanna and Attalage (2016)	Sri Lanka; tropical savanna climate (Köppen Aw)	Roof forms	Mathematical, and experimental	6 buildings with different roof designs	Numerical model using finite volume method to calculate roof enclosure factors, and experimental measurements of six buildings for validation
Ayoub and Elseragy (2018)	Aswan, Egypt; hot desert climate (Köppen BWh)	Domed roof forms	Simulation	N/D	Grasshopper parametric model for domed-roof forms and Ecotect simulations of the daily solar irradiance
Moustafa et al. (2018)	–Cairo, Egypt; Riyadh, Saudi Arabia; hot desert (Köppen BWh) –Istanbul, Turkey; hot-summer Mediterranean (Köppen Csa)	Domed and vaulted-roof forms	Simulation	A domed-roof space 13 m^2, a vaulted-roof and a double-domed roof space with an area of 26 m^2	Autodesk Simulation CFD thermal simulations of 14 roof cases to calculate the distribution of heat flux and heat transfer
Elnokaly et al. (2019)	Aswan, Egypt; hot desert climate (Köppen BWh)	Vaulted roofs	Simulation	A domestic residential space located in the ground floor	Grasshopper parametric model and simulations of solar irradiance and energy consumption for 2310 cases using Radiance and Energyplus
Mazzeo and Kontoleon (2020)	Italy and Greece; hot-summer Mediterranean climate (Köppen Csa)	Pitched roof forms	Mathematical	N/D	Numerical dynamic model in the steady periodic regime, taking into account the heat balance on the outer surface and the inner surface of the roof
Akbarpoor et al. (2022)	Hot regions in Iran; hot desert climate (Köppen BWh)	Domed roof forms	Simulation	A 64 m^2 test building with four domes	CFD simulations of airflow inside and around were carried out using ANSYS FLUENT software

(continued)

of Cairo, Egypt. Simulation-based approach was applied to a living room in a typical house design, withing 37 roof design possibilities. Findings indicated that the vaulted-roof form with high albedo coating material achieved a 53% improvement in discomfort hours and 826 kWh energy savings compared to the flat non-insulated roof. Sirimanna and Attalage (2016) conducted a mathematical and experimental study to assess the roof thermal performance for different roof configurations and local materials in Sri Lanka. The comparison was implemented for four pitched roof buildings and one flat roof building, where measurements were recorded for several days. Findings showed a difference of 5 °C between the cases. Moreover, the best material performance was realized by clay tiles roof. Akbarpoor et al. (2022) assessed a novel system involving a domed-roof system and evaporative cooling channel using 3D CFD simulations of airflow. Findings implied that the performance of the system is considerably influenced by the dome's diameter and the cooling channel's measurements, providing thermal comfort under a maximum cooling demand of about 112 W/m^2.

Using simulation software, changing an input parameter and analysing the effects on the performance output is a common method to demonstrate the impact of a parameter on building energy use in what is known as a sensitivity analysis (Elhadad & Orban, 2021). Performing a sensitivity analysis can identify the most influential parameters, the insignificant parameters that can be eliminated, how much each parameter contributes to output variability, how input and output parameters are correlated and how changing an input will affect the model's output (Hamby, 1994). In order to perform a sensitivity analysis, typical steps include determining input and output variables, creating building energy models, running and collecting simulation results, conducting sensitivity analysis and finally presenting the sensitivity analysis results (Tian, 2013).

As early as possible in the design stages, sensitivity analyses can provide important information about the significant design parameters to concentrate on as well as which parameters are unimportant and with little influence on the outputs (Heiselberg et al., 2009). Sensitivity analysis is generally categorized into local and global methods. Local sensitivity analysis explains the influence of uncertain inputs around a specific local point, while global sensitivity analysis explains the effect of uncertain inputs onto the entire input space. Global sensitivity analysis is preferred in many design problems, but it also requires higher computational efforts. In building performance analysis, both local and global sensitivity methods are widely used (Tian, 2013). The Morris's method, Sobol's method as well as regression-based sensitivity coefficients are the most used measures of sensitivity in building energy research (Nguyen et al., 2014).

The sensitivity of the parameters in the study is measured using the standardized regression coefficient (SRC) of the simulation input variables, i.e. the independent variables, and the simulation output variables, i.e. the dependent variables. On the other hand, exploratory factor analysis explores the underlying structure of the input variables. It demonstrates the correlations between them, providing further insightful information about the nature and interrelations of the independent variables. One of the most common application methods is the unsupervised machine learning technique known as Principal Component Analysis (PCA). This method also reduces the dimensionality of large datasets by loading each group of related variables onto a principal component (PC). The analysis is employed in the study as a following step to the statistical regression analysis.

This study aims to explore the impact of roof design variables on both energy use intensity (EUI) and thermal comfort in residential buildings in the hot desert climate of Alexandria, Egypt. It focusses on the geometrical configurations possibilities of the roof form, ranging from flat-to-pitched-to-curved roof forms such as vaults and domes. The developed parametric model explores thousands of form variations, and subsequently the most crucial geometrical design factors in terms of building performance are identified by the sensitivity analysis, using regression-based global sensitivity analysis and PCA analysis. The study's findings can assist researchers and energy modellers at the early stage of the design process, with an emphasis on improving the roof thermal performance in Alexandria, Egypt.

2 Methodology

In addition to the studies previously reviewed, the study involved more in-depth investigations of the relationship between roof geometry and energy use by examining a broader range of geometric principles controlling the shape. A parametric approach is adopted in this research. Figure 1 illustrates the methodological tools and plan of work considered in the research. The framework can be divided into three main steps. First, the base geometric model is created using the software Rhinoceros v7.2.2 and Grasshopper graphical interface for a prototype house located in Alexandria, Egypt. For the parameterization of the roof, 11 variables shaping the various geometrical configurations are defined. Consequently, the energy model is developed using Ladybug and Honeybee plugins v1.4.0. Thermal zone parameters, schedules, setpoints, construction materials and weather files are set up in the model. The output variables are then defined to test the energy use intensity (EUI) as well as the total thermal discomfort hours (TDH).

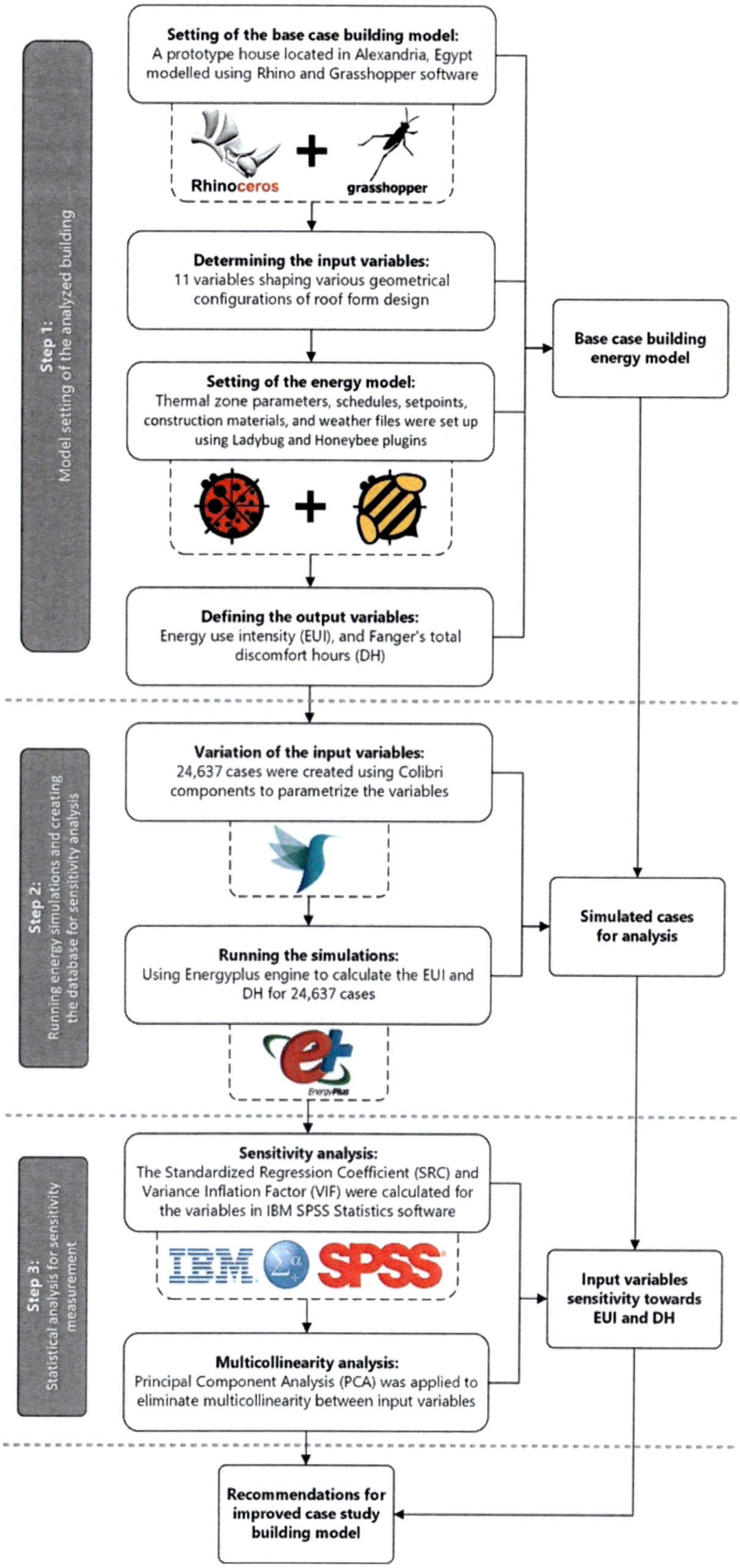

Fig. 1 The methodological tools and plan of work adopted in the research

The second step includes the automation process for the simulations to create the database for sensitivity analysis. By employing the plugin TT Toolbox v1.9, Colibri components are used to generate variations of the geometrical configurations, and 24,637 cases are simulated in EnergyPlus v9.2.2. The third step is the statistical analysis for sensitivity measurement. Global sensitivity analysis using the Standardized Regression Coefficients (SRCs) method is implemented to measure the sensitivity of the input variables towards the output variables through the software IBM SPSS Statistics v26.0.0. Furthermore, PCA is applied to explain the structure of the variables with one another. The extracted Principal Components (PCs) are rotated, and their factor loadings are calculated using Varimax orthogonal rotation. Then, the emergent PCs are interpreted to relevant design factors. The paper eventually explains the impact of the tested roof geometrical configurations on the performance metrics and provides architectural interpretations for an optimized roof model design for low-rise residential buildings in Alexandria, Egypt.

3 Case Study

The analysis is performed on a hypothetical residential building prototype, which is a single-floor building located in Alexandria, Egypt (Fig. 2). The footprint of the building is fixed with a floor area of 100 m^2 imitating a typical middle-income house. The volume of the building is fixed to be 500 m^3 in order to orientate the thermal calculations to be solely affected by the geometry and not the volume. Building height is normalized depending on the geometrical configuration to maintain the fixed volume. Tables 2 and 3 summarize the materials used in the building construction. Moreover, Table 4 compares the thermal-physical properties of the proposed prototype model and the ranges of design values recommended by the Egyptian Code for Improving Energy Efficiency in Buildings (MHUC, 2006). The properties of the building were chosen to fall within the recommended ranges from the design standard. The four external walls accommodate two windows, each with a glazing ratio of 20%, and the building orientation is 0 °C towards the north.

The number of occupants in the residence is four people, which represents the average family size in Egypt, according to the Central Agency for Public Mobilization and Statistics (CAPMAS, 2017). Based on regular daily patterns assumptions, schedules for occupancy and lighting were developed and defined in Ladybug throughout the year, as shown in Fig. 3. It is necessary to set the weather data in EnergyPlus in advance in order to calculate the model's energy consumption. In this paper, the climate of Alexandria, Egypt (31° 12′ N, 29° 55′ E) was used for the test simulations. The weather data used in the simulation process were obtained from the weather source "Alexandria Nozha Intl AP 623,180 ISD-TMYx". The TMYx are typical meteorological files acquired from hourly weather station meteorology data through 2021 in the ISD (US NOAA/NCEI's Integrated Surface Database). The climate of Alexandria is classified as a hot desert climate (BWh) by the Köppen–Geiger system. The warmest month of the year is August, with an average temperature of 27 °C, while coldest month is February with an average temperature of 14 °C. The average relative humidity throughout the year ranges from 63 to 72%.

4 Parametric Model Definition

4.1 Design Variables

The parametric model was developed in order to create various geometrical configurations for the roof shape design, ranging from pitched roof types such as gable roofs, hip roofs, shed roofs and pyramid roofs. Moreover, the parametric definition covers curved geometry transformations as well, such as vaulted-roof and domed-roof transformations. A custom form-finding definition

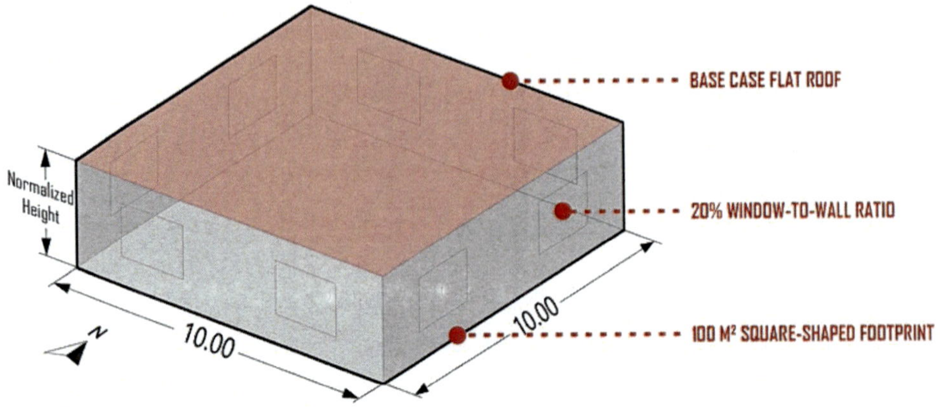

Fig. 2 The geometry of the tested house prototype

Table 2 The materials used in the construction of the tested house prototype

Type	Material	Thickness (mm)	K-value (W/m K)	Specific heat (J/kg K)	Density (kg m^3)	Solar absorption	R-value (m^2 K/W)	System R-value (m^2 K/W)
Wall	Stucco	25	0.72	839	1856	0.7	0.035	2.12
	Brick wall	100	0.9	720	1920	0.65	0.111	
	Polyurethane foam	50	0.028	1470	30	0.6	1.786	
	Brick wall	100	0.9	720	1920	0.65	0.111	
	Gypsum	12.5	0.15	829	784	0.4	0.079	
Roof	Clay tiles	25	1	800	2000	0.7	0.025	3.03
	Bitumen felt	3	0.5	1000	1700	0.87	0.006	
	Polyurethane foam	50	0.028	1470	30	0.6	1.786	
	Cast concrete	150	2.27	837	2321	0.7	0.066	
	Gypsum	12.5	0.15	829	784	0.4	0.079	
Floor	Porcelain tiles	30	1.3	840	2300	0.4	0.023	0.26
	Floor screed	70	0.41	840	1200	0.73	0.17	
	Bitumen felt	3	0.5	1000	1700	0.87	0.006	
	Cast concrete	150	2.27	837	2321	0.7	0.066	

Table 3 Glazing parameters in the tested house prototype

Type	Solar heat gain coefficient (SHGC)	T_{sol}, solar transmittance	T_{vis}, visible transmittance	Glazing U-factor (W/m^2 K)	R-value (m^2 K/W)
Glazing	0.45	0.367	0.6	2.214	0.279

Table 4 Design values of the reference building and recommend value ranges by the design standard

Design values	Building prototype	Values recommended by the design standard (MHUC, 2006)
Thermal resistance of the external walls (m^{2}K/W)	R-value = 2.12	East and West: R-value ≥ 1.15 South: R-value ≥ 0.69 North: R-value ≥ 0.47
Thermal resistance of the roof (m^{2}K/W)	R-value = 3.03	R-value ≥ 2.15
Window-to-wall ratio	WWR = 20%	East and West: WWR ≤ 30% South: WWR ≤ 30%; North: N/D
Glazing solar heat gain coefficient	SHGC = 0.45	East and West: SHGC ≤ 0.5 South: SHGC ≤ 0.71; North: N/D

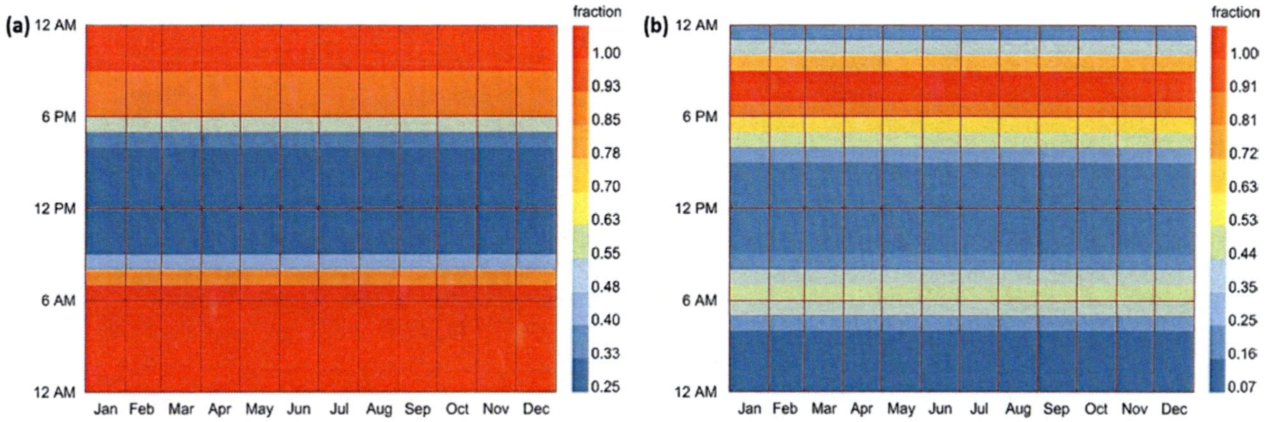

Fig. 3 Schedules for the tested thermal zone throughout the year; **a** Occupancy schedule, **b** Lighting schedule

Table 5 Design input variables and their parameter ranges

Input variable	Symbol	Unit	Range	No. of steps
Corner1 height	Cr1_Ht	m	[3–5]	2
Middle1 depth	Mid1_D	m	[0–5]	3
Middle1 height	Mid1_Ht	m	[3–5]	3
Corner2 height	Cr2_Ht	m	[3–5]	2
Middle2 depth	Mid2_D	m	[5–10]	3
Middle2 height	Mid2_Ht	m	[3–5]	3
Dome height	D_Ht	m	[3–7]	3
Arc1 segments	Arc1_Seg	None	[2–10]	2
Arc2 segments	Arc2_Seg	None	[2–10]	2
Repetition count	Rep	None	[1–2]	2
Orientation angle	N	Deg	[0–270]	4

was developed in order to facilitate the modelling and adjustment of roof geometry parametrically. The building footprint and walls structure are fixed, while the roof geometrical parameters act as input variables, whereas 11 input variables were defined to shape the different forms (Table 5).

The definition works by controlling specific points in the roof form to generate simple to complex shapes. The base case is a flat roof, and the position of the points can be defined by their height or depth in reference to their base case. The control points form linear geometry (or arcs), of which the number of segments can be controlled as well. Furthermore, the repetition of the resulting roof geometry is controlled in the form of a planar array. And finally, the orientation angle of the resultant roof form is controlled as well. Since this occurs parametrically, these parameters can be automated during the simulation stage to generate all possible forms and analyse the variables' sensitivity. Figure 4 illustrates a few iterations of the tested cases while pointing out the geometry input variables and the impact of their parameterization.

4.2 Objective Performance Metrics

As previously mentioned, the research aims at evaluating the energy performance in the residential building prototype under the various tested roof forms. The goal is to minimize energy consumption while maximizing indoor thermal comfort. The annual energy consumption of the building is calculated by Energyplus within a heating setpoint of 18.3 °C and a cooling setpoint of 25 °C throughout the year, based on the Egyptian Code for Improving Energy Efficiency in Buildings (MHUC, 2006). The metric used to assess energy loads is the Energy Use Intensity (EUI), which is a measure of the total energy consumed by the building in 1 year per unit area of the building. EUI metrics can be calculated as follows:

$$EUI = \frac{E_c + E_h + E_l + E_q + E_w}{A} \tag{1}$$

where E_c is the energy load intensity for space cooling (kWh), E_h is the energy load intensity for space heating (kWh), E_l is the energy load intensity of lighting (kWh), E_q is the energy load intensity for equipment (kWh), E_w is the energy load intensity for service hot water (kWh) and A is the total gross floor area of the building (m^2).

On the other hand, the second objective considered in this research is thermal comfort. For assessment, the study uses Fanger thermal comfort, the most widely used thermal comfort measurement, whereas the predicted percentage of dissatisfaction (PPD) and the predicted mean vote (PMV) are calculated. Developed by Fanger (1970), the PMV is a seven-point scale ranging from+3 to −3, referring to the expected mean vote for the thermal sensations of a large group of people in a particular environment. It depends on six factors, which are metabolic rate, clothing insulation, air temperature, mean radiant temperature, relative humidity and air velocity, while PPD describes the percentage of dissatisfied people in the thermal environment. For the tested cases, the metabolic rate is assumed to be 1.1 met for a seated person, while the clothing insulation is set to 0.7 clo for long sleeve shirts and pants. The output variable metric for thermal comfort is the number of thermal discomfort hours throughout the year (TDH), represented by the number of hours above or below the indoor temperature between 18.3 and 25 °C.

5 Results and Discussion

5.1 Sensitivity Analysis

The first step of the analysis was the performance of Energyplus simulations to obtain the energy use intensity and discomfort hours results. 24,637 cases were simulated and reported to form a database for the sensitivity analysis. The simulations were parallelized over five virtual machines, each allocated with 5 GB of RAM clocked at 3200 MHz and four CPU cores out of an Intel Core i9-10850K processor clocked at 5 GHz. Due to the high dimensionality of the dataset, a parallel coordinate plot is used to illustrate the results (Fig. 5). In this graph, a polyline maps each simulation run. Each variable is represented by its own vertical axis with a different scale. Polylines intersect all vertical axes at a single point, and

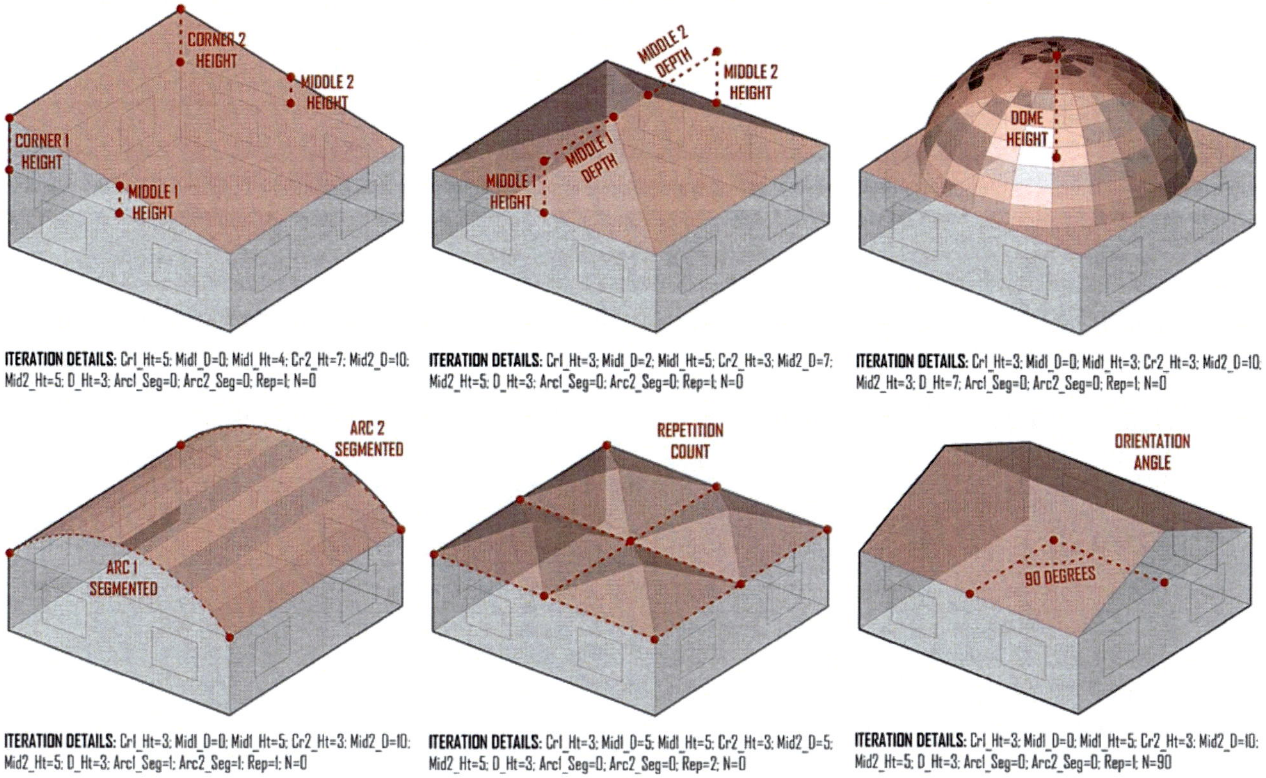

Fig. 4 Examples of the generated iterations from the parametric definition whose design input variables are indicated

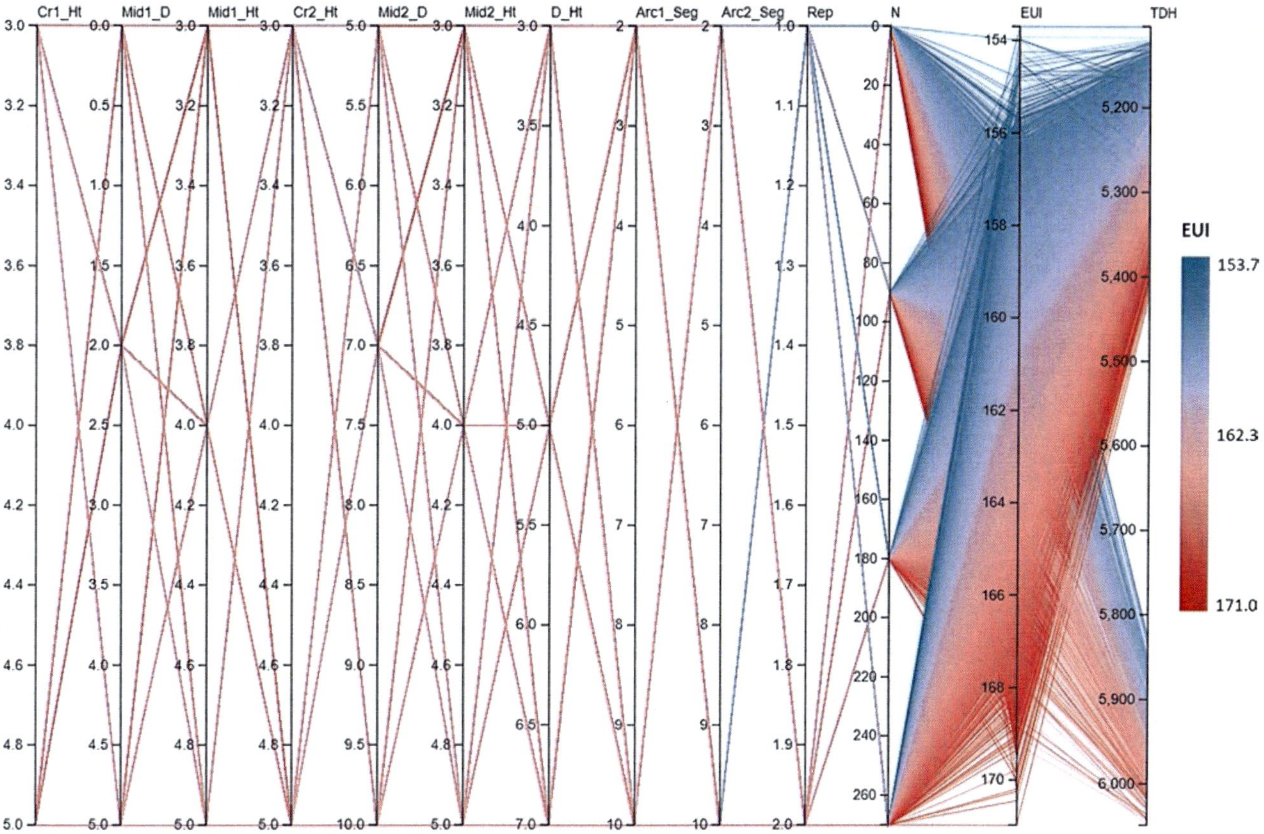

Fig. 5 Parallel coordinate plot for the input design variables and output performance metrics

the output metric or input variable value can be measured at each point of intersection. The plot's left side represents the combinations of tested input design variable values, while the left side extends to visualize the resulted output performance metrics.

Moreover, another 12th variable in terms of the dome definition (D_def) is added to represent whether the dome form category is generated in the iteration. To achieve this, the one-hot encoding technique is applied. It is a popular encoding method employed when processing datasets containing categorical variables (Al-Shehari & Alsowail, 2021), turning the variable into numerical values (0 or 1). In this case, D_def, which has a categorical data type, was implemented based on the if function logical test used in the parametric definition to generate the dome state, which requires the condition of 'Cr1_Ht=3', 'Cr2_Ht=3', 'Mid1_Ht=3', 'Mid2_Ht=3' and 'D_Ht>3', rendering 'D_def=1'.

The goal of this part of the statistical analysis is to measure the sensitivity of the input variables, i.e. the independent variables, towards the simulation output variables, i.e. the dependent variables. There is difficulty in directly comparing regression coefficients since they may be expressed in different measurement units. This problem is eliminated by standardized regression coefficients (SRCs), which express the coefficients in a common, statistically reasonable set of units, and allow comparisons to be performed (Siegel, 2016). In this paper, for each output variable, regression analysis is performed, and the standardized regression coefficients are calculated as follows:

$$SRC_i = \frac{s_{x_i}}{s_y} * B_i \tag{2}$$

where, for each ith independent variable, s_{x_i} and s_y are the estimated standard deviations of x_i and y, respectively, and B_i is the regression coefficient resulting from a linear regression between y and x_i.

Moreover, a multicollinearity check is performed on the data by calculating the Variance Inflation Factor (VIF) for each input variable. Small VIF values lower than 5 indicate a low correlation among the variables, which implies a more robust model (Sheather, 2009). The results of the VIF of independent variables, presented in Table 6, show acceptable collinearity in the data, which supports the SRCs results in the case study. VIF is calculated as follows:

$$VIF_i = \frac{1}{1 - R_i^2} \tag{3}$$

where R_i^2 represents the unadjusted coefficient of determination for regressing the ith independent variable on the remaining ones.

Simulation results varied with a difference of 17.3 kWhm^2 in EUI and 11% in indoor thermal comfort (945 discomfortable hours). Considerable variations were implied in the SRCs between the parameters. Figure 6 shows the SRCs of the input variables representing the change in each parameter in the EUI and TDH. The SRCs of the repetition variable (Rep), first corner height variable (Cr1_Ht) and second corner height variable (Cr2_Ht) represent the highest positive values of 0.567, 0.367 and 0.317 for EUI, respectively. Furthermore, the SRCs of the second middle point height variable (Mid2_Ht), first middle point height variable (Mid1_Ht) and the dome definition variable (D_def) yielded the highest negative values of −0.325, −0.297 and −0.154 for EUI, respectively. Moreover, the SRCs of the first arc segments variable (Arc1_Seg) and the second arc segments variable (Arc2_Seg) yielded small negative values of −0.018 and −0.013 for EUI, respectively, while the SRC of the dome height variable (D_Ht) was 0.01 for EUI. On the other hand, the second middle point depth variable (Mid2_D), the first middle point depth variable (Mid1_D) and the orientation variable (N) had the lowest sensitivity and least impact on EUI.

The sensitivity of variables towards the TDH output was analysed as well, which had shown an overall less sensitivity than towards the EUI output (Fig. 6). The SRCs of orientation (N), second arc segments (Arc2_Seg), repetition (Rep), first corner height (Cr1_Ht) and second corner height (Cr2_Ht) variables represented the highest positive values of 0.239, 0.182, 0.139, 0.135 and 0.112 on TDH, respectively. On the other hand, the SRCs of the first arc segments (Arc1_Seg), the second middle point height (Mid2_Ht) and the first middle point height (Mid1_Ht) variables yielded the highest negative values of −0.125, −0.113 and −0.098 on TDH, respectively. Dome height (D_Ht) and dome definition (D_def) variables have yielded negative SRC values of -0.06 and −0.04, respectively. On the other hand, first middle point depth (Mid1_D) and second middle point depth (Mid2_D) variables had the least sensitivity towards TDH.

Table 6 VIF values for the independent input variables

Variable	Cr1_Ht	Mid1_D	Mid1_Ht	Cr2_Ht	Mid2_D	Mid2_Ht
VIF	1.027	1.003	1.037	1.028	1.002	1.036
Variable	D_Ht	Arc1_Seg	Arc2_Seg	Rep	N	D_def
VIF	1.009	1.003	1.002	1.009	1.001	1.122

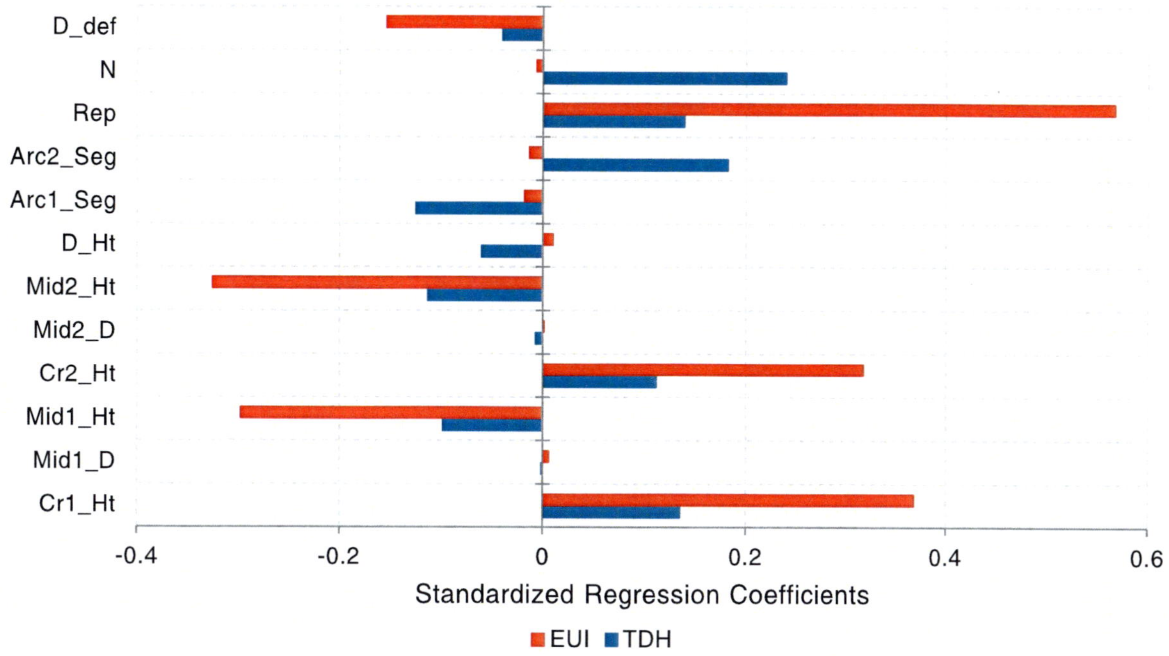

Fig. 6 SRCs of the design input variables for EUI and TDH

The results depict that the most beneficial parameters on the overall building performance for this case study are the roof middle point heights as well as the generation of the dome form state. Conversely, the parameters of the roof corners' heights and the roof geometry repetition contributed the worst to the building performance for both energy consumption and thermal comfort. These findings indicate that forms such as gable roofs, domes, vaults and geometries whose corners tend to slope downwards and midpoints slope upwards would perform better than the others in similar design conditions. However, to provide a better explanation for the variables' correspondence to the test, further analysis is performed in the next section.

5.2 Principal Component Analysis

The second goal of the statistical analysis is exploratory factor analysis in order to explore the interrelations between the design variables, i.e. the independent variables, to form not only a mathematical interpretation but also an architectural interpretation of the results. The unsupervised machine learning data reduction technique of Principal Components Analysis (PCA) is employed to describe the variability of the multivariate dataset. To achieve this, all variables were analysed by the PCA, and 12 PCs were generated. A correlation matrix was calculated and decomposed into magnitudes known as eigenvalues (Fig. 7), and directions known

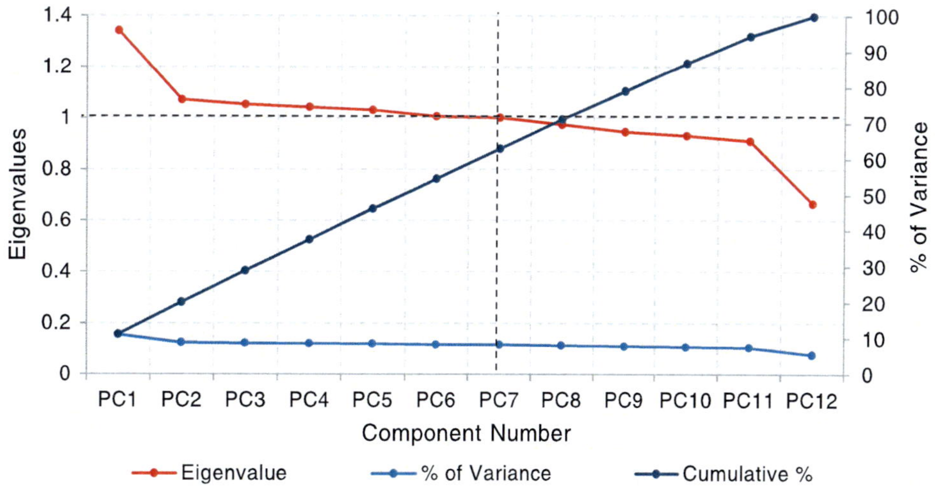

Fig. 7 Eigenvalues and the explained variability of the PCs produced by the correlation matrix

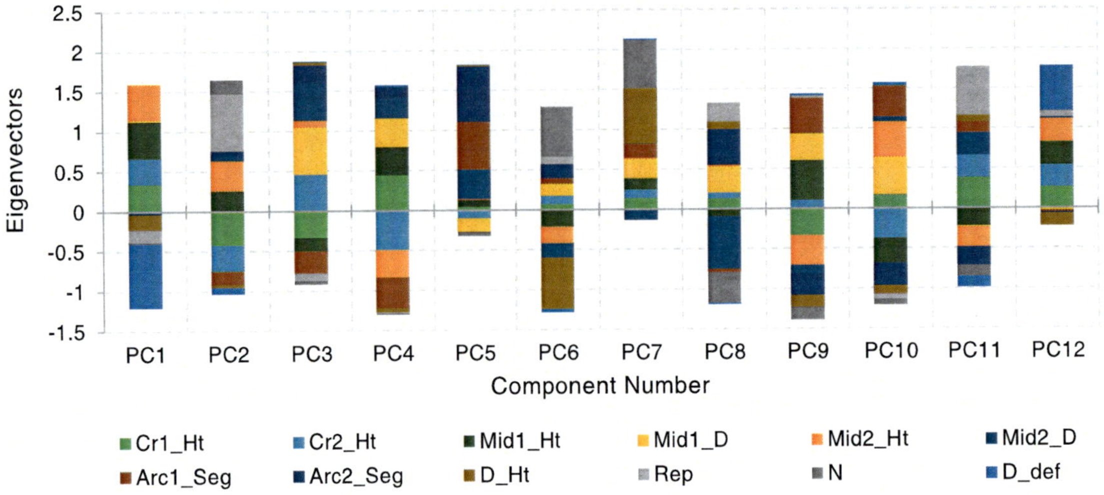

Fig. 8 Eigenvectors of the PCs produced by the correlation matrix

as eigenvectors (Fig. 8). The eigenvalue is a measure of the variance accounted by the corresponding PC, while the eigenvector represents the relative weight of each variable in the new PCs.

Furthermore, to reduce the dimensionality of the 12-dimensional dataset, the least significant PCs should be eliminated, which is known as factor retention. Factors with small eigenvalues indicate little variances and should be eliminated from the analysis. The most common retention method is known as the Kaiser (1960) criterion, which suggests the retention of all the PCs with eigenvalues above 1. Moreover, the line plot of the eigenvalues, known as the scree plot (Fig. 7), is often used to determine the number of PCs to extract in an approach called the scree test (Cattell, 1966). It suggests that the "elbow" of the graph, where it tends to level off, is usually where the important PCs to retain are located. Consequently, 7 PCs have been extracted for further analysis, representing 63% of the cumulative percentage of variance.

As soon as the number of components is determined, it is necessary to interpret the loading of the individual variables. In order to produce a factor matrix that is easy to interpret, the factor solution must be rotated. Several methods exist, which can be categorized into orthogonal and oblique. Factors resulting from orthogonal rotation are not correlated with each other, while those resulting from oblique rotation are allowed to correlate (Howard, 2016). Both methods were performed on the extracted PCs, and the results were nearly identical. Also, the correlation matrix resulting from the oblique rotation confirmed that the PCs are orthogonal, i.e. not correlated. Therefore, the oblique rotation was not required.

Varimax rotation, created by Kaiser (1958), is the type of orthogonal rotation employed in the study. It generates both large and small factor loadings, as it aims to increase factor loadings variances. It is the most widely used form of orthogonal rotation that converts vectors from principal component analysis or factor analysis into simple structures (Jackson, 2005). The analysis produces the distribution and loadings of each variable within the emergent PCs. Furthermore, factor analysis also determines the extent to which loading values can be representative of the associated PC, known as factor loading cutoff. Research has indicated that a "good" factor loading cutoff in a primary factor is 0.40 (Hinkin, 1995), while other authors suggested values of 0.30 (Costello & Osborne, 2005) and 0.20 (Knekta et al., 2019). Consequently, three approaches for loadings cutoff have been followed, where loadings <0.2, 0.3 and 0.4 have been suppressed, respectively (Fig. 9).

The emerged PCs reveal new features out of the original variables. The dominantly correlated variables with PC1 were the middle point heights and the dome form definition, which were also the most beneficial parameters for the thermal performance. PC2 is highly correlated with the first corner depth as well as the repetition array. While on the contrary, PC3 is highly correlated with the second corner height as well as the repetition array. PC2 and PC3 refer to the variables with the worst impact on the energy performance from the previous analysis. Furthermore, PC4 is correlated with the first arc number of segments and the first middle point depth, while PC5 is correlated with the second arc number of segments and the second middle point depth. Finally, each of PC6 and PC7 represents one dominant parameter, the dome form height and the roof orientation angle, respectively.

The findings are in good agreement with the sensitivity analysis results. The use of PCA is beneficial to decrease the number of dimensions in the dataset while still allowing it to represent each variable in the optimization problem. From an architectural observation, PC1 can be linked

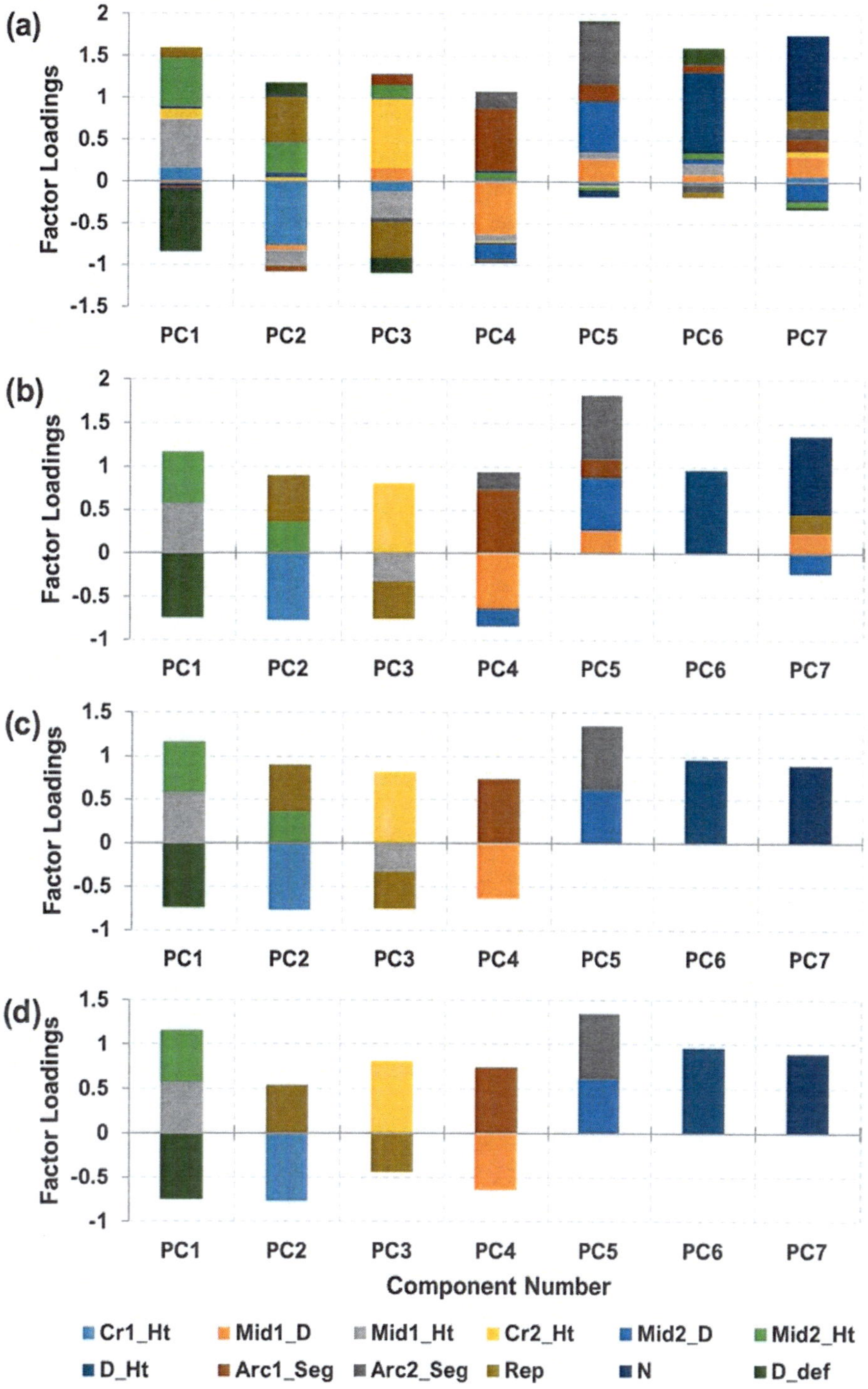

Fig. 9 Factor loadings onto the extracted 7 PCs using Varimax orthogonal rotation. **a** Initial full loadings without cutoff. **b** 0.2 factor loadings cutoff. **c** 0.3 factor loadings cutoff. **d** 0.4 factor loadings cutoff

with the generation of gable, domed and vaulted-roof forms. On the other hand, PC2 and PC3 represent producing shed roofs as well as duplicated geometries like saw-toothed roofs. PC4 and PC5 can be associated with hipped roof forms and curve-edged roofs like vaulted-roof types. The orientation factor, represented by PC7, is not related to the other variables, which is understandable as the change in it can produce all sorts of geometry, and it does not depend on any other factor. Additionally, the dome height variable is dominantly represented by PC6 solely, which is supposedly due to the formula for the dome generation. Therefore, the dome definition variable loading on PC1 provides a better interpretation for the dome form state.

6 Conclusion

Building performance simulation is one of the powerful steps towards energy optimization and efficient early design. However, simulation results are susceptible to uncertainties due to the lack of information about the effect of each parameter. Therefore, sensitivity analysis is typically coupled with building performance simulation problems. The paper investigated the key features of a roof geometrical configuration influencing the energy performance of residential buildings. It was implemented within a case study of a hypothetical house prototype in Alexandria, Egypt. Following a parametric approach, the test input variables were defined to set up a custom form-finding definition that imitates realizable residential building roofs such as flat, pitched, gable, vaulted and domed roofs, as well as other transformations between them. Subsequently, the energy model was set up with parameters including thermal zone parameters, construction materials, schedules and setpoints, which were selected to comply with the conditions of the case study and the Egyptian code standards. The following step included creating the database for sensitivity analysis, where 24,637 cases were simulated, and objective outputs, including the EUI and TDH, were calculated.

The study provides an insightful overview of the important and insignificant design parameters, assisting energy modellers with limited time and computational effort in the early design stages. The analysis results suggested that the most significant variables on the overall performance with positive influence were the roof middle point heights and the definition of the dome state. Conversely, the most significant variables with negative influence were the roof corners' heights and the repetition array of the roof geometry. On the other hand, the least significant variables were roof middle points' depth. This indicates that forms in which corners tend to slope downwards and midpoints slope upwards, such as gable roofs, domes and vaults, would achieve better energy and thermal comfort performance. Moreover, the PCA analysis was able to form relationships between the roof geometrical configurations referring to the forms that are related and would perform in a similar manner, which was in compliance with the sensitivity analysis results. Future studies are recommended to implicate a larger design space in the test as well as alterations in construction materials and climatic conditions in order to assess the variation of the results in diverse design problems.

Lastly, based on the results of the study, the following recommendations are suggested:

- Encouraging building designers to rethink the design possibilities towards sloping and curved roofs regardless of the aesthetic and financial aspects.
- Setting regulations for newly built contexts addressing the use of sloped and curved roof systems in low-rise residential buildings.
- Driving decision-makers to recognize and pursue potential tactics and pathways for future developments.
- Extending the exploitation of thermal simulation procedures to assess the energy performance in case-specific project designs.
- Training architects to expand the use of parametric approaches in design exploration to transfer the gained knowledge during the assessment of passive techniques.

References

Akbarpoor, A. M., Gilvaei, Z. M., Poshtiri, A. H., & Zhong, L. (2022). A hybrid domed roof and evaporative cooling system: Thermal comfort and building energy evaluation. *Sustainable Cities and Society, 80*, 103756.

Al-Shehari, T., & Alsowail, R. A. (2021). An insider data leakage detection using one-hot encoding, synthetic minority oversampling and machine learning techniques. *Entropy, 23*(10), 1258.

Awwad, B. A., Suliman, M. O., & Safran, M. (2018). Study of different pitched roof types. *Civil and Environmental Research, 10*(3), 98–113.

Ayoub, M., & Elseragy, A. (2018). Parameterization of traditional domed-roofs insolation in hot-arid climates in Aswan, Egypt. *Energy and Environment, 29*(1), 109–130.

Basu, R. (2009). High ambient temperature and mortality: A review of epidemiologic studies from 2001 to 2008. *Environmental Health, 8*(1), 40. https://doi.org/10.1186/1476-069X-8-40

CAPMAS. (2017). *Census—population for 2017 (Governorate)*. Central Agency for Public Mobilization and Statistics.

Cattell, R. B. (1966). The scree test for the number of factors. *Multivariate Behavioral Research, 1*(2), 245–276.

Costello, A. B., & Osborne, J. (2005). Best practices in exploratory factor analysis: Four recommendations for getting the most from your analysis. *Practical Assessment, Research, and Evaluation, 10*(1), 7.

Dabaieh, M., Wanas, O., Hegazy, M. A., & Johansson, E. (2015). Reducing cooling demands in a hot dry climate: A simulation study for non-insulated passive cool roof thermal performance in residential buildings. *Energy and Buildings, 89*, 142–152.

Elhadad, S., & Orban, Z. (2021). A sensitivity analysis for thermal performance of building envelope design parameters. *Sustainability, 13*(24), 14018.

Elnokaly, A., Ayoub, M., & Elseragy, A. (2019). Parametric investigation of traditional vaulted roofs in hot-arid climates. *Renewable Energy, 138*, 250–262.

Elnokaly, A., & Elseragy, A. (2006). *Revitalisation of traditional curved roofs for indoor thermal comfort in hot climates*. In Paper presented at the international conference on architecture.

Elseragy, A., & Elnokaly, A. (2007). *Assessment criteria for form environmental performance of building envelope in hot arid climates*. In *Paper presented at the 24th conference on passive and low energy architecture*.

Elseragy, A. A. (2003). *Architectural and solar potential of curved and flat roofs in hot arid regions (with reference to Egypt)*. University of Nottingham.

Fanger, P. O. (1970). *Thermal comfort: Analysis and applications in environmental engineering*. Mcgraw-Hill.

Fathy, H. (1986). *Natural energy and vernacular architecture*.

Ghaedi, A., Ghaedi, H., & Ghaedi, H. (2012). The role of roof shapes in design of green building systems (Case Study: Iran, BandarAbbas). *APCBEE Procedia, 1*, 335–340.

Hadavand, M., & Yaghoubi, M. (2008). Thermal behavior of curved roof buildings exposed to solar radiation and wind flow for various orientations. *Applied Energy, 85*(8), 663–679.

Hamby, D. M. (1994). A review of techniques for parameter sensitivity analysis of environmental models. *Environmental Monitoring and Assessment, 32*(2), 135–154.

Heiselberg, P., Brohus, H., Hesselholt, A., Rasmussen, H., Seinre, E., & Thomas, S. (2009). Application of sensitivity analysis in design of sustainable buildings. *Renewable Energy, 34*(9), 2030–2036.

Hinkin, T. R. (1995). A review of scale development practices in the study of organizations. *Journal of Management, 21*(5), 967–988.

Howard, M. C. (2016). A review of exploratory factor analysis decisions and overview of current practices: What we are doing and how can we improve? *International Journal of Human-Computer Interaction, 32*(1), 51–62.

IEA. (2021). *World Energy Outlook 2021*. Retrieved from International Energy Agency—Publications: https://iea.blob.core.windows.net/assets/888004cf-1a38-4716-9e0c-3b0e3fdbf609/WorldEnergyOutlook2021.pdf

Jackson, J. E. (2005). Varimax rotation. *Encyclopedia of Biostatistics, 8*.

Kaiser, H. F. (1958). The varimax criterion for analytic rotation in factor analysis. *Psychometrika, 23*(3), 187–200.

Kaiser, H. F. (1960). The application of electronic computers to factor analysis. *Educational and Psychological Measurement, 20*(1), 141–151.

Head, K., Clarke, M., Bailey, M., Livinski, A., Ludolph, R., & Singh, A. (2018). *Report of the systematic review on the effect of indoor heat on health*. Retrieved from Geneva: https://www.ncbi.nlm.nih.gov/books/NBK535282/

Knekta, E., Runyon, C., & Eddy, S. (2019). One size doesn't fit all: Using factor analysis to gather validity evidence when using surveys in your research. *CBE—Life Sciences Education, 18*(1), rm1.

Lin, B., Chen, H., Liu, Y., He, Q., & Li, Z. (2021). A preference-based multi-objective building performance optimization method for early design stage. *Building Simulation, 14*(3), 477–494.

Mazzeo, D., & Kontoleon, K. J. (2020). The role of inclination and orientation of different building roof typologies on indoor and outdoor environment thermal comfort in Italy and Greece. *Sustainable Cities and Society, 60*, 102111.

MHUC. (2006). Egyptian code for improving energy efficiency in buildings: Code No. 306–2005 ECP—Part one: Residential buildings. In *Ministry of housing, utilities and urban communities*. National Center for Housing and Building Research.

MOEE. (2013). *Egyptian electricity holding company annual reports*. Retrieved from The Ministry of Electricity and Renewable Energy: http://www.moee.gov.eg/english_new/EEHC_Rep/2013-2014en.pdf

Moustafa, W. S., Hegazy, I. R., & Eldabousy, M. M. (2018). Roof geometry as a factor of thermal behavior: Simulation based study of using vaults and domes in the Middle East zone. *International Journal of Low-Carbon Technologies, 13*(3), 204–211.

Nguyen, A.-T., Reiter, S., & Rigo, P. (2014). A review on simulation-based optimization methods applied to building performance analysis. *Applied Energy, 113*, 1043–1058.

Sachindra, D. A., Ng, A. W. M., Muthukumaran, S., & Perera, B. J. C. (2016). Impact of climate change on urban heat island effect and extreme temperatures: A case-study. *Quarterly Journal of the Royal Meteorological Society, 142*(694), 172–186. https://doi.org/10.1002/qj.2642

Sheather, S. (2009). *A modern approach to regression with R*. Springer Science & Business Media.

Siegel, A. F. (2016). *Practical business statistics* (7th ed.). Academic Press.

Sirimanna, M., & Attalage, R. (2016). A model for analyzing the thermal performance of roof configurations with flat inclined surfaces. *Energy and Buildings, 116*, 122–132.

Tian, W. (2013). A review of sensitivity analysis methods in building energy analysis. *Renewable and Sustainable Energy Reviews, 20*, 411–419.

Vijaykumar, K., Srinivasan, P., & Dhandapani, S. (2007). A performance of hollow clay tile (HCT) laid reinforced cement concrete (RCC) roof for tropical summer climates. *Energy and Buildings, 39*(8), 886–892.

Visser, F., & Yeretzian, A. (2013). *Energy Efficient Building Guidelines for MENA Region: Energy Efficiency in the Construction Sector in the Mediterranean*. Retrieved from MED-ENEC: https://www.climamed.eu/wp-content/uploads/files/Energy-Efficient-Building_Guideline-for-MENA-Region-NOV2014.pdf

Investigating the Impacts of Biochar Addition to Substrate on Mitigating Carbon Dioxide Emissions in Green Roofs

Gaochuan Zhang, Bin Tang, Haohao Bian, Yangfei Huang, Hexian Jin and Bao-Jie He

Abstract

The utilization of biochar is often heralded as a salient approach to curtailing emissions and sequestering carbon. Agricultural and forestry soils have demonstrated its efficaciousness in reducing carbon emissions; however, the urban greening industry has yet to be thoroughly researched in this regard. To address this lacuna, the present study undertook the application of biochar to the soil of an urban green roof, with the aim of ascertaining its potential to mitigate carbon dioxide emissions. By studying the effects of applying BS-0%, BS-1%, BS-5%, and BS-10% biochar on green roof substrate, the study examined the carbon dioxide value over a period of 36 h encompassing both sunny and nocturnal environments. The results indicated that the addition of biochar to the soil was effective in reducing carbon emission, with all of the biochar additions demonstrating an aptitude for carbon dioxide fixation. Notably, the application of BS-1% biochar evinced a statistically significant reduction in the carbon dioxide value. Furthermore, the addition of biochar to the roof substrate was found to be crucial in controlling rainwater runoff and alleviating urban eutrophication, indirectly reducing carbon emissions. This study provides a practical and theoretical framework for the application of biochar to mitigate the urban heat island effect.

Keywords

Green roof · Biochar · Carbon dioxide reduction · Climate change

G. Zhang · Y. Huang
School of Civil Engineering and Architecture, Zhejiang University of Science & Technology, Hangzhou 310023, China
e-mail: Zjhz_zgc@sina.com

Y. Huang
e-mail: 9543938@qq.com

B. Tang
Huzhou Spatial Planning and Research Center, Huzhou 310000, China
e-mail: 529669644@qq.com

H. Bian
Zhejiang Province Key Laboratory of Recycling and Ecological Treatment of Waste Biomass, School of Environment and Natural Resources, Zhejiang University of Science and Technology, Hangzhou 310023, China
e-mail: bianhaohao@vip.qq.com

H. Jin
College of Landscape Architecture, Zhejiang Agriculture & Forestry University, Hangzhou 311300, China
e-mail: 529669644@qq.com

B.-J. He (✉)
Centre for Climate-Resilient and Low-Carbon Cities, School of Architecture and Urban Planning, Chongqing University, Chongqing 400045, China
e-mail: baojie.unsw@gmail.com

Smart City Research Institute in Liyang, Chongqing University, Liyang, Jiangsu 213300, China

Key Laboratory of New Technology for Construction of Cities in Mountain Area, Ministry of Education, Chongqing University, Chongqing 400045, China

State Key Laboratory of Subtropical Building Science, South China University of Technology, Guangzhou 510640, Guangdong, China

Nomenclature

CEC	means cation exchange capacity;
TN	means Total N; TP means Total P;TK means Total K; SSA means specific surface area;
BS-0%	means The mass of biochar accounts for 5% of the total mass of substrate;
BS-10%	means The mass of biochar accounts for 10% of the total mass of substrate.

B.-J. He et al. (eds.), *Resilient Horizons: Building Sustainable Environments for Climate Adaptation and Health*, Advances in Science, Technology & Innovation, https://doi.org/10.1007/978-3-031-46109-5_11

1 Introduction

As the world has become more urbanized and industrialized, there have been clear improvements in the lives of those who reside in cities. Unfortunately, this progress has come at a significant cost to the natural environment and climate of urban areas (Chen & You, 2020). A prime example of this is the urban heat island effect, which is a direct result of the changes that have taken place in the impervious surfaces of urban regions, along with the creation of densely packed building groups and the generation of copious amounts of greenhouse gases through human activities (Li et al., 2022; Rajapaksha et al., 2022; Wan et al., 2022). The negative effects of the heat island effect are felt across cities of all sizes, as evidenced by the findings of various studies (Li et al., 2022; Rajapaksha et al., 2022; Wan et al., 2022). Not only can the heat island effect pose a direct threat to human health through thermal hazards, it can also indirectly contribute to an increased incidence of respiratory diseases, as well as cardiovascular and cerebrovascular diseases by worsening air pollution (Song et al., 2022). However, in recent years, there has been a growing body of research that focuses on mitigating the heat island effect, and by extension, improving the overall quality of life for city dwellers It is important to note that carbon dioxide is the main greenhouse gas responsible for climate change in cities, with its content being significantly higher than that of other greenhouse gases such as CH_4, N_xO_y (Wood et al., 2012). Furthermore, soil properties also play a key role in determining the amount of carbon dioxide emissions (Jones et al., 2011). Urban greening has been identified as a particularly effective means of alleviating the heat island effect, promoting carbon absorption, and releasing oxygen. In addition, it can enhance the aesthetic value of the environment and improve the overall quality of life for urban residents (Du et al., 2022). Despite the obvious benefits of urban greening, however, there are significant challenges that need to be addressed. For example, urban buildings are divided into different regions, which makes it difficult to achieve a balance of green spaces across urban areas. As a result, the cooling effect of urban greening can be minimal, rendering it less effective in mitigating the heat island effect (Xu et al., 2022). Therefore, it is essential that urban planners and policymakers work together to find creative solutions that address these issues, so that the benefits of urban greening can be maximized, and the heat island effect can be minimized.

Green roof, a form of greening where green plants are placed on a roof, has been in use since more than 2000 BC (Oberndorfer et al., 2007). With the increasing diversity and number of buildings in urban areas, there has been a significant increase in rooftop space, which presents a ripe opportunity for green roof development to balance the insufficiency of urban greening in space (Sitzenfrei et al., 2020). Previous studies have identified a multitude of benefits of green roofs, including reduction of air temperature, rainwater runoff, and building energy consumption (Alim et al., 2022). However, the carbon sequestration and decarbonization effects of green roofs have been paid less attention to. Moreover, the reckless use of green roofs can become a source of pollution. This is because the substrate of green roofs is typically soil, which causes nutrients to be lost easily, and after rainfall, leads to a large amount of runoff that causes eutrophication of urban water bodies. This can, in turn, release large amounts of carbon dioxide, the main greenhouse gas, from water bodies, aggravating the heat island effect (Cai et al., 2011). Therefore, it is important to explore and understand the potential of green roofs for reducing carbon emissions, as well as the proper management and maintenance of green roofs to ensure their overall sustainability and effectiveness in mitigating the urban heat island effect.

Biochar, a solid material produced through anaerobic and high temperature processes utilizing waste, possesses exceptional properties such as high specific surface area, high ion exchange capacity, and carbon negativity, and has found broad applications in the fields of energy, construction, and environment (Lehmann et al., 2021). The environmental sector has specifically leveraged the advantages of biochar as an adsorbent, catalyst, or soil conditioner to improve substrate characteristics (He et al., 2022; Lee et al., 2019; Yang et al., 2020). Recent researches have shown that biochar can effectively reduce carbon emissions from soils in certain conditions and mitigate climate change, hence making it a promising solution in the fight against global warming (Tan & Wang, 2023). Although studies in the past have mostly focused on the application of biochar in agricultural or forest soil, applying the conclusions of such studies in the urban heat island is difficult due to the different conditions (Li et al., 2018). Previous researches on biochar combined with green roof substrates have been limited to examining their soil substrate properties and hydrological properties (Gan et al., 2021; Huang et al., 2020). However, studies that specifically investigate the impact of biochar addition to green roof substrates on carbon emissions are still scarce. Interestingly, under specific conditions, the application of biochar to soil may even increase carbon emissions, thereby countering the mitigation of the heat island effect (Hawthorne et al., 2017). Therefore, it is crucial to conduct research on the effects of biochar application in green roof substrates on carbon emissions, especially as rooftop spaces endure much more extreme environmental conditions such as drought, high temperature, and weathering, compared to agriculture and forests on land.

Table 1 Basic physicochemical properties of soil and biochar

Parameter	TN	TP	TK	SSA (cm^2/g)
Soil	645.5	268.30 mg/kg	6.2 mg/kg	–
Biochar	0.98%	–	–	19.42
Biogas slurry	2724.37 mg/L	13.56 mg/L		–

The utilization of biochar substrate for green roof implementation is a highly intriguing and multifaceted subject that warrants close attention. In particular, the mitigation of the urban heat island phenomenon can be achieved through two distinct but interrelated pathways. First, the green roof has the capacity to directly absorb carbon emissions, thereby alleviating the negative impact of this harmful gas on the environment. Secondly, the green roof can adjust runoff for water eutrophication, thereby indirectly reducing carbon emissions. In order to fully explore the intricacies of this topic, a comprehensive study was conducted which involved the application of four distinct weight rates of biochar (accounting for the mass of 0, 1, 5, and 10%) across a range of weather conditions. This investigation had three primary objectives, namely to: (i) assess the potential of biochar in reducing carbon emissions in green roofs under different weather conditions, (ii) examine the value of adding biochar at appropriate proportions for fixing carbon dioxide, and (iii) categorize the water-retaining and fertilizer-retention properties of biochar. The findings of this study are expected to provide valuable insights and recommendations for the scientific application of biochar in green roof implementation. This includes new reference points and weight ratios that can be utilized to optimize the effectiveness of biochar in mitigating the negative environmental impacts of urbanization. Overall, this study highlights the immense potential of biochar in promoting sustainable urban development and reducing the deleterious effects of the urban heat island phenomenon.

2 Materials and Methodologies

2.1 Experimental Materials

The study in question availed itself of a panoply of four distinct materials, to wit: soil, biogas slurry, biochar, and vegetation. Both soil and biogas slurry were meticulously gleaned from the campus farm of the esteemed Zhejiang University of Science and Technology. The soil, once procured, was subjected to a natural air-drying process in the laboratory, whereas the plant residues were methodically jettisoned and subsequently ground through a 0.5 mm nylon sieve. Table 1 adumbrates the multifarious physical and chemical properties of the soil. With regard to the biochar, it was meticulously prepared by initially drying corn stalks, followed by subjecting them to pyrolysis under 500 °C using the OTF-1200X-V-VT five-stage temperature-controlled tube furnace, which is available through Kejing Material Technology Co., Ltd. located in the city of Hefei, China. In terms of vegetation, the study in question adopted the ubiquitous and commercially available sedum lineare, which is a plant species that is commonly employed in rooftop space, and was dutifully acquired from the reputable Hangzhou Flower Market.

2.2 Experimental Site and Design

The locus of experimentation was meticulously selected to be none other than the illustrious Zhejiang University of Science and Technology, situated in the scintillating city of Hangzhou, which happens to be the largest city in the Zhejiang province, with a burgeoning population that has been estimated to be no less than 12.204 million, resulting in a burgeoning urbanization rate that currently stands at an impressive 83.6%. Moreover, the urban weather station data collected from this bustling metropolis reveals that Hangzhou is situated in an area that can be described as "hot and humid," with the average temperature hovering around a balmy 17.8 °C, and the maximum temperature reaching a blistering 42.9 °C, a fact which is corroborated by Fig. 1, which is appended hereto.

In this particular study, the experimental design is divided into two tests, one being the outdoor test, and the other being the indoor test. The outdoor experiment was further classified into four distinct treatments. These treatments include the soil control treatment, without the addition of biochar, and three mixed treatments of soil and biochar. The biochar was mixed with soil in varying proportions, with biochar accounting for 0%, 1%, 5%, and 10% of the total mass of substrate, respectively. For the outdoor experiment, two liters of biogas slurry were used as a nutrient source to irrigate and penetrate the soil substrate until it was suitable for planting. Then, Sedum lineare was screened and transplanted into the green roof substrate. It was essential to maintain the vegetation for seven days before the outdoor test, and during this time, daily watering was necessary to ensure root growth. Once the Sedum had fully recovered, the carbon dioxide experiment was set

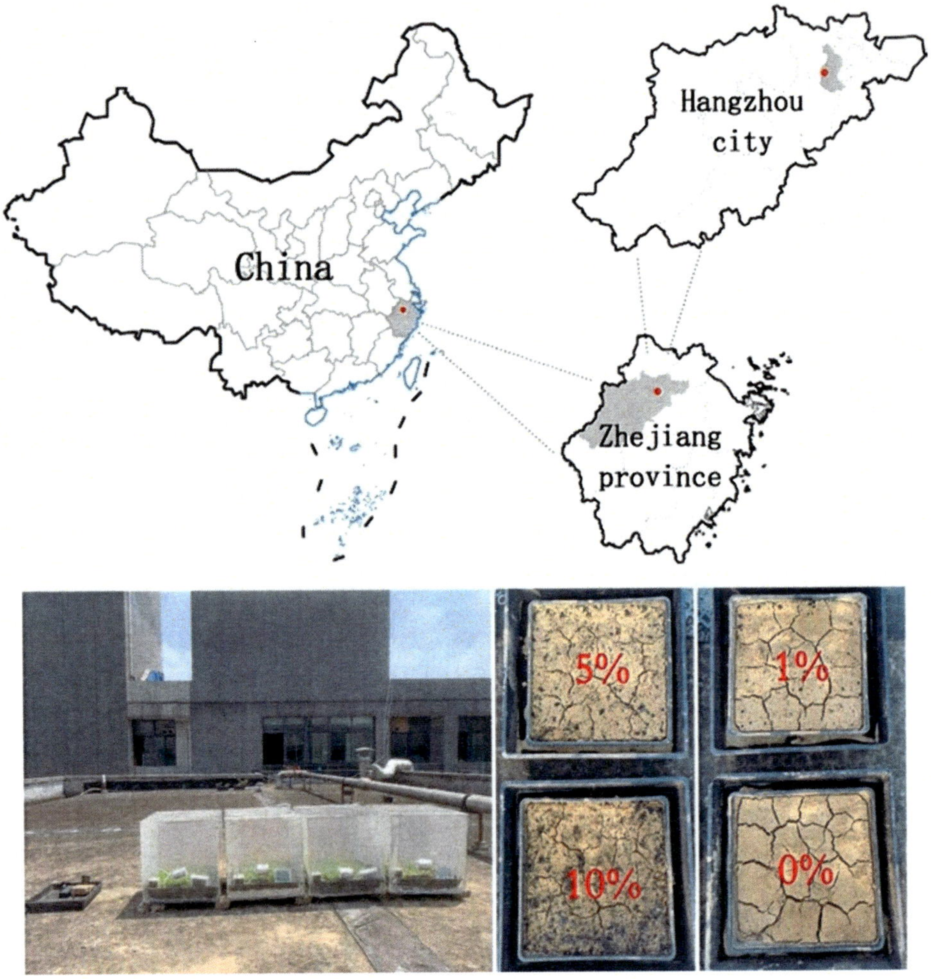

Fig. 1 Biochar substrate model with four proportions in site

up. To create a closed microenvironment, the module of the green roof was sealed with an impermeable acrylic sheet. The interface was then fixed with waterproof tape to ensure the module's airtightness. As illustrated in Fig. 2, the carbon dioxide value in the microenvironment was recorded by an automatic recorder (AirNow, UK). All the equipment was calibrated in advance, and the accuracies and ranges of the test instruments can be seen in Table 2. Additionally, the meteorological parameters were collected by the weather station on campus. The data collected was exported for analysis after 36 continuous hours from May 16, 8 A.M. to May 17, 8 P.M. Soil samples of each treatment were collected using the five-point soil sampling method (Zhang et al., 2021). In the indoor test, the ring knife method was used as international standards to test the performance of water and fertilizer retention and analyze the soil's physical and chemical properties.

3 Results and Discussion

In accordance with the multifarious findings elucidated in Fig. 2, biochar substrate manifests a remarkable proclivity for carbon emissions reduction, particularly in the treatments of BS-1%, BS-5%, and BS-10%. This observation aligns with the notion that biochar is capable of modulating soil properties and sequestering carbon dioxide (Lehmann et al., 2021). Notably, the BS-1% treatment exhibits the most superior performance in carbon fixation. Upon a closer inspection, although the organic carbon input of the BS-5% and BS-10% treatments is five times and ten times greater than that of BS-1%, the extent of carbon dioxide emission reduction is somewhat limited, failing to correspond with the proportional increase in organic carbon input. This could be attributed to the organic carbon component of biochar, which increases the organic carbon content

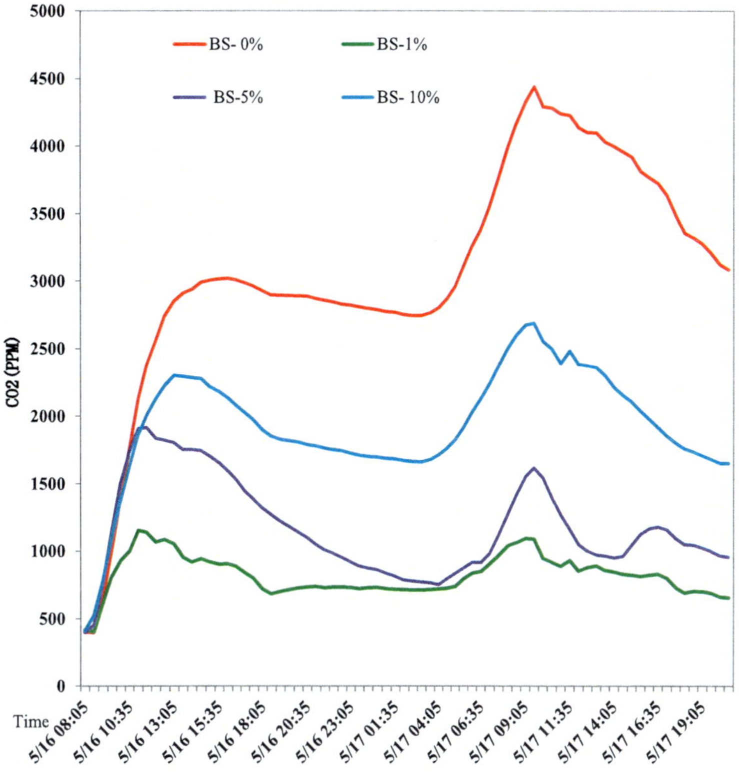

Fig. 2 The amount of carbon dioxide emissions in four experimental module

Table 2 Parameters and instruments used during the field measurements

Item	Instrument	Parameter	Resolution/range	Frequency (min)
Temperature	Sensirion-STS21	Air temperature	±0.2 °C (−40 to 125 °C)	5
Humidity		Relative humidity	±1.5% (0–100%)	5
CO_2	Senseair S8	PPM	±200 ppm ± 10% of reading	10
Radiation	TES-1333R	Solar radiation	0.1 W/m^2(0–2000 W/m^2)	10
Illumination	TES-1336A	Lux	0.01 lx (0 ~ 20,000 lx)	10

in the substrate, and the additional biochar, which enhances the soil pore structure, thereby stimulating microbial content and activity, hastening soil organic carbon decomposition, and carbon dioxide release in the soil, leading to an uptick in carbon emissions (Zhao et al., 2022). Thus, it is evident that the adsorption capacity of biochar substrate is far superior to the organic carbon release capacity. The former is known to absorb a greater quantity of carbon dioxide than the latter and is known to release during decomposition. As the proportion of biochar in the substrate escalates, the carbon dioxide content in the confined microenvironment also increases, substantiating the capacity of biochar in reducing emissions from green roofs. Nevertheless, the threshold quantity of biochar application that yields the most desirable outcomes necessitates further investigation, which will be undertaken in the ensuing research.

3.1 Directly Effect of Carbon Dioxide Emission from Meteorological Environment

3.1.1 Relationship Between Illumination Intensity and Carbon Dioxide Emission

In the context of the enclosed microenvironment, with the increase of time under the same illumination intensity, the concentration of carbon dioxide in the box tends to increase. However, it is worth noting that the inflection points of this increase in concentration are markedly distinct and dependent on the proportion of additional biochar. Specifically, for BS-1%, BS-5%, and BS-10%, the peak value of carbon dioxide emission is observed at 11 A.M. on May 16th and at 10:30 A.M. on May 17th under an illumination intensity of 3500 lx. It is noteworthy that the release of carbon dioxide in the substrate of these biochar additions tends to decrease at night, indicating that the influence of illumination intensity is quite helpful for the release of carbon dioxide. As per the same condition of illumination intensity during the daytime, it can be concluded that all the three additional biochar proportions, namely, BS-1%, BS-5%, and BS-10%, can indeed mitigate carbon emissions, as shown in Fig. 3. However, in the case of a high illumination intensity, the module containing a high proportion of biochar is seen to release a greater amount of carbon dioxide, which is in line with the fact that the illumination can improve the respiration of vegetation and soil (Hawthorne et al., 2017).

3.1.2 Relationship Between Atmospheric Temperature and Carbon Dioxide Emission

As the data presented in Fig. 4, it could discern a noteworthy pattern: the carbon dioxide value in the modules mirrors the atmospheric temperature change curve. When the outdoor temperature rises, the carbon dioxide value released increases, which suggests that the microbial activity of carbon dioxide released in the substrate is accelerating organic-matter degradation. It turns out that biochar contains unstable organic carbon which contributes to the unstable organic carbon pool in the soil after its application, resulting in a boost to soil carbon dioxide emissions (Spokas, 2013). Additionally, the increase in soil moisture content, porosity, aggregation, and CEC may also lead to increased carbon dioxide emissions (Liang et al., 2010). In fact, the diversity and activity of microbial taxa involved in carbon dioxide production can also increase by the application of biochar (Wang et al., 2016). The intricacy and complexity of these processes make it a truly perplexing subject. All of this underscores the importance of maintaining proper hydration levels by adding water or spraying the roof to help regulate surface temperature for green roof, as well as for the substrate. Doing so not only improves indoor energy consumption and reduces carbon dioxide but can also help to curb the carbon dioxide emission of green roof itself. It's a lot to take in, but it's clear that these factors must be considered when implementing green roofs.

3.1.3 Relationship Between Solar Radiation and Carbon Dioxide Emission

Based on the data shown in Fig. 5, it can be observed that when solar radiation is present, all of the modules tested have generated carbon dioxide, with differing results depending on the biochar proportion added. Specifically, the modules with BS-1% and BS-5% have shown a tendency to reduce the amount of carbon dioxide generated. The carbon dioxide emissions from the modules correspondingly increased or decreased as the solar radiation levels fluctuated. This means that there is a direct correlation between the release of carbon dioxide and the level of solar radiation, and as the radiation value increases, the amount of carbon dioxide released also increases. This finding is in agreement with previous research that indicates that solar radiation can promote the activity of microbial organisms,

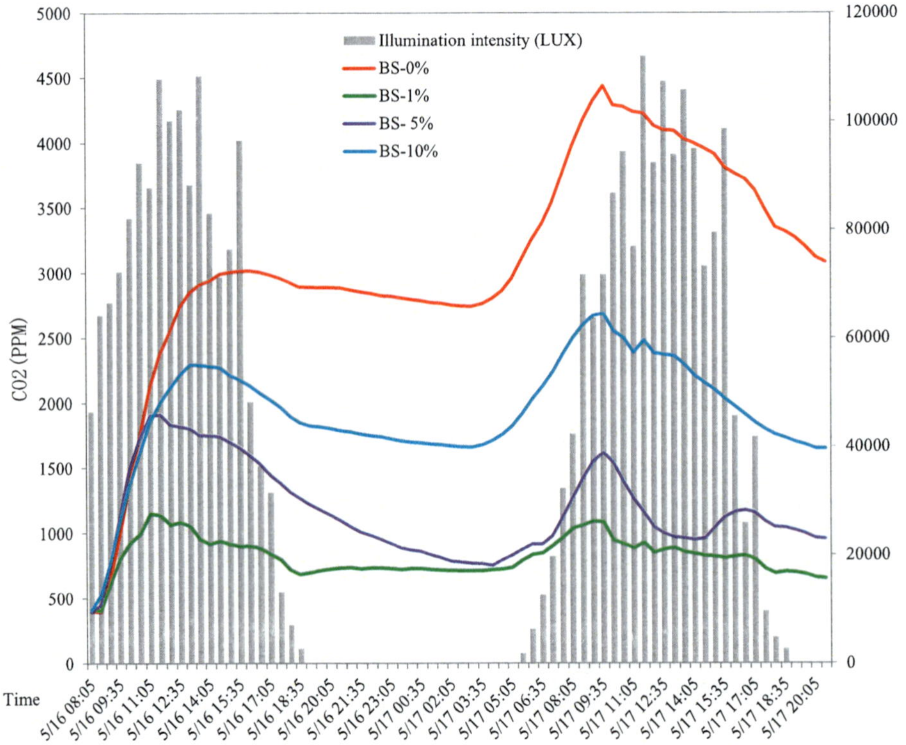

Fig. 3 Relationship between Illumination intensity and carbon dioxide emission

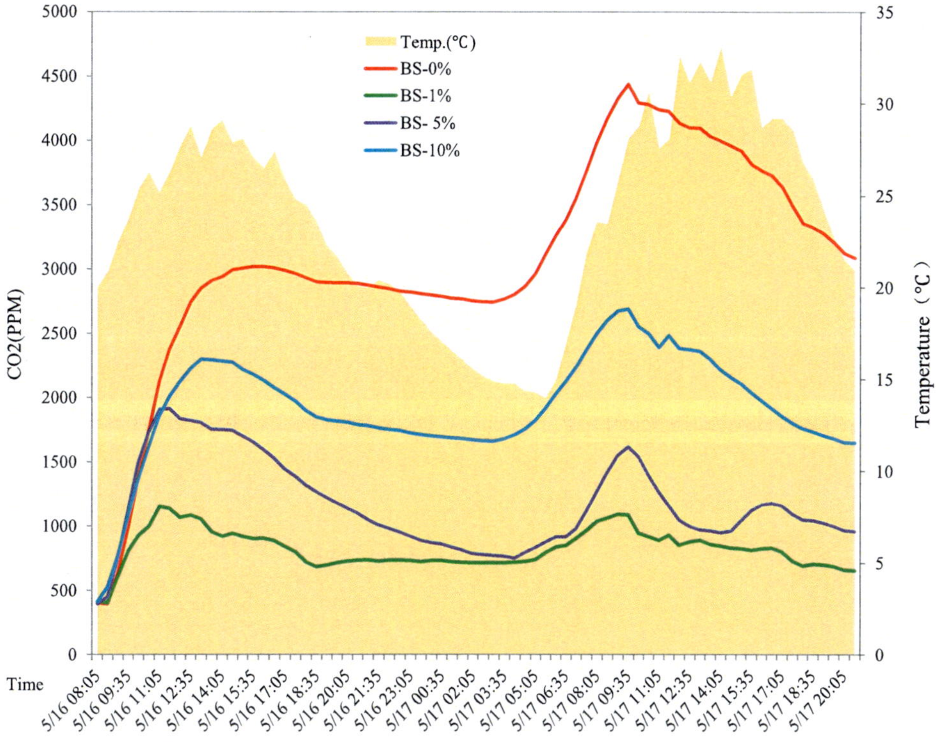

Fig. 4 Relationship between atmospheric temperature and carbon dioxide emission

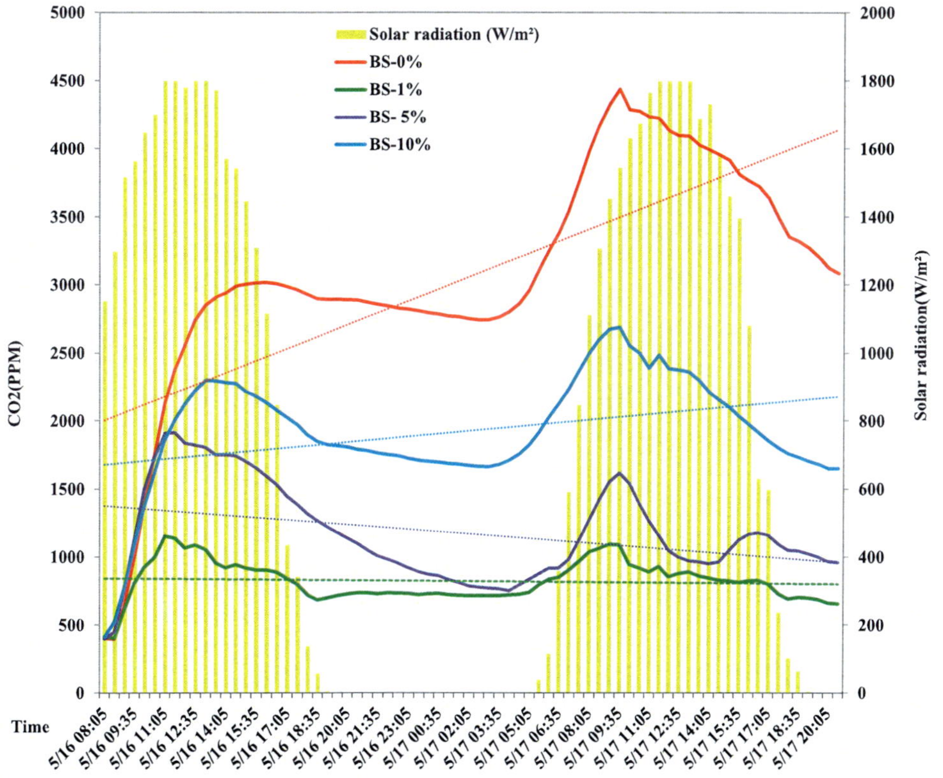

Fig. 5 Relationship between solar radiation and carbon dioxide emission

resulting in increased carbon dioxide release (Wang et al., 2016). Interestingly, the results have also revealed that when the proportion of biochar is excessive, such as in the case of BS-10%, it may even lead to the release of additional carbon dioxide. This highlights the importance of carefully considering the appropriate amount of biochar to be added, as excessive amounts may have unintended consequences.

3.2 Indirectly Effects of Biochar Addition on Variations of Hydrological Properties

3.2.1 Water-Retaining Property

The hydrological and fertilizer retention properties of green roof substrates are critical factors that influence soil carbon dioxide emissions. The hydrological properties are especially significant as they play a crucial role in reducing flooding during heavy rainfall in urban areas. These properties, including soil bulk density and soil water retention, have potential impacts on carbon dioxide emissions. Table 3 provides a detailed view of the soil properties after cultivation for each treatment, which indicates the compactness of soil as the biochar application rate increases. The bulk density of the soil decreased with the increase in biochar application rate, where BS-0%, 1%, 5%, and 10% showed compactness, compactness, suitability, and suitability, respectively. The less compact the soil is, the more porous it becomes, thereby storing more water. High soil moisture helps to reduce carbon dioxide emissions and reduce the possibility of flooding during extreme precipitation (Norberg et al., 2018), thus promoting the development of a 'sponge city.' The soil water retention of different treatment groups was also tested, and it was observed that the soils with a higher biochar application rate retained more water. For instance, the BS-0%, 1%, 5%, and 10% soils with a moisture content of 50% lost 25.1%, 20.4%, 16.3%, and 10.5% of their mass in 1 h at 50 °C, respectively. This indicates that the higher the biochar application rate, the better the water retention of the soil. Interestingly, the BS-1% treatment group had the maximum variation of water retention extent from 2.6 to 10.6, which shows that it has excellent water retention Thus, good water retention properties are not only conducive to the survival of green roofs in extreme environments but also reduce the number of artificial water supplies needed, enabling more photosynthesis for carbon fixation.

3.2.2 Fertilizer-Retention Property

In urban areas, eutrophication caused by runoff from green roofs can lead to serious consequences. An increase in carbon dioxide emissions from water bodies due to eutrophication can exacerbate the heat island effect. That's why it's critical to consider the fertilizer retention properties of green roof substrates. And the soil cation exchange capacity (CEC) is a key factor in this regard. In Table 4, it could display how the CEC of different green roof substrates changed based on the biochar application amount. The BS-0%, 1%, 5%, and 10% CEC showed different grades of performance, such as grade 5, grade 3, grade 3, and grade 2, respectively. Notably, the more biochar was applied, the better the fertilizer retention performance became. It's also worth mentioning that residual nitrogen and phosphorus were detected in the soil, and the team calculated the fertilizer retention rate using element conservation. In Table 1, you can see that the nitrogen content of the original soil and biogas liquid. Based on this information, the team calculated the nitrogen loss rate for different biochar application amounts. For instance, the nitrogen loss rates of BS-0%, BS-1%, BS-5% and BS-10% were 40.9%, 20.7%, 16.3% and 10.9%, respectively. These results show that the nitrogen loss rate decreased with the increase of biochar application, indicating that the fertilizer retention performance of the soil was improved. Moreover, biochar was the main source of phosphorus in the soil, which is consistent with the results of previous studies (He et al., 2021).

Interestingly, when 1% of the total mass of biochar was added to the soil, it resulted in more phosphorus in the soil. But when the biochar application amount was increased, there was no significant increase in the phosphorus content in the soil substrate. This indicates that the phosphorus in the soil substrate had reached the upper limit due to the addition of biochar. However, an increased phosphorus content in the soil substrate could promote plant growth. The study also found a significant increase in soil pH from 4.84 to 6.64. Based on the national soil fertility grade standard in China, the BS-0%, 1%, 5% and 10% soil pH showed that the soil was acidic, acidic, subacidic and neutral, respectively. This suggests that biochar can alleviate the trend

Table 3 Soil properties after different treatments

Parameter	pH	Bulk density (g/cm³)	CEC (cmol/kg)	TN (mg/kg)	TP (mg/kg)
BS-0%	4.84	1.39	2.6	1.20×10^3	286
BS-1%	5.32	1.26	10.6	1.36×10^3	353
BS-5%	5.85	1.15	12.4	1.61×10^3	430
BS-10%	6.64	1.02	16.3	1.81×10^3	422

Table 4 Soil fertility grading standard

Parameter	pH[a]	Bulk density[b] (g/cm^3)	CEC (cmol/kg)
Grade 1	<4.5	<1.00	>20.0
Grade 2	4.5–5.5	1.00–1.25	15.4–20.0
Grade 3	5.5–6.5	1.25–1.35	10.5–15.4
Grade 4	6.5–7.5	1.35–1.45	6.2–10.5
Grade 5	>7.5	1.45–1.55	<6.2
Grade 6	–	>1.55	–

[a] The grading standard of pH ranges from Grade 1 to 5: strong acid, acidic, subacidity, neutral, alkaline

[b] The grading standard of Bulk density ranges from Grade 1 to 6: too loose, suitable, subcompact, compact, too compact, solid

of soil acidification, which is a serious problem in southeastern China. By mitigating soil acidification, biochar can improve soil health and increase the abundance of soil microbial communities, which in turn can promote better plant growth. Soil acidification also reduces soil fertility, as nitrogen and phosphorus are more active in acidic conditions and are more easily lost. Therefore, the application of biochar in green roof substrates can not only improve the hydrological properties of the substrates, but also reduce nutrient runoff from green roofs. Overall, these findings suggest that biochar is a promising substrate additive for green roofs.

4 Conclusions and Prospects

This study investigates the use of biochar, a widely utilized carbon-friendly material in agriculture, in the green roof substrate to study the chrematistics of carbon fixation. Through a comprehensive analysis of four groups of comparative experiments, the study uncovers fascinating findings, suggesting that (1) the addition of biochar in the substrate can effectively reduce carbon dioxide emissions from green roofs, (2) while all the factors including illumination, radiation, and temperature can accelerate the emission of carbon dioxide, it is not always the case that the higher the biochar added, the better the carbon sequestration and decarbonization effect of the green roof, (3) BS-1% exhibits the best effect of carbon fixation and maximum variation of water retention extent, (4) the value ranging from BS-1% to BS-5% has the control ability of eutrophication runoff, and (5) it is not always the case that the more biochar added, the better the carbon reduction effect. This study stands out from the traditional research on carbon fixation performance of green roofs and opens up new avenues for reducing urban carbon dioxide. With biochar as a substrate additive, the study proposes a more comprehensive and effective strategy for carbon sequestration and decarbonization, which will contribute significantly to the development of sustainable and eco-friendly urban environments.

Acknowledgements This research was supported by National Natural Science Foundation of China (E51978623; 42301339) and Zhejiang University of Science & Technology (NO. F701104K02; 2021QN056). Project NO. 2023CDJXY-008 supported by the Fundamental Research Funds for the Central Universities. State Key Laboratory of Subtropical Building Science, South China University of Technology (No. 2022ZA01). Bao-Jie He appreciates the Travel Grant by the Buildings and Sustainability journals under MDPI Publisher.

References

Alim, M. A., Rahman, A., Tao, Z., Garner, B., Griffith, R., & Liebman, M. (2022). Green roof as an effective tool for sustainable urban development: An Australian perspective in relation to stormwater and building energy management. *Journal of Cleaner Production, 362*, 132561. https://doi.org/10.1016/j.jclepro.2022.132561

Cai, W.-J., Hu, X., Huang, W.-J., Murrell, M. C., Lehrter, J. C., Lohrenz, S. E., Chou, W.-C., Zhai, W., Hollibaugh, J. T., Wang, Y., Zhao, P., Guo, X., Gundersen, K., Dai, M., & Gong, G.-C. (2011). Acidification of subsurface coastal waters enhanced by eutrophication. *Nature Geosci, 4*(11), 766–770. https://doi.org/10.1038/ngeo1297

Chen, R., & You, X. (2020). Reduction of urban heat island and associated greenhouse gas emissions. *Mitigation and Adaptation Strategies for Global Change, 25*(4), 689–711. https://doi.org/10.1007/s11027-019-09886-1

Du, C., Jia, W., Chen, M., Yan, L., & Wang, K. (2022). How can urban parks be planned to maximize cooling effect in hot extremes? Linking maximum and accumulative perspectives. *Journal of Environmental Management, 317*, 115346. https://doi.org/10.1016/j.jenvman.2022.115346

Gan, L., Garg, A., Wang, H., Mei, G., & Liu, J. (2021). Influence of biochar amendment on stormwater management in green roofs: Experiment with numerical investigation. *Acta Geophysica, 69*(6), 2417–2426. https://doi.org/10.1007/s11600-021-00685-4

Hawthorne, I., Johnson, M. S., Jassal, R. S., Black, T. A., Grant, N. J., & Smukler, S. M. (2017). Application of biochar and nitrogen influences fluxes of CARBON DIOXIDE, CH4 and N2O in a forest soil. *Journal of Environmental Management, 192*, 203–214. https://doi.org/10.1016/j.jenvman.2016.12.066

He, M., Xiong, X., Wang, L., Hou, D., Bolan, N. S., Ok, Y. S., Rinklebe, J., & Tsang, D. C. W. (2021). A critical review on performance indicators for evaluating soil biota and soil health of biochar-amended soils. *Journal of Hazardous Materials, 414*, 125378. https://doi.org/10.1016/j.jhazmat.2021.125378

He, Q., Li, X., & Ren, Y. (2022). Analysis of the simultaneous adsorption mechanism of ammonium and phosphate on magnesium-modified biochar and the slow release effect of fertiliser. *Biochar, 4*(1), 25. https://doi.org/10.1007/s42773-022-00150-5

Huang, S., Huang, D., Garg, A., Jiang, M., Mei, G., & Pekkat, S. (2020). Stormwater management of biochar-amended green roofs: Peak flow and hydraulic parameters using combined experimental and numerical investigation. *Biomass Conversion Biorefinery*. https://doi.org/10.1007/s13399-020-01109-x

Jones, D. L., Murphy, D. V., Khalid, M., Ahmad, W., Edwards-Jones, G., & DeLuca, T. H. (2011). Short-term biochar-induced increase in soil CARBON DIOXIDE release is both biotically and abiotically mediated. *Soil Biology and Biochemistry, 43*(8), 1723–1731. https://doi.org/10.1016/j.soilbio.2011.04.018

Lee, S. Y., Choi, J.-W., Song, K. G., Choi, K., Lee, Y. J., & Jung, K.-W. (2019). Adsorption and mechanistic study for phosphate removal by rice husk-derived biochar functionalized with Mg/Al-calcined layered double hydroxides via co-pyrolysis. *Composites Part B: Engineering, 176*, 107209. https://doi.org/10.1016/j.compositesb.2019.107209

Lehmann, J., Cowie, A., Masiello, C. A., Kammann, C., Woolf, D., Amonette, J. E., Cayuela, M. L., Camps-Arbestain, M., & Whitman, T. (2021). Biochar in climate change mitigation. *Nature Geoscience, 14*(12), 883–892. https://doi.org/10.1038/s41561-021-00852-8

Li, K., Chen, Y., & Gao, S. (2022). Uncertainty of city-based urban heat island intensity across 1112 global cities: Background reference and cloud coverage. *Remote Sensing of Environment, 271*, 112898. https://doi.org/10.1016/j.rse.2022.112898

Li, Y., Hu, S., Chen, J., Müller, K., Li, Y., Fu, W., Lin, Z., & Wang, H. (2018). Effects of biochar application in forest ecosystems on soil properties and greenhouse gas emissions: A review. *Journal of Soils and Sediments, 18*(2), 546–563. https://doi.org/10.1007/s11368-017-1906-y

Liang, B., Lehmann, J., Sohi, S. P., Thies, J. E., O'Neill, B., Trujillo, L., Gaunt, J., Solomon, D., Grossman, J., Neves, E. G., & Luizão, F. J. (2010). Black carbon affects the cycling of non-black carbon in soil. *Organic Geochemistry, 41*(2), 206–213. https://doi.org/10.1016/j.orggeochem.2009.09.007

Norberg, L., Berglund, Ö., Berglund, K. (2018). Impact of drainage and soil properties on carbon dioxide emissions from intact cores of cultivated peat soils. *Mires and Peat,* (21), 1–14. https://doi.org/10.19189/MaP.2017.OMB.284

Oberndorfer, E., Lundholm, J., Bass, B., Coffman, R. R., Doshi, H., Dunnett, N., Gaffin, S., Köhler, M., Liu, K. K. Y., & Rowe, B. (2007). Green Roofs as urban ecosystems: Ecological structures, functions, and services. *BioScience, 57*(10), 823–833. https://doi.org/10.1641/B571005

Rajapaksha, S., Nnachi, R. C., Tariq, M. A. U. R., Ng, A. W. M., Abid, M. M., Sidiqui, P., Rais, M. F., Aamir, E., Herrera Diaz, L., Kimiaei, S., & Mehdizadeh-Rad, H. (2022). An estimation of the anthropogenic heat emissions in Darwin city using urban microclimate simulations. *Sustainability, 14*(9), 5218. https://doi.org/10.3390/su14095218

Sitzenfrei, R., Kleidorfer, M., Bach, P. M., & Bacchin, T. K. (2020). Green infrastructures for urban water system: Balance between cities and nature. *Water, 12*(5), 1456. https://doi.org/10.3390/w12051456

Song, J., Lu, Y., Zhao, Q., Zhang, Y., Yang, X., Chen, Q., Guo, Y., & Hu, K. (2022). Effect modifications of green space and blue space on heat–mortality association in Hong Kong, 2008–2017. *Science of the Total Environment, 838*, 156127. https://doi.org/10.1016/j.scitotenv.2022.156127

Spokas, K. A. (2013). Impact of biochar field aging on laboratory greenhouse gas production potentials. *GCB Bioenergy, 5*(2), 165–176. https://doi.org/10.1111/gcbb.12005

Tan, K., Wang, J. (2023). Substrate modified with biochar improves the hydrothermal properties of green roofs. *Environmental Research, 216*(Pt 1), 114405. https://www.sciencedirect.com/science/article/abs/pii/S0013935122017327?via%3Dihub

Wan, J., Yong, B., & Zhou, X. (2022). Spatial and temporal analysis of the increasing effects of large-scale infrastructure construction on the surface urban heat island. *Ecotoxicology and Environmental Safety, 237*, 113521. https://doi.org/10.1016/j.ecoenv.2022.113521

Wang, J., Xiong, Z., & Kuzyakov, Y. (2016). Biochar stability in soil: Meta-analysis of decomposition and priming effects. *GCB Bioenergy, 8*(3), 512–523. https://doi.org/10.1111/gcbb.12266

Wood, T. E., Cavaleri, M. A., & Reed, S. C. (2012). Tropical forest carbon balance in a warmer world: A critical review spanning microbial- to ecosystem-scale processes. *Biological Reviews, 87*(4), 912–927. https://doi.org/10.1111/j.1469-185X.2012.00232.x

Xu, C., Chen, G., Huang, Q., Su, M., Rong, Q., Yue, W., & Haase, D. (2022). Can improving the spatial equity of urban green space mitigate the effect of urban heat islands? An empirical study. *Science of the Total Environment, 841*, 156687. https://doi.org/10.1016/j.scitotenv.2022.156687

Yang, Z., Zhang, C., Zeng, G., Tan, X., Wang, H., Huang, D., Yang, K., Wei, J., Ma, C., & Nie, K. (2020). Design and engineering of layered double hydroxide based catalysts for water depollution by advanced oxidation processes: A review. *J Mater Chem A, 8*(8), 4141–4173. https://doi.org/10.1039/C9TA13522G

Zhang, Y., Hu, Y., You, Z., Li, Z., Kong, M., Han, M., Liu, Z., Zhang, J., & Yao, Y. (2021). Soil ventilation benefited strawberry growth via microbial communities and nutrient cycling under high-density planting. *Frontiers in Microbiology, 12*, 666982. https://doi.org/10.3389/fmicb.2021.666982

Zhao, Z., Wang, B., Theng, B. K. G., Lee, X., Zhang, X., Chen, M., & Xu, P. (2022). Removal performance, mechanisms, and influencing factors of biochar for air pollutants: A critical review. *Biochar, 4*(1), 30. https://doi.org/10.1007/s42773-022-00156-z

Influence of Human Footprint on Eco-System: The Necessity of Creating Sustainable Environment

Adaptation of Plant Ecosystems to Rapid Climate Change in the Ural Region

Irina V. Petrova, Alena A. Kochubey, Svetlana O. Medvedeva, Elena A. Sharova and Olga E. Cherepanova

Abstract

There is a direct linear relationship between the total amount of carbon emitted by human activities and the level of warming on the Earth's surface. At the same time, warming in the northern regions of Eurasia is faster than in the southern ones. In this work, we considered changes in atmospheric temperature and accumulation of precipitation in the Urals and their influence on the number and area of fires in the region. We assessed the combined effect of elevated atmospheric temperatures and high humidity on the pollen fertility of *P. sylvestris* L. The Ural region has faced the trend of climate warming and fixing of new records of high temperatures. Thus, the average annual temperature in the region has increased by 3 °C since 2020. A decrease in the thickness of the snow cover in early spring leads to dryness in May, which increases forests burning. The most dangerous weather in the Urals was in 2010, which led to growth in the number of forest fires. The decrease in spring humidity has led to almost 10 times increase in the number of fires. High atmospheric temperature led to a sharp decrease in the proportion of fertile pollen, which was almost three times lower compared to climatically normal years. Sharp changes in climatic factors lead to the transformation of the ecosystems. In the long term, broad-leaved tree species have the greatest adaptive potential in the Ural region.

Keywords

Carbon fields · Scotch pine · Ural · Transformation · Climate change

1 Introduction

The natural landscape of the Ural region is rapidly changing. The territories of the Polar and Subpolar Urals, the northernmost parts of the Russian Federation, harbor glaciers up to 3000 years old (Grigor'ev et al., 2019; Sannikov et al., 2020a). Scientists have been actively monitoring these glaciers since the middle of the twentieth century: today the area and number of glaciers are significantly decreasing. According to forecasts, by 2040 most glaciers will cease to exist, and only the largest ones will remain (Solly et al., 2017). Due to glaciers retreat in the Subpolar Urals, vertical and horizontal shifts of larch sparse forests and closed forests boundaries can be noted. The average upper boundary of light forests rose by 26 m, and of closed forests by 35 m. The average horizontal shift of the boundaries is 290 and 520 m, respectively (Fomin et al., 2020).

Not only the territory of the northern Urals is undergoing significant changes but the transformations also affected the Middle Urals, where we can observe the shift of the forest border higher into the mountains along the entire length of the Ural Range. These processes lead to changes in phytocenoses structure and redistribution of resources in ecosystems—typical tundra species disappear (endemic species—blueberries, crowberries), and they are replaced by the tree species (scotch pine, cedar pine, common

I. V. Petrova · A. A. Kochubey · S. O. Medvedeva · E. A. Sharova · O. E. Cherepanova (✉)
Botanical Garden of the Ural Branch of the Russian Academy of Sciences, Ekaterinburg, Russian Federation
e-mail: botgarden.olga@gmail.com

I. V. Petrova
e-mail: irina.petrova@botgard.uran.ru

A. A. Kochubey
e-mail: 79326010873@yandex.ru

A. A. Kochubey · S. O. Medvedeva · E. A. Sharova · O. E. Cherepanova
Laboratory of Population Biology and Forest Dynamics, Botanical Garden Ural Branch of the Russian Academy of Sciences, Yekaterinburg, Russian Federation
e-mail: 79326010873@yandex.ru

B.-J. He et al. (eds.), *Resilient Horizons: Building Sustainable Environments for Climate Adaptation and Health*, Advances in Science, Technology & Innovation, https://doi.org/10.1007/978-3-031-46109-5_12

spruce, larch) (Hagedorn et al., 2020; Sannikov et al., 2020b).

Within 1966–2018 observation period there was a shift in the average annual temperature ranging from 1.6 to 2.0 °C. The most serious transformation is noted in winter period (growth on 2.3–3.0 °C) and to a lesser extent in spring period (growth on 0.8–1.7 °C). The entire territory of the Ural region is characterized by an increase in the amount of spring precipitation. The changes in temperature, precipitation, conditions of growing season are extremely uneven in various regions of the Urals (Vasiliev et al., 2021).

The transformation of climatic conditions is also observed in the Southern Urals. The trends revealed in meteorological values determine the directions of climate change: an increase in temperature and amount of solid precipitation during the cold period leads to some mitigation of winter conditions; a significant increase in air temperature and a decrease in the amount of precipitation during the warm period, provokes raising of drier conditions.

During last century, the average January temperature was −17 °C, now January temperatures are on average 2.4 °C warmer, which is −14.6 °C. The average annual temperature is also rising. The identified trends in temperature and humidity climatic values are one of the consequences of global warming. With sustainable development of this process, the territory of the Southern Urals can be subject to very unfavorable natural and climatic processes and phenomena: growth of heat waves and extreme fire periods, development of atmospheric and soil droughts with further aridization, raising of snow cover depth, increased water erosion in spring, etc. In turn, changes will also occur in phenological processes, which are very dependent on climatic conditions (Galimova & Silantyev, 2019).

On average, in the Ural Federal District (UFD), as well as throughout Russia, for the period 1976–2020 in all seasons and in general for the year, surface air temperature trends are positive. The increase in the average annual air temperature is 0.49 °C/10 years. The strongest warming in the Ural Federal District is observed in spring (0.74 °C/10 years), and the weakest in summer (0.31 °C/10 years). Estimates of temperature trends have a high degree of reliability only in spring and on average for the year. Precipitation per year in the district is growing on average by 2.3%/10 years. The greatest increase in precipitation is observed in spring (8.4%/10 years). The degree of reliability of precipitation and temperature trends is approximately the same. During last thirty years, dangerous hydrometeorological phenomena associated with high wind speeds, extreme precipitation, and fire hazards have most often caused damage in the Ural Federal District (Climate Center of Roshydromet, 2021).

Natural/undisturbed forest ecosystems play an important role in solving many global issues, including climate regulation. More than 45% of the territory of the Urals is covered with boreal forests, which are destroyed by fires (Sannikov et al., 2017; Sannikova et al., 2019). Climate change creates conditions, but is not the main cause of forest fires. Due to atmospheric temperature rising, small amount of snow, and dry weather, it is easier for plant objects to ignite, which leads to stronger and more uncontrollable fires. At the same time, the consequences of fires make an additional contribution to climate change, by releasing a large amount of free carbon. A fire-damaged forest recovers more slowly, carbon dioxide absorption and oxygen emission decreases, and as a result, its function of slowing down climate change is reduced. Fire destroys beneficial soil microorganisms, reduces the supply of nutrients, causes soil erosion and desertification begins. Degraded lands do not absorb carbon dioxide, and it accumulates in the atmosphere. Black carbon, or soot, affects the albedo of the surface, contributes to the heating of ice caps, which reduces the natural cooling of the Earth (Kuznetsova, 2022).

The pyrological features of the climate and vegetation are one of the main natural factors affecting the patterns of fires. They determine the processes of transition of forest combustible materials (FCM) to the state of fire maturity and can lead to fires in the presence of fire sources of both natural and anthropogenic origin. The spread of forest fires depends on several natural factors—weather conditions, stocks of FCM and relief (Turner et al., 1994; Agee, 1996; Ryan, 2002). A fire that occurs in conditions of severe drought, with high wind strength, is catastrophic and has practically no natural barriers, it can even pass through extremely wet forest areas (Román-Cuesta et al., 2009; Sannikov et al., 2017; Sannikova et al., 2019).

A forest fire smokes large spaces, changing the composition of atmospheric air. Large forest fires are not only accompanied by peak emissions of CO_2 into the atmosphere, but also reduce atmospheric CO_2 sinks. Both the first and the second lead to growth of atmospheric CO_2 content and, accordingly, to increase in the greenhouse effect, thus closing the "fire-warming" cycle.

After European Union adopted the law on achieving carbon neutrality by 2050, that is, reducing dangerous emissions into the atmosphere to a minimum, Russia joined the project. Our country is making its contribution to solving the climate problem. Thus, in several regions of the country, including the Urals, a project of carbon polygons was launched—land plots established in order to implement measures for climate active gases control. One of the aims of climatic polygons is the creation of carbon farms in the Urals.

The transformation of forest ecosystems, being the result of global warming, will lead to a change of the main forest-forming coniferous species to broad-leaved species, which are distinguished by high adaptive potential in conditions of constant climatic fluctuations. Coniferous tree species which form Russia's boreal forests are reducing their productivity due to abnormal heat in summer, low atmospheric humidity in spring, increasing diseases, insect infestations and fire activity (Schaphoff et al., 2016). In this regard, it is necessary to pay attention to promising highly productive broad-leaved species that can participate in the fixation of free carbon during the recultivation of disturbed areas.

The purpose of this study: to consider regional changes in temperature, precipitation accumulation and their impact on fires as a source of CO_2 emissions; assess the impact of elevated atmospheric temperatures on the restoration of coniferous species natural populations in the region; to recommend highly productive broad-leaved species capable of fixing free carbon for a long time in the Ural region in unstable climatic conditions.

2 Materials and Methods

The assessment of the main climatic parameters (average annual atmospheric temperature, atmospheric humidity in spring, snow depth, the sum of negative and positive annual temperatures, maximum and minimum monthly average temperatures) was carried out using open data from meteorological databases.

Location of the meteorological station in Yekaterinburg (Sverdlovsk region, Russia): latitude 56.83 longitude 60.63 altitude 281 m. The information was obtained for a 30-year period from 1990 to 2020, and an additional assessment of atmospheric temperature changes over a 100-year period was made (Website of Weather and climate, 2022).

The assessment of the number of fires and the area of forest land covered by fires was carried out based on the data from the Unified Interdepartmental Information and Statistical System (EMISS). The information is generated on a quarterly cumulative basis based on the on-line reporting "Information on the burning of forests on the Russian Federation territory" submitted by federal budget institution "Avialesookhrana" (Website of the EMISS State Statistics, 2022).

The study of pollen vitality of the main forest-forming species of the region was carried out to assess the resistance of coniferous tree species under the conditions of climate transformation. Works on the fixation of Scots pine pollen grains were carried out in 2009–2010 in the subzone of the southern taiga of the foothill strip of the Middle Urals in the lingonberry-broom upland pine forest and in the adjacent flowing swamp.

The age of forest stands typical for the region in both plots is 60–80 years, the composition is 10 C. The study of pollen grains vitality was carried out right before the start of microstrobil pollination phenophases (at the stage of microsporophylls separation), in the phases of pollen peak and at the end of pollination. Three to five microstrobiles were collected from a tree on each trial plot ($n=30$). The strobiles were thoroughly dried at room temperature until pollen was completely precipitated. The viability of pollen grains was assessed by germinating pollen in Petri dishes on a nutrient medium (1% agar, 2.5% sucrose) in a thermostat for five days at +27 °C (Cherepanova & Mishchikhina, 2012; Petrova et al., 2013). Those grains whose pollen tube size was equal to two pollen grain diameters were considered germinated. Pollen tubes were measured in five view fields in each Petri dish (150 view fields for one trial plot, five trial plots in total). Pollen grains from each tree were studied separately.

To identify the links between climatic factors and fires in the Ural region, a correlation statistical analysis was carried out using STATISTICA 10 portable software package.

The transformation of ecosystems caused by global warming will lead to a gradual change from coniferous tree species to cool temperate broadleaf species. Broad-leaved species appear to be more adaptive to higher atmospheric temperatures and precipitation (Hiura et al., 2019). In order to assess the potential of tree species resistance to abrupt climatic changes, as well as to assess the ability of these species to reduce the level of free carbon in the conditions of the Middle Urals, we set up a small experimental site in 2022. The site is an anthropogenically disturbed piece of land, which is divided into 3 separate plots planted with representatives of woody broad-leaved species growing in the Urals: *Acer pensylvanicum* L., *Salix schwerinii* E.L. Wolf, *Tília cordáta* Mill. Each fragment represents a separate type of community, the purpose of which is to study the strategy of fixing free carbon and its subsequent deposition. On each plot, we estimated the growth rate of the vegetative shoot, the stock of biomass at the end of the growing season, and the productivity of the tree species.

The stand phytomass was determined in each plot based on allometric models of tree phytomass dependence on trunk diameter, developed for the forests of the Ural region (Usoltsev et al., 2021, 2022).

3 Results

3.1 Dynamics of Average Annual Temperatures in the Middle Urals

We have analyzed the linear trends of temporal changes of the average annual temperature over 30 years in the Middle Urals. According to the graph (Fig. 1), the growth trend for winter months' sum of temperatures is confirmed, which contributes to the average annual temperatures in the

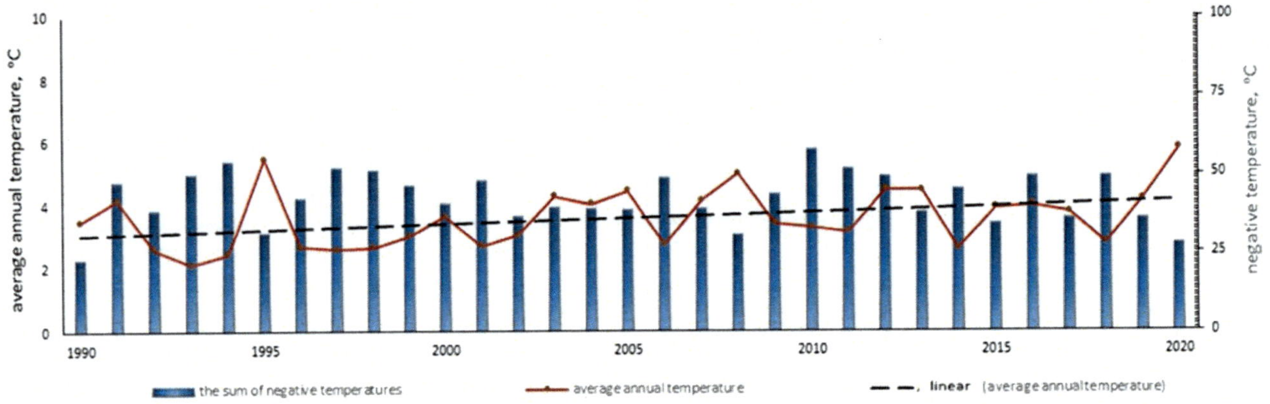

Fig. 1 Average annual curves of atmospheric temperature in the middle Urals

region. Thus, the minimum sum of negative atmospheric air temperatures (−30.7 °C) was recorded by us in 2008, which also affected the average annual temperature, which amounted to 5 °C, which is 18.2% higher than the average annual temperature in the previous ten years (4.08 °C).

Periods of winter warming are accompanied by periods of increased cold, which indicates the instability of climatic phenomena. One of the coldest winter periods occurred in 2010–2011. The sum of negative temperatures was −57.9 °C, which is lower than the average for ten years on 20.1% (−46.5 °C). It should be noted that one of the coldest winters in the Urals over the past thirty years was preceded by the warmest year, when the sum of negative temperatures was −30.7 °C, which is higher than the ten-year average (−34.0 °C) (Fig. 1). The positive temperatures of November–December, which have occurred twice over the past thirty years, since 2008, have become more frequent.

The sum of monthly summer temperatures over the past thirty years has crossed the threshold of 20 degrees thirteen times, the frequency exceeding the twenty-degree threshold in the summer months has been increasing since 2010. The maximum indicator of the sum of atmospheric air temperature in summer fell on July 2016 (23 °C), and it was also preceded by a hot June (20.1 °C). After 2018, the situation in the Urals region continues to be difficult, so the spring of 2021 was marked by a sharp warming in April (7.0 °C) and May (18.5 °C), which has not been observed for April temperatures since 2000, and for May, the entire period of meteorological observations. The years after 2015 were the warmest on record, and 2011–2020 is the warmest decade on record (Climate center of Roshydromet, 2021). The average temperature, according to Roshydromet (Climate center of Roshydromet, 2021), increased on 1.2–1.5 °C from 1900 to 2000, but starting from 2020 (Website of Weather and climate), according to the weather station, there was a sharp increase—by 3 °C.

3.2 Fires in Forest Areas and Reduction of Air Humidity

Wildfires released 1.76 billion tons of carbon worldwide in 2021, with fires in Yakutia, Western Siberia, Turkey and the United States contributing the most (Hu et al., 2023).

In Russia in 1991–2004 both the number of forest fires and the forest area covered by fires increased in general (Sherstyukov & Sherstyukov, 2007). In the Middle Urals, the general increase in winter temperatures led to an increase in early spring aridity, which contributed to growth of the frequency and the area of fires (Fig. 2).

Figure 2 clearly shows sharp peaks which reflect the strongest fires in 2004, 2010, and 2021. For example, in summer 2010, an extreme blocking anticyclonic regime was established over the European territory of Russia. As a result, the summer of 2010 was a record year in terms of weather and climate: abnormally high temperatures combined with low humidity in the atmosphere led to the formation of an extensive drought and, as a result, fires (Mokhov, 2011). The sum of positive annual temperatures in these extremely dry years was: 2004—81.6 °C and the area of fires was 30.5 thousand hectares, 2010—94.5 °C and 253.7 thousand hectares, 2021—97.5 °C and 58.1 thousand hectares. For the Middle Urals, the sum of positive annual temperatures during periods of minimum fire activity did not exceed 78 °C (1990—76.1 °C, 2015—77.9 °C). The sum of positive annual temperatures is not a determining factor for the fire hazard of the season. High summer temperatures were recorded in 2016 (sum of positive annual temperatures 93.6 °C, the sum of July temperatures 23.0 °C). At the same time, the area of fires did not exceed 3.5 thousand hectares, which is 78.7% less than the average value for a 30-year period (16.45 thousand hectares). According to Roshydromet (Climate center of Roshydromet, 2021), 2016 was a warm year, but with little snow in winter. We assume that it was the sum of climatic

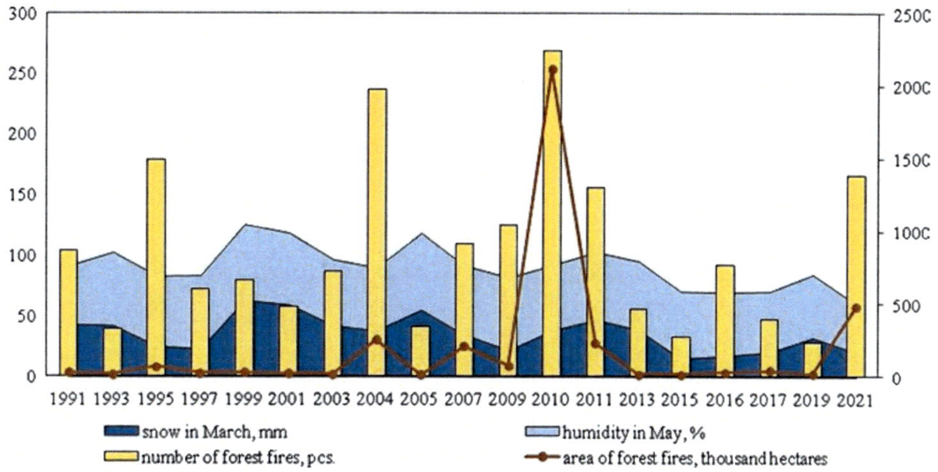

Fig. 2 Correlations of climatic factors (snow cover, atmospheric humidity) with the area and number of forest fires in the Urals in 1991–2021

factors that limited the number of fires in the region: air humidity in spring was 52.5%, which is only 7% lower than the average value for the study period (56.7%).

Along with sharp fluctuations in atmospheric temperature (Fig. 1), changes are also detected in the amount of snow cover, which has a primary effect on the atmospheric humidity in spring (Fig. 2). According to the analysis of the data, the fire hazard situation in the region develops with the obligatory presence of an anticyclone in the spring period and the preceding warm and little snowy winter. Thus, in the climatic conditions prevailing in 2004: the sum of negative winter temperatures and the amount of snow cover were close to the minimum values (−39.1 °C, 38.33 mm), the high average monthly temperature in May (+15 °C) and the anticyclone which came in the first ten-day period of May caused catastrophic forest fires, despite the high humidity of the spring period (62.5% on average).

According to the results of the correlation analysis, the highest negative relationship is observed between the following characteristics: the area of fires—the average monthly atmospheric humidity in May (−0.46, $p=0.05$) and the number of fires—the sum of negative annual temperatures (−0.51, $p=0, 05$). It can be concluded that with a decrease in atmospheric humidity, the area of fires in forest areas increases, and with a decrease in the sum of winter temperatures, a decrease in the number of fires in the region is observed.

It is known that the number and area of fires can be calculated quite easily using satellites. The calculation of CO_2 emissions from fires in a particular region is difficult. At the moment, mathematical models are still being developed that can estimate CO_2 emissions immediately after the fire and the extinction of plant species, as well as after the soil erosion (Hu et al., 2023). A long-term calculation of CO_2 emissions after the forest fires on a regional scale will make it possible to specify the factors that have the greatest impact on the transformation of climatic conditions.

3.3 Pollen Grain Anomalies as a Reproductive Response of Coniferous Species to Temperature Changes in the Region

The main forest-forming species of the region, the Scots pine, during sharp climatic fluctuations, forms a reduced amount of healthy pollen. The years 2009 and 2010 were marked with an early dry spring and higher temperatures in May. Woody plants quickly launched reproductive processes, not having time to accumulate a sufficient amount of water and microelements, which are extremely important for the last division occurring in microstrobiles when pine pollen grains are formed. Simultaneously with the formation of reproductive structures in April–May, root-forming processes also take place, for which, due to insufficient heating of the soil upper layers, organic and ash elements which are kept in well-heated wood and, probably, in last year's needles can be consumed (Solodovnikova, 2009).

The optimal conditions for the formation of fertile or "healthy" pollen are considered to be atmospheric temperature from 5 to 25 °C and relative humidity not more than 60%. With an increase in the air relative humidity, we noted various morphocytological anomalies of pollen grains and their frequency among which the adhesion and partial underdevelopment can be considered the most frequent (Cherepanova & Mishchikhina, 2012). The combined effect of the abnormally high for April–May average daily air temperature (28–30 °C), which we observed in 2009–2010, and relative humidity (over 90%) for several days, leads to the formation of almost 100% sterile pollen. Thus, the proportion of germinated pollen grains in samples from the southern taiga subzone of the foothill zone of the Middle Urals in 2009 was less than 5%. In 2010, it was marked by a heat wave and short periods of heavy rainfall during pollination, where the production of healthy pollen was only

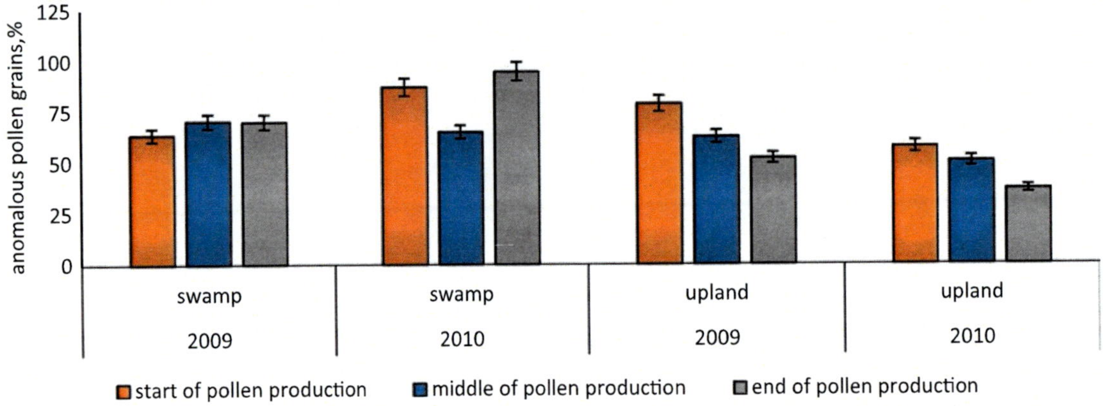

Fig. 3 The occurrence of anomalous pollen grains during the period of high humidity in two adjacent ecotopes

15%, which is almost three times lower than the optimal temperature (Demezhko et al., 2020).

Pollen collected from pines growing in swamps turned out to be less adaptive. In the conditions of the Ural swamps, pine begins dusting processes even before the complete disappearance of the snow cover. The proportion of pollen grains anomalies in 2010 in pines growing in swamps in the middle of pollination is almost 25% higher than in pines growing on elevations (Fig. 3). In the same year, but at the end of dusting, this indicator for pines from swamps was more than three times higher than for pines growing on elevations.

3.4 Carbon Polygon Model as an Arena for Studying the Productivity of Broad-Leaved Tree Species

The most productive growth of the vegetative shoot of woody plants was recorded by us in *S. schwerinii*, where the gain in vegetative shoot growth was 88.67 ± 1.25 cm. *A. pensylvanicum* seedlings turned out to be the most adaptive to increased solar insolation. In all variants, the appearance of lateral branching was recorded, and the maximum growth was 19 ± 1.08 cm. The maximum value of dry biomass at the end of the growing season was also noted by us in representatives of *S. schwerinii* (347.8 ± 19.87 g). Representatives of *S. schwerinii* also accumulated the largest dry leaf biomass (0.05 ± 0.03 g).

The maximum values of phytomass and organic carbon stock were noted for *S. schwerinii* plants (Table 1). In general, *S. schwerinii* plants showed the best results in terms of organic carbon storage and accumulation of aboveground phytomass. In the first year, aboveground phytomass values were the lowest in *T. cordáta*. The aboveground phytomass productivity was slightly higher in *A. pensylvanicum* plants.

The rapidly changing climatic conditions of the Ural region lead to a change of tree species in ecosystems. Coniferous boreal forests, being forced out by the increasing aridity of the climate to mountainous regions, will most likely be replaced by broad-leaved species or herbaceous species.

S. schwerinii is the most promising tree species for the formation of carbon polygons, which has a high growth rate of vegetative shoots, accumulation of aboveground phytomass, and fixation of organic carbon.

4 Discussion

There is a direct linear relationship between the total amount of carbon emitted by human activities and the level of warming on the Earth's surface. Over the past 70 years, free carbon emissions have multiplied, and for many years the main share of CO_2 came from the processing of minerals. The increase in the number of forest fires and deforestation, as well as the intensification of agricultural land use over the past 5 years, have a negative impact on the

Table 1 Calculation of productivity of aboveground phytomass (Pa, t/ha), biomass (g) and stock of organic carbon (C org, t/ha) of tree crops

Parameters	*Tílìa cordáta*	*Salix schwerinii*	*Acer pensylvanicum*
Pa, t/ha	0,82	1,82	0,52
Total plant biomass, g	162,7 ± 13,72	347.8 ± 19.87	264,5 ± 10,93
C org, t/ha	0,41	0,91	0,26

processes of free carbon sequestration (Sannikov et al., 2020b).

According to Fig. 2, the total number of fires, although it shows growth in a linear trend, still changes slightly and even has a periodicity. It is possible to single out periods of a decrease in the number of fires, which usually follow periods of mass burnout of coniferous forests. In a number of Ural regions, including the middle Urals, there is a shortage of spring water, which is partly due to winters with little snow and indicates an uneven course of climate change.

Forest fires are not only a disaster for the local population, but also create an additional source of emissions of greenhouse gases and aerosols into the atmosphere. About 30% of tropospheric ozone, carbon monoxide and carbon dioxide content in the atmosphere is due to the contribution of forest fires. Associated with forest fires, aerosol emissions into the atmosphere can have a significant impact on the physical and optical characteristics of the cloud cover and, consequently, on the climate (Kondratiev & Grigoriev, 2004).

It is noted that the overall global warming on the planet does not occur evenly. Different regions are warming at different rates. There is a loss of snow cover and reduction of ice caps, which significantly accelerates the warming of boreal regions in northern latitudes (NOAA, 2023). The snow cover of Eurasia does not change evenly. It is noted that the greatest loss of snow cover occurs in spring. The period of preservation of snow cover has significantly decreased in the west of the Tibetan Plateau, as well as in parts of Central Asia and northwestern Russia (Zhang & Ma, 2018). The Ural region belongs to those regions of the globe where warming processes are most active. This is evidenced by a decrease in the quality of pollen from woody plants, an increase in the average annual temperature, a decrease in snow cover and air humidity in spring, and, of course, growth in the number and area of forest fires.

The decrease in the quality of pollen grains and their viability was noted not only for coniferous woody species of the Urals. A similar trend was noted in the Mediterranean region for *Olea europaea* L., where at high temperatures the olive blossom phases are reduced (Iovane et al., 2022). The same effect, a reduction in pollination time, was also observed in our work for *P. sylvestris* (Cherepanova & Mishchikhina, 2012). High humidity results in less heat-tolerant pollen. In the olive, the destruction of pollen grains was observed at the stage of the final division in the anthers (Iovane et al., 2022). It was noted that three days of the combined effect of high temperature and humidity is sufficient for the destruction of all pollen.

The pollen of cistus (*Cistus incanus*) and myrtle (*Myrtus communis*) is able to survive short-term periods of atmospheric temperatures rise to critical values. However, at high humidity and atmospheric temperatures, the fertility of pollen grains simultaneously decreases on 100% (Aronne, 1999). The cumulative effect of humidity and temperature is also indicated for species inhabiting the territory of South America. Black Locust (*Acacia mearnsii*) pollen grains formed abnormal pollen tubes that reduced the total number of healthy seeds (Beck-Pay, 2012).

Anomalies and a significant decrease in the quality of germination of pollen tubes are associated with violation of the last stage of reproductive cells division, which occurs in spring. If the air warms up too quickly, then the microstrobiles are released from the cover of the bud scales earlier. Extreme hygrothermal factors have a significant, sometimes radical impact on the processes of microsporogenesis, and as a result, on the viability of pollen, and finally on production of a new seed crop. According to paleontological data, the appearance of anomalous pollen grains has previously been associated with serious fluctuations in climate change. Paleontological sections show high content of teratomorphic pollen coinciding with sharp decrease in atmospheric air temperature during the periods of late Holocene. Most often, abnormal development is associated with the formation of various variations of air sac mutations, which was also noted by us in 2009–2010. Thus, the appearance of a large number of anomalous pollen grains of spore-pollen plants is associated with a sharp cooling on the border between Subboreal and Subatlantic periods (Melnikova, 2004). For this reason, we believe that the optimal species for planting in Ural carbon farms may be broad-leaved woody highly productive species, the pollen of which is less sensitive to temperature rise.

Depending on the ecological situation, composition, structure and condition of forests, tree species can be both carbon storages and sources of its emission into the atmosphere. And also, when reaching a mature age, the productivity of forest-forming species begins to decline, decreasing more actively in coniferous species, and less actively in broad-leaved species. With an increase in continentality, the productivity of forest stands also decreases. It falls most actively in pine forests, and less actively in broad-leaved species. The main task of carbon farms establishment in the Urals is to find a sustainable ecosystem that stores the maximum amount of free carbon at all possible levels.

Taking into account the decrease in the resistance of indigenous forest-forming species in the Urals—Scotch pine, spruce, larch—it is necessary to test species growing in more southern regions. It is known that reforestation in treeless areas contributes to additional CO_2 sequestration (Koutika, 2022). At the same time, it is important to test certain species in the conditions of carbon polygons, which will allow selection of the most adaptive ones.

The northern regions of Eurasia, including the Northern Urals, show the following trends: change in the main forest-forming species, increase in plant productivity, as well as increase in forest cover. The southern regions of Eurasia show more diverse response, but with a trend towards reduced productivity and loss of forest cover (Schaphoff et al., 2016). It is necessary to test broad-leaved species that can adapt to disturbed areas and abrupt changes of climatic conditions.

Among the proposed broad-leaved species of woody plants, special attention should be paid to the species of the genus *Salix*, among which there are species bred by scientists of the Botanical Garden, Ural Branch of the Russian Academy of Sciences. Thus, *S. schwerinii* showed high growth rate, which, in general, is typical for representatives of this taxon. Willows also responded positively to the decrease in soil acidity. This species can also be recommended for experimental work in disturbed and depleted areas, since it maintains satisfactory growth rates with sufficient moisture. Attention should also be paid to *A. pensylvanicum*, whose representatives are characterized by efficient biomass accumulation, but are demanding to growing conditions.

5 Conclusions

Summarizing many years of domestic and foreign experience in studying the processes of carbon sequestration, fire activity, forest productivity, soil respiration, we can conclude that there is no consensus or template for the formation of systems capable of sequestering free carbon permanently and in large quantities. So, each region, including the Middle Urals, must develop its own unique strategy for preserving natural landscapes, transforming disturbed wastelands, and restoring post-fire areas to reduce the carbon footprint. Our work is aimed at identifying highly productive species that can survive in conditions of sharp climatic fluctuations, deposit free carbon and grow on technogenic and post-fire substrates. The experimental site, which includes the main type of anthropogenic landscape in the Ural region, needs to be expanded with areas simulating agricultural areas, restored industrial areas, and aquatic ecosystems. According to primary results, in order to increase the storage capacity of forests, multilayered communities should be formed, the main species of which are broad-leaved tree species, for example, representatives of the genus *Salix*.

Acknowledgements *The work was financially supported by the ministry of science and higher education of the Russian federation within the State Assignment FEUZ-2023-0023.*

The work was carried out as part of the implementation of the state task of the Botanical Garden of the Ural Branch of the Russian Academy of Sciences.

References

Agee, J. K. (1996). The influence of forest structure on fire behavior. *Forest Vegetation Management Committee* (pp. 52–68). Sacramento.

Aronne, G. (1999). Effects of relative humidity and temperature stress on pollen viability of *Cistus incanus* and *Myrtus communis*. *Grana, 38*, 364–367. https://www.tandfonline.com/doi/pdf/10.1080/00173130050136154

Beck-Pay, S. L. (2012). The effect of temperature and relative humidity on Acacia mearnsii polyad viability and pollen tube development. *South African Journal of Botany, 83*, 165–171. https://doi.org/10.1016/j.sajb.2012.08.011

Cherepanova, O. E. & Mishchikhina, Y. D. (2012). Influence of environmental factors (temperature and air humidity) on the quality of Scots pine pollen (Pinus sylvestris L.). *Agrarian Bulletin of the Urals, 7*(99), 72–73.

Climate center of Roshydromet. (2021). *Global climate change and the Ural Federal District. On the way to adaptation.* St. Petersburg: High technologies. https://cc.voeikovmgo.ru/images/dokumenty/2022/%D0%A3%D1%80%D0%B0%D0%BB%D1%8C%D1%81%D0%BA%D0%B8%D0%B9%20%D0%A4%D0%9E.pdf

Demezhko, D. Y., Gornostaeva, A. A., & Antipin, A. N. (2020). The Fennoscandian ice sheet during the late weichselian: Geothermal evidence. *International Journal of Earth Sciences, 109*(6), 1941–1955.

Fomin, V. V., Shiyatov, S. G., & Mikhailovich, A. P. (2020). Trees in the upper treeline ecotone in the Polar Urals: Centuries-old change and spatial patterns. *Mountain Research and Development, 40*(2), 32–40. https://doi.org/10.1659/MRD-JOURNAL-D-20-00002.1

Galimova, R., & Silantyev, K. (2019). The analysis of the long-term dynamic of the amount of atmospheric precipitation on the territory of the Republic of Bashkortostan. *IOP Conference Series: Earth and Environmental Science,* (–), 12–20.

Grigor'ev, A. A., Devi, N. M., Kukarskikh, V. V., V'yukhin, S. O., Galimova, A. A., Moiseev, P. A. & Fomin V. V. (2019). Structure and dynamics of tree stands at the upper timberline in the western part of the Putorana plateau. *Russian Journal of Ecology, 50*(4), 311–322. https://doi.org/10.1134/S1067413619040076

Hagedorn, F., Dawes, M. A., Bubnov, M. O., Devi, N. M., Grigoriev, A. A., Mazepa, V. S., Shiyatov, S. G., Moiseev, P. A., & Nagimov, Z. Y. (2020). Latitudinal decline in stand biomass and productivity at the elevational treeline in the Ural mountains despite a common thermal growth limit. *Journal of Biogeography, 47*(8), 1827–1842. https://doi.org/10.1111/jbi.13867

Hiura, T., Go, S., & Iijima, H. (2019). Long-term forest dynamics in response to climate change in northern mixed forests in

Japan: A 38-year individual-based approach. *Forest Ecology and Management., 449*, 117469. https://doi.org/10.1016/j.foreco.2019.117469

Hu, X., Chen, D., Hu, L., Li, B., Li, X., & Fang, X. (2023). Global methyl halide emissions from biomass burning during 2003–2021. *Environmental Science and Ecotechnology, 14*, 100228. https://doi.org/10.1016/j.ese.2022.100228

Iovane, M., Cirillo, A., Izzo, L. G., Di, V., & Aronne, G. (2022). High temperature and humidity affect pollen viability and longevity in Olea Europaea L. *Agronomy, 12*(1), 1. https://doi.org/10.3390/agronomy12010001

Kondratiev, K. Y., & Grigoriev, A. A. (2004). Forest fires as a component of natural ecodynamics. *Atmospheric and Ocean Optics, 17*(4), 279–292.

Koutika, L. S. (2022). Boosting C sequestration and land restoration through forest management in tropical ecosystems: A mini-review. *Ecologies, 3*(1), 13–29. https://doi.org/10.3390/ecologies3010003

Kuznetsova, E. (2022). Impact of environmental consequences of forest fires on climate change. In *Fire safety: Modern challenges, problems and solutions: Conference* (pp. 188–190). St. Petersburg. https://elibrary.ru/item.asp?id=48933540

Melnikova, T. A. (2004). Anomalous pollen of the genus *Pinus* L. as an indicator of paleoclimatic fluctuations in the Late Holocene. *Bulletin of the Far Eastern Branch of the Russian Academy of Sciences, 3*, 178–182.

Mokhov, I. I. (2011). Specific features of the 2010 summer heat formation in the European territory of Russia in the context of general climate changes and climate anomalies. *Izvestiya. Atmospheric and Oceanic Physics, 47*(6), 653–660.

NOAA National Centers for Environmental Information (2023). State of the Climate: Global Climate Report for 2022. https://www.ncei.noaa.gov/access/monitoring/monthly-report/global/202213

Petrova, I. V., Sannikov, S. N., Cherepanova, O. E., & Sannikova, N. S. (2013). Reproductive isolation and disruptive selection as factors of genetic divergence between Pinus sylvestris L. populations. *Russian Journal of Ecology, 44*(4), 296–302. https://doi.org/10.1134/S1067413613040103

Román-Cuesta, R. M., Gracia, M., & Retana, J. (2009). Factors influencing the formation of unburned forest islands within the perimeter of a large forest fire. *Forest Ecology and Management, 258*, 71–80.

Ryan, M. G. (2002). Canopy processes research. *Tree Physiology, 22*, 1035–1043.

Sannikov, S. N., Petrova, I. V., Sannikova, N. S., Kochybei, A. A., & Sannikov, D. S. (2017). Divergence of biogeocenoses within pine forest types. *Russian Journal of Ecology, 48*(4), 340–349. https://doi.org/10.1134/S1067413617040142

Sannikov, S. N., Petrova, I. V., Egorov, E. V., & Sannikova, N. S. (2020a). Searching for and revealing the system of pleistocene refugia for the species *Pinus sylvestris* L. *Russian Journal of Ecology., 51*(3), 215–223. https://doi.org/10.1134/S1067413620030133

Sannikov, S. N., Sannikova, N. S., Petrova, I. V., & Cherepanova, O. E. (2020b). The forecast of fire impact on *Pinus sylvestris* renewal in Southwestern Siberia. *Journal of Forestry Research,* (–), 1–9. https://doi.org/10.1007/s11676-020-01260-1.

Sannikova, N. S., Sannikov, S. N., Kochubei, A. A., & Petrova, I. V. (2019). Natural pine regeneration on burns in forest-steppe of Western Siberia. *Forest Science, 5*, 22.

Schaphoff, S., Reyer, C. P. O., Schepaschenko, D., Gerten, D., & Shvidenko, A. (2016). Observed and projected climate change impacts on Russia's forests and its carbon balance. *Forest Ecology and Management, 361*, 432–444. https://doi.org/10.1016/j.foreco.2015.11.043

Sherstyukov, B. G., & Sherstyukov, A. B. (2007). Climatic conditions of potential forest fire in Russia in the 20th and 21st centuries. *Proc VNIIGMI WDC, 173*, 137–151.

Solly, E. F., Djukic, I., Schweingruber, F. H., Hagedorn, F., Moiseev, P. A., Andreyashkina, N. I., Devi, N. M., Mazepa, V. S., Shiyatov, S. G., Trubina, M. R., Göransson, H., & Wilmking, M. (2017). Treeline advances and associated shifts in the ground vegetation alter fine root dynamics and mycelia production in the south and polar Urals. *Oecologia, 183*(2), 571–586. https://doi.org/10.1007/s00442-016-3785-0

Solodovnikova, M. P. (2009). Daily dynamics of plant transpiration intensity in arid environmental conditions of the Burtinskaya Steppe site of the Orenburgsky State Reserve. *Bulletin of the Orenburg State University, 6*(100), 351–353.

Turner, M. G., Romme, W. H., & Gardner, R. H. (1994). Landscape disturbance models and the long-term dynamics of natural areas. *Nat Areas, 14*, 3–11.

Usoltsev, V. A., Tsepordei, I. S., & Noritsin, D. V. (2022). Allometric models of the biomass of forest-forming trees in the Urals. *Forests of Russia and the Economy in Themata, 1*, 4–14.

Usoltsev, V. A., Tsepordey, I. S., & Zukow, W. (2021). Climatically determined spatial and temporal changes in the biomass of Pinus sp. of Eurasia in the context of the law of the limiting factor. *Ecological Questions, 33*(1), 24–32. https://doi.org/10.12775/EQ.2022.007

Vasiliev, A. A., Glaz, N.V. & Nokhrin, D.Yu. (2021). Climate change in the Urals. *Problems of Agrochemistry and Ecology, 3–4*, 45–50. https://doi.org/10.26178/AE.2021.43.59.004

Website of the EMISS State Statistics. (2022). Date of application. https://www.fedstat.ru/

Website of Weather and climate. (2022). Date of application. http://www.pogodaiklimat.ru/history/28440.htm

Zhang, Y., & Ma, N. (2018). Spatiotemporal variability of snow cover and snow water equivalent in the last three decades over Eurasia. *Journal of Hydrology, 559*, 238–251. https://doi.org/10.1016/j.jhydrol.2018.02.031

Climate Determinants of Health

Emine Didem Evci Kiraz and Aylin Sönmez Gün

Abstract

Climate change is identified as the biggest problem of the 21st century. Governments must be more concerned than ever with the health dimension of climate change, and low-carbon economic recovery paths must be pursued reduce inequalities and improve human health by implementing policies around the world. However, climate change speed is faster than people's attempts to adapt and mitigate. There is a need for the most accurate, most concise, most accessible, and clearest data-based indicators and/or determinants in the relationship between climate and health due to limited resources and time. In this sense, the most appropriate approach is to call the determinants that are likely to reveal the health hazards and effects of climate change signals and act as a guide in monitoring and evaluation work, "climate determinants of health".

Keywords

Public health · Climate change · Social determinants of health

E. D. Evci Kiraz (✉) · A. Sönmez Gün
Faculty of Medicine, Department of Public Health, Aydın Adnan Menderes University, Aydın, Türkiye
e-mail: devci@yahoo.com

A. Sönmez Gün
e-mail: aylin.sonmez@adu.edu.tr

A. Sönmez Gün
Iğdır Provincial Health Directorate, Iğdır, Türkiye

1 Introduction

Studies dealing with the health effects of climate change are augmenting rapidly. Apart from many groups concerned about the issue, the World Health Organization (WHO) also started to study the connection between climate and health. The milestone in the relationship between climate and health is the Health and Climate Change Special Report published by the WHO at the 24th Conference of the Parties (COP24). The report pointed to the high number of failures, emphasizing that the lack of language and system unity among disciplines and sectors has played a major role in this failure, and defined climate change as the biggest problem of the 21st century. The report made seven recommendations (COP24 Special Report: Health and Climate Change, n.d.);

1. Actions to be taken for reducing both air pollution and carbon emissions should be identified and encouraged in the Nationally Determined Contributions in accordance with the Paris Agreement.
2. The health effects of adaptation and mitigation measures should be included in the designment of financial policies.
3. Health concerns should be included in Nationally Determined Contributions, National Adaptation Plans and National Communications to the United Nations Framework Convention on Climate Change (UNFCCC).
4. In order to adapt the health system to the climate-related considerations, obstacles to relevant investments should be removed.
5. The health community should be able to take part in climate action easily as reliable and committed advocates.
6. Mayors must be mobilized to increase resilience, reduce carbon emissions, and improve health. Because mayors are leaders of intersectoral action.
7. Health progress resulting to climate change mitigation and adaptation activities should be systematically monitored and reported.

B.-J. He et al. (eds.), *Resilient Horizons: Building Sustainable Environments for Climate Adaptation and Health*, Advances in Science, Technology & Innovation, https://doi.org/10.1007/978-3-031-46109-5_13

WHO facilitated the creation of the Clean Air Fund at COP25 conference and publication of health and climate change indicators for policy makers in a report (COP25 Climate Change Conference, n.d.). In COP26, policy makers were called to take urgent action on the current climate and health crises, and the health sector was defined as a reliable and effective stakeholder, but often an overlooked climate actor (COP26 Special Report on Climate Change and Health: The Health Argument for Climate Action, 2021). Global health and climate change research reports published by the WHO at COP25 and COP26 revealed that the management of the relationship between climate and health was weak in terms of twin actions (mitigation and adaptation) (2021 WHO Health and Climate Change Global Survey Report, n.d.). According to the data of the countries participating in the 2021 research:

1. All around the world, 67% of the countries have conducted or are currently conducting assessments in terms of both climate change and health vulnerability, and the compatibility of health-related practices with them.
2. National health and climate change strategies have been or are being developed by 77% of the countries.
3. Climate change and health troubles have been involved in the recovery plans after COVID-19 by 33% of the countries.
4. While the stakeholders focusing on the structural and social determinants of health are known to be less commonly represented, progress has been made in developing cooperation focused on the environmental determinants of health (40–50% of coordination mechanisms).
5. Less than 40% of the countries involved weather and climate-related information in their health surveillance systems in consideration of climate-sensitive diseases.
6. Around 30% of the countries have a climate-informed health early warning system for heat-related morbidity or mortality due to extreme weather events.
7. Health workforce is becoming increasingly knowledgeable about climate change and health, yet more effort is needed.
8. Countries are making assessments regarding the climate resilience of healthcare facilities at an increasing rate (27%).
9. Only a small portion (28%) of low- and low-middle-income countries receive international funding from their ministries of health to support climate change and health-related work.
10. Countries are significantly raising health-related awareness through their nationally determined contributions.

The Intergovernmental Panel on Climate Change (IPCC) reports included exposure, vulnerability (sensitivity and adaptability), and risk definitions in hazard, and impact analysis (Field et al., 2014). In the IPCC Sixth Assessment Report, risk was assessed through the effects of climate change on ecosystems, biodiversity, and human systems (Langsdorf et al., 2022). Impacts on human systems are examined under the subheadings of water security, and food production, health and well-being, cities, settlements, and infrastructure (Langsdorf et al., 2022).

Rapid urbanization makes people more susceptible to climate change (UNFCC, 2017). Despite the fact that cities in different geographical areas are affected by climate change in different ways, those protected from the extremes of climate change will also come under increasing pressure to host climate refugees (While & Whitehead, 2013).

Conducting independent international studies, the Lancet Countdown is working on indicators to monitor the health effects of climate change. There are 44 indicators under five main headings (Romanello et al., 2021). According to the 2021 report, efforts to build resilience have been slow and fraught with inequities (Romanello et al., 2021). It must be ensured that government leaders are more concerned than ever with the health dimension of climate change and that low-carbon economic recovery paths are pursued by implementing policies around the world in such a way as to reduce inequalities and improve human health (Romanello et al., 2021). In this connection, the Lancet Countdown indicators show relevant evidence supporting the urgency and opportunity of this transition, indicating that no one is safe until everyone is safe (Romanello et al., 2021).

The climate change speed is faster than people's attempts to adapt and mitigate. Thus, there is a need to determine the most accurate, most concise, most accessible, and clearest data-based indicators and/or determinants in the relationship between climate and health since relevant resources and time are limited. For this purpose, the health hazards of climate change should be addressed under six headings and the health impact of these hazards under eight headings (Fig. 1). Other hazards and impacts may also arise in a way that has become much more critical for countries, regions, cities and societies. Any new development can be included in the analysis.

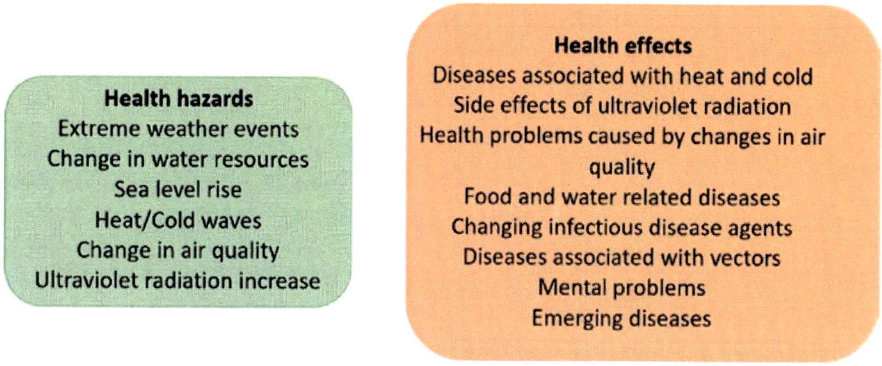

Fig. 1 Health effects of climate change, Prof. Dr. Evci Kiraz (2021) (Evci Kiraz, 2019a, 2019b)

2 Method

The study is a classical review type research carried out by examining scientific articles and articles on the subject in academic databases and web pages. COP, UNFCC, WHO reports have been reviewed.

Based on all the studies highlighted in the introduction, the most appropriate approach is to call the determinants in Fig. 1, which can reveal the health hazards and effects of climate change signals on health, and can guide monitoring and evaluation, as "climate determinants of health" (Evci Kiraz, 2019a, 2019b). The climate determinants of health are presented in Fig. 2.

Risk is defined as a function of hazard (H), exposure (E), and vulnerability (V) [$R = f(\text{H, E, V})$], whereas vulnerability is defined as a function of exposure (E), sensitivity (S), and adaptive capacity (AC) [$\text{V} = f(\text{E, S, AC})$] (Field et al., 2014). All these formulas should be used in preparing the chain of climate and health impact in order to reveal the true extent of the health impacts of climate change (Fig. 2). Based on all the studies underlined above, the climate determinants of health should guide the health impact chains. The determinants that can reveal the hazards and impacts of climate change signals on health and that can guide monitoring and evaluation activities are called "climate determinants of health" (Fig. 2) (Evci Kiraz, 2019a, 2019b).

3 Findings

The "climate determinants of health" in Fig. 2, proposed to be used to identify and analyse the relationship between climate and health, are described below.

4 Climate Determinants of Health

1. Social determinants of health: High levels of social determinants of health are required to build and maintain good health and complete well-being. The 17 Sustainable Development Goals play a guiding role in promoting health (Evci Kiraz, 2019a, 2019b).
2. Individual level of health: According to the WHO, "health is a state of complete physical, mental and social well-being and not merely the absence of disease or infirmity". Social factors such as living in a corrupt

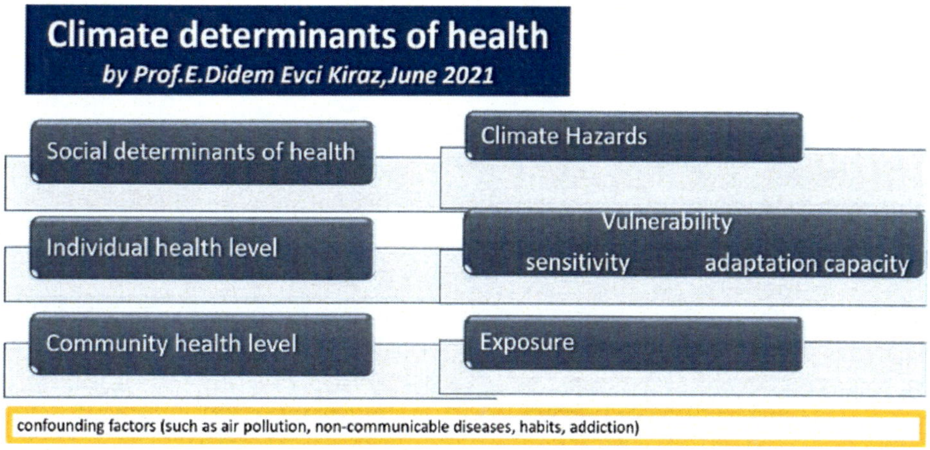

Fig. 2 Climate determinants of health, Prof. Dr. Evci Kiraz (2021) (Evci Kiraz, 2021)

environment with wrong beliefs, lack of education, and poor nutrition as a result of economic inadequacies, as well as the inability to benefit from health services can be counted among the main causes of many diseases (Öztek, 2020).

3. Community health level: It can be tracked with fertility, morbidity and mortality statistics. Information that can explain the health level and problems of the society, such as the types of diseases seen, whether or not there are epidemics and the main causes of mortality by age group play a major role at this level (Öztek, 2020).
4. Climate hazards:
   - Extreme weather events
   - Alteration in water resources
   - Sea level rise
   - Heat/cold waves
   - Alteration in air quality
   - Increased ultraviolet radiation
5. Vulnerability: It is defined as a lack of ability to cope with and adapt to a level of vulnerability or susceptibility to harm (Glossary of İklimiDuy, 2021).
6. Exposure: It is defined as the presence of human, household, species or ecosystem in places and environments where environmental functions, services and resources, infrastructure or economic, social or cultural assets are likely to be adversely affected (Glossary of İklimiDuy, 2021).

5 Conclusions

"Social determinants of health" cannot be abstracted from the demographic, environmental and social determinants of health. However, it should be used in theoretical and mathematical analyses in order to know the real level of the effects of climate change on health and to calculate the real risk.

6 What Else Do the Climate Determinants of Health Do?

6.1 They Reveal the True Extent of the Relationship Among Climate and Health

Research conducted on the relationship between environment and health is actually complex since such studies include many confounding factors, and are interdisciplinary and inter-sectoral, which also require the shared use of data and the joint interpretation of results. Climate and health studies have similar characteristics. Further ecological studies (correlation studies) are needed to analyse the relationship between climate and health at the global, regional, country, city, neighbourhood and household level. Climate determinants of health should be made use of in ecology research to reveal the extent of the relationship. Once the relationship is determined, the significance and size of the relationship should be analysed with analytical studies (cross-sectional, cohort, case study etc.).

6.2 The Determinants also Guide Academic Studies and Practices for the Protection of People, Communities, Cities, and Global Health from the Effects of Climate Change

Climate determinants of health are based on measurable, standardized data, and can be supported by quantitative and qualitative data. In addition to those obtained from the field by epidemiological methods, data can also be collected from national and international observation platforms. Climate determinants of health are key to ensuring a unified language and close cooperation with other sectors in the adaptation efforts of the health sector in relation to the effects of climate change. It seems that the existing disease surveillance systems in the countries remain insufficient to monitor and evaluate the health effects of climate change, prevent the effects, protect those who will be affected and to operate early warning systems. Climate-sensitive diseases should be incorporated into the surveillance systems and should also be addressed along with the climate determinants of health. Until all these developments occur, it will be beneficial for countries to create lists of climate-sensitive diseases at the national, local, city-neighbourhood-household level. Consequently, real risk maps will emerge in particular, thanks to the climate-sensitive disease data integrated with geographical information systems and the climate determinants of health.

6.3 The Climate Determinants of Health Prove That the Health Impacts of Climate Change are Not Under the Ultimate Responsibility and Administration of Only One Sector

Climate determinants of health should be considered as a separate indicator in this regard. This approach has some merits with new contributions, but the level of evidence available is rather low. Further studies are needed to reveal the relationship between climate and health and to calculate the real risk. Planning the future with climate change adaptation actions focused on real risks will not only save manpower but also budget and time. Monitoring at the local level and determining and monitoring health compliance policies is critical. Local governments should take the existing research on climate change and health as a reference, and in case of insufficient data, they should determine their working areas. Indicators form the basis for demonstrating the situation and making comparisons with other communities (Evci Kiraz & Çelik, 2022).

References

2021 WHO health and climate change global survey report. (n.d.).

COP24 special report: health and climate change. (n.d.). CBY-NC-SA3.0IGO.

COP25 Climate Change Conference. (n.d.). Retrieved March 27, 2022, from https://www.who.int/news-room/events/detail/2019/12/02/default-calendar/cop25-climate-change-conference

COP26 special report on climate change and health: The health argument for climate action. (2021). CCBY-NC-SA3.0IGO.

Evci Kiraz, D. (2019a, November). Social determinants of health. https://www.sdplatform.com/Dergi/1228/Sagligin-sosyal-belirleyicileri.aspx

Evci Kiraz, E. D. (2019b). Effects of climate change on human health. Climate Change Training Modules Series.

Evci Kiraz, E. D. (2021, December 15). Climate and health from a public health approach.

Evci Kiraz, E. D., & Çelik, E. (2022). *Guide to protecting health from climate change through local climate change planning*. Alter Publications.

Field, C. B., Barros, V. R., Dokken, D. J., Mach, K. J., Mastrandrea, M. D., Bilir, T. E., Chatterjee, M., Ebi, K. L., Estrada, Y. O., Genova, R. C., Girma, B., Kissel, E. S., Levy, A. N., MacCracken, S., Mastrandrea, P. R., White, L. L., (Eds.). (2014). *Climate change 2014: Impacts, adaptation, and vulnerability*. Part A: Global and Sectoral Aspects. Contribution of Working Group II to the Fifth Assessment Report of the Intergovernmental Panel on Climate Change.

ClimateHear Glossary of Terms/Glossary. (2021). https://www.iklim-iduy.org/terimler-sozlugu/

Langsdorf, S., Löschke, S., Möller, V., & Okem, A. (2022). Climate change 2022 impacts, Adaptation and Vulnerability Working Group II Contribution to the Sixth Assessment Report of the Intergovernmental Panel on Climate Change. www.ipcc.ch

Öztek, A. Z. (2020). Public health theories and practices. *Health and Social Assistance Foundation.*

Romanello, M., Mcgushin, A., Napoli, C. di, Drummond, P., Hughes, N., Jamart, L., Kennard, H., Lampard, P., Solano Rodriguez, B., Arnell, N., Ayeb-Karlsson, S., Belesova, K., Cai, W., Campbell-Lendrum, D., Capstick, S., Chambers, J., Chu, L., Ciampi, L., Dalin, C., et al. (2021). The 2021 report of the Lancet Countdown on health and climate change: Code red for a healthy future. *The Lancet, 398*, 1619–1662. https://doi.org/10.1016/S0140-6736(21)01787-6

UNFCC. (2017). Rapid urbanization increases climate risk for billions of people. https://unfccc.int/news/rapid-urbanization-increases-climate-risk-for-billions-of-people

While, A., & Whitehead, M. (2013). Cities, urbanisation and climate change. *Urban Studies, 50*(7), 1325–1331. https://doi.org/10.1177/0042098013480963

Effects of Climate Change on Human Security and Sustainable Development

Kwadwo Addo Tuffuor and Emmanuel Yamoah Tenkorang

Abstract

There has been growing international concerns about the severe impact of climate change which has led to global warming and other challenges with respect to the environment. Climate change also poses a threat to human security. Added to the above, it is also a development issue in view of the fact that it has a direct bearing on the incidence of poverty. Moreover, climate issues have inter and intra-generational equity since it affects human security. Human security emphasises the worth and dignity of the human person in addition to the protection and safety of all persons, environmental security, the provision of healthcare, sustainable economic development and the sustainability of communities in tandem with international law. Developing an integrative approach towards the protection of these in relation to climate change is key to sustainable development. It is worth mentioning here that economic security help ensures a secured income, health security is also dependent on the environment and it is the environment that supports human livelihood in its entirety. The success and failure of all these hinge on the climate. This paper presents the findings of the review of literature of academic research across the world to highlight the effects of climate change on human security and aims to reveal the insecurities that result from it. It then concludes with recommendations for sustainable environmental governance at the international level and policy reform to ensure sustainable development.

K. A. Tuffuor (✉)
Department of Peace Studies, School for Development Studies, University of Cape Coast, Cape Coast, Ghana
e-mail: kaddotuffour@ucc.edu.gh

E. Y. Tenkorang
Department of Environment, Governance and Sustainable Development, School for Development Studies, University of Cape Coast, Cape Coast, Ghana
e-mail: eytenkorang@ucc.edu.gh

Keywords

Climate change · Human security · Environmental governance · Sustainable development · Economic security

1 Introduction

Climate change is one of the most significant and complex issue that poses a challenge to the international community in contemporary times and in the future. In the view of Dokos et al. (2008), the medium and long-term impacts of persistent environmental degradation have been a source of concern for global environmentalists. Environmental challenges invariably impact socio-economic and political developments worldwide. Hadžić (2022) has intimated that climate change tends to cause security challenges such as food insecurity, violent extremism, social and political instability in societies. The author continued that, for the vulnerable and marginalized in society, the limited access to resources will continue to impact negatively on their security as the climate changes. The impact of climate change poses a serious threat to the world's poorest and most vulnerable groups that face human security challenges such as intractable conflict, poverty and uneven distribution of resources, weak governance institutions and disease burden (Dokos et al., 2008). This has been reinforced by the International Institute for Environment and Development (IIED) (2015), which has opined that the multiplier effects of climate change will be most felt by 'vulnerable regions and vulnerable groups' because of existential threats to their human security.

It is no denying of the fact that, changes occurring in the world's climate has 'no precedent in the history of human civilisation' (IPCC, 2007; Stern, 2007). It is also seen as contemporary age's most disastrous environmental challenge (Yasmeen, 2021). Changes in the climate system also poses challenges on the "the economic, social and

B.-J. He et al. (eds.), *Resilient Horizons: Building Sustainable Environments for Climate Adaptation and Health*, Advances in Science, Technology & Innovation, https://doi.org/10.1007/978-3-031-46109-5_14

political systems" that underly the global system, and this becomes manifest on the health front in countries making it a national security issue (IPCC, 2021). Data from the World Health Organisation (WHO), suggests that regions and populations that are currently experiencing a high incidence of diseases that are due to changes in temperature are the people who contribute the least to greenhouse gas that warm the earth (Patz et al., 2007). Statistics indicate that, global carbon emissions on average are approximately about one metric ton per year (tC/yearr) per person. That of "developing countries per capita emissions are about 0.6 tC/year, and more than fifty countries have emissions that are below 0.2 tC/year" (Patz et al., 2007).

Even though, there are differences between the rich industrial world and developing economies, environmental diseases, as well as climate related natural disasters occur commonly in the developing economies as compared to the highly industrialised ones (Patz et al., 2007). Baer et al. (2000) have opined that "greenhouse gas emissions should be based on every person's equal right to the atmospheric commons." Human, industrial and agricultural activities have led to expanded emission of these gases into the atmosphere thereby resulting in an expanded greenhouse effect which brings about increase in mean atmospheric temperature otherwise known as global warming (Ani et al., 2022). Large-scale changes to the climate system occur, which results in the transformation of the human condition as communities become urban (Forest et al., 2010). There is every reason for humanity to be disturbed by the adverse impacts of changes in human systems due to climate change in view of the fact that, the rate of change is "unprecedented" in the last 10,000 years, and its multiplier effects has caused large-scale social upheavals (Lautensach & Lautensach, 2020).

Current scenario data points to the fact that climate change will lead to food insecurity, invariably causing hunger globally and in particular Africa with its attendant public health challenges. Environmental disasters triggered by climate change will also affect food security. Dokos et al. (2008) have opined that developing countries are at risk to variability in climate due to a host of complex factors and absence of shocks to adapt to in times of emergencies. The devastating consequences of climate change will, in general have a myriad of effects in sub-Saharan Africa as compared to other regions in spite of low emission rate from African citizens (Ramin & McMichael, 2009). Added to the above, several African countries are bearing the brunt with combined pressures of poverty, weak public health as well as weak physical institutions (Reid & Cogel, 2006; UNDP, 2007).

Climate change is highly complex and multi-faceted which affects all aspects of life including "international politics, economics, migration, human rights, development, trade, health and environmental systems" and can easily serve as a trigger for the myriad of issues that undermine human security (Dokos et al., 2008). In the view of the European Union (EU) as cited by Dokos et al. (2008), climate change is seen as 'a threat multiplier' with the potential to inflame existing tensions and political instability. The most significant challenge about climate change is that it threatens the world's regions that have large incidence of poverty and are vulnerable to natural disasters. The potential dangers are not only humanitarian, but are also political. Moreover, many of the issues that confront human security affect international security that demands an integrative response (Lautensach & Lautensach, 2020). Hadžić (2022) has emphasized that climate change in the Western Balkans impacted negatively on the health of people, security and quality of life inclusive of the agricultural sector, water management and the tourism industry and this acts as a challenge to the security of the individual.

The United Nations (2017) stipulates that the human security framework should be used to build and strengthen states capabilities to withstand climate change and natural disasters that occur in order to maintain peaceful societies, while at the same time addressing humanitarian crisis and sustainable development. The human security methodology is therefore used to realise the UN 2030 transformative Agenda on the Sustainable Development Goals (SDGs). The consensus reached by the General Assembly in resolution 66/290 on human security highlighted the need for people to live in freedom, live in dignity, free from deprivation and called for a people-centred approach to strengthen measures to address disasters when they occur (United Nations, 2017). The human security framework addresses five key fundamental principles. First, all security related activities must be "human centred, comprehensive, context-specific, prevention-oriented and lastly, but not the least, ability to empower the people".

This paper focuses on the threats posed by climate change to human sustainability in relation to the self-worth of the individual, freedom from fear, politics and international relations. In addition, it addresses the intricate relationship between the environment and the basic necessities of life such as sustainable economic livelihood, healthcare, food security and forced migrations that result from public emergencies such as the earthquake that happened in Syria and Turkey recently. The paper then presents the findings of literature of academic research across the world to highlight the effects of climate change on human safety and sustainable development and aims to reveal the insecurities that result from it. This is achieved through a desk review of relevant literature in the area of climate change and human sustainability.

2 Climate Change: What is It?

There is strong and overwhelming scientific evidence to show that it is human activities that cause climate change. It occurs largely due to the burning of fossil fuels, which causes a rise in global temperatures that affect precipitation patterns leading to extreme weather (United Nations, 1992). SOAS University of London (2018), sees climate change as changes that occur over a long period of time with respect to variability in the atmosphere as a result of natural activities or human action. The most significant attributes of climate are 'temperature, precipitation (rainfall, snow, hail), clouds, humidity and sunlight" (SOAS University of London, 2018). In the view of Dalby (2013), climate change is about "extensive set of scientific investigations" that specifically measures changes that happens in the environment mostly temperature and rainfall patterns. Moreover, it is also about changes in the climate spanning some period as a result of natural phenomena or human activity (Dalby, 2013). The Intergovernmental Panel on Climate Change (IPCC) define "climate change" as:

> a change of climate which is attributed directly or indirectly to human activity that alters the composition of the global atmosphere and which is in addition to natural climate variability observed over comparable time periods (IPCC, 2001).

Climate change has been part of mankind over years and it is part of "the natural geological history of the world" according to Uitto et al. (2017) and Mathez and Smerdon (2018). Large scale natural disasters in the world are attributable to changes in climate. Threats to human security such as instability in the politics of a country, low economic development, food insecurity and forced migration intersects with climate change according to Goodman et al. (2008). The IPCC (2007) has also intimated that human activities inclusive of transportation, industry, agriculture and land use and consumer choices that occur daily are the elementary factors that causes a change to the earth's climate. Changes in weather patterns, rise in water in the seas and changes in weather conditions are specific examples of climate change. The three main major factors that causes climate change are astrological causes which include:

> variations in the obliquity of the ecliptic; the eccentricity of the earth's orbit around the sun; precession of the equinoxes caused by the previous ones mentioned above; and the bombardment of the earth by extra-terrestrial objects, volcanic eruptions and environmental changes due to social, economic and political factors as deforestation, damming of rivers to create artificial lakes, production of energy to the atmosphere such as combustion of fossil fuels such as petrol, diesel, coal and changes in the composition and land - use activities (Anyadike, 2009, p. 13).

These lead to a pile-up of "carbon monoxide (CO), methane (CH_4), sulphur (IV) oxide (SO_2)" called greenhouse gases due to their ability to absorb terrestrial radiation from the surface of the earth and redirecting the heat back, invariably leading to global warming (Anyadike, 2009). It is estimated that black carbon causes the loss of about two million lives a year and of these, a sizable number are women and children resident in the regions of the world where majority of the people are vulnerable and deprived (WHO, 2014). Climate change affects every facet of human life, inclusive of public health and national security concerns. This spans from water in the wells used at home, land used to cultivate food on the dining table and vital for the myriad of challenges faced by humanity globally (United Nations, 2011).

3 Climate Change and Human Security

Globally, climate change has increasingly become one of the issues that poses risk to all the indicators of human security. This has been echoed by the IPCC report that indicated that climate change is having enormous effect worldwide through rise in sea levels and dynamics in rainfall patterns. Resolution 66/290 by the UN General Assembly opined that "human security is an approach to assist member states in identifying and addressing widespread and crosscutting challenges to the survival, livelihood and dignity of their people" (United Nations, 2017). Climate change consequently threatens national security, which invariably unleashes a spiral of insecure situations which change the social order, reduce revenue inflows to the state and eventually cause instability in the political system.

The impact of climate change on the security of the state as well as international order can better be appraised from a more integrative human security perspective. This has placed enormous challenge on social, economic, political as well as institutions in the international arena that protect the environment. Climate change has significantly become a development issue affecting public health and posing a security risk to both human and national security (Nanthini & Nair, 2021). Most often, activities that threaten or boost the security of the individual are often outside the communities in which those affected lives. The immediate effects could be seen from the melting of ice glaciers, extreme weather conditions unleashing tsunamis, hurricanes which consequently results in forced migration, resource scarcity and protracted conflict with other cumulative effects (Nanthini & Nair, 2021). The EU Council on climate diplomacy has concluded that, "climate change multiplies threats to international stability and security" especially those in weak and vulnerable communities with their attendant pressures on the environment predisposing them to disasters which leads to the loss of livelihood through forced displacements in various countries (Council of the European Union, 2020).

It is imperative to frame climate change beyond the usual 'environmental, scientific and securitisation' which for some time now has occupied academic and policy discourses to one of 'humanistic,' with crucial considerations to vulnerable and low-income economies (Nanthini & Nair, 2021). According to the UNDP (2020), as well as the IPCC assessment Report (2021), which all emphasise human security and emerging issues in Anthropocene, it is crucial not to focus only on "human-induced climate change", but also climate variability on regions, taking cognisance of vulnerable people who suffer the most during natural disasters. Nanthini and Nair (2021) have emphasised that climate change has gradually become "the front and centre" when it comes to understanding international security and peace. Yasmeen (2021) has buttressed this and indicated that changes in the climate system is a challenge to international security and poses an existential threat to the health of people due to rise in temperature, degradation of the environment, food insecurity and increase in disease burden.

With mounting evidence from scientific data on the negative effects of climate change, it is increasingly seen as a security issue. A number of sessions on discussions about climate change have been held by the United Nations Security Council (UNSC) on the emerging security threats posed by climate change, and also highlighting it in political discussions regarding peace and security in the international sphere. Some developed countries have indeed classified climate change as one of the biggest threats to national security. Indeed, there is overwhelming evidence to conclude that there is a synergy between climate change and human security. Huntjens and Nachbar (2015) have emphasised that minimal notions of climate change which threaten national security, inclusive of the security of people affirms certain measures that are taken politically, as compared to wider notions of security.

Establishing the linkage between climate variability and human security involves dealing with vulnerable communities and the socio-political milieu in which it occurs as well as the geographical areas in which they find themselves. In the view of Dalby (2013), the responsibility to protect under international law has become more complex due to climate related disasters and inter-generational equity. Environmental justice concerns regarding obligations to support and protect vulnerable populations are also increasingly being pursued by institutions at the international level. Human security evolves around potential activities that will be triggered in times of public emergency. It is imperative that attention is focused on measures to deal with emerging threats. All these require a critical look at actions taken to protect vulnerable populations which is now part of the broad human security architecture regarding the right to be free from natural disasters (Dalby, 2013).

There are several notions of human security depending on the particular discipline. Further, human security has been discussed under many disciplines where differing methods are used (Brauch, 2009; Norris & Inglehart, 2012). Adger et al. (2014) sees human security regarding climate change as, "a condition that exists when the vital core of human lives is protected, and when people have the freedom and capacity to live in dignity." In this context, the main factors that underpin human security involve the "universal and culturally specific, material and non-material elements necessary for people to act on behalf of their interests" (Adger et al., 2014). The Commission on Human Security (2003) also took the opportunity to highlight the key issues under the concept. Human security is all encompassing and a broad inter-relational concept that brings to the fore emerging vulnerabilities in economic, social and the political sphere in which they manifest. From this perspective, this conceptualization sees vulnerability as the one which is man-made and hinges on what happens externally.

Adger et al. (2014) have again noted that several factors influence the dynamics in human security, especially happenings within the economy, in the state and activities of civil society. Low economic development, selectivity based on one's background and the devastating consequences of natural disasters impact adversely on human security (Adger et al., 2014). According to Nanthini and Nair (2021), there are a myriad of key factors that must be considered in having any assessment on human security including developing coping strategies in times of natural disasters. Neocleous (2000, p. 9) citing Wilhelm von Humboldt's conceptualisation of security in the later part of the eighteenth century has emphasised that 'security is a matter of the legal assurance of freedom' which encapsulates legal as well as economic security and the minimum assurance of better living standards. This has been reinforced by Rothschild (1995) who has also intimated that human security is multifaceted encompassing socio-economic and environmental issues. The international community having the legitimate task to protect people in times of violent extremism and other issues that threaten human survival, is what the former UN Secretary—General call "a larger freedom" (United Nations, 2011).

The whole concept of human security was first and foremost defined by the UNDP report in 1994, within the provisions of Ullman's designation as:

> first, safety from such chronic threats as hunger, disease and repression. And second, protection from sudden and hurtful disruptions in the patterns of daily life – whether at home, at work or in the society (UNDP, 1994, p. 23).

Human security consequently connotes complex international political economy together with events that occur in times of disaster. These intersect with human security in the general context of climate. The UNDP report in 1994 regarding human security gave four main essential features of the concept. These are:

> First, it is a universal concern relevant to people everywhere. Second, the components of security are interdependent. Third, human security is easier to ensure through early prevention, lastly, but not the least, and perhaps the crucial innovation in this formulation, is the shift of the referent object of security from states to people. This in turn raises issues of rights, justice, vulnerability, power relations, and empowerment.

Other international threats to the security of the individual include, population growth, inequity in the distribution of resources, forced migration, environmental degradation, narcotics and terrorism (Dalby, 2013). Climate change did not feature initially, but has become a major factor that threatens human security. The Commission on Human Security (2003, p. 4) defines human security as:

> protecting the vital core of all human lives in ways that enhance human freedoms and human fulfillment. Human security means protecting fundamental freedoms—freedoms that are the essence of life. It means protecting people from critical (severe) and pervasive (widespread) threats and situations. It means using processes that build on people's strengths and aspirations. It means creating political, social, environmental, economic, military and cultural systems that together give people the building blocks of survival, livelihood and dignity.

Human security in the wider context encapsulates respect for the fundamental human rights of people, good governance, availability and access to free quality education as well as a robust public health care system and ensuring that individuals have opportunities and freedoms to fulfill their potential. Each step towards the attainment of these objects leads to a reduction in poverty, promoting economic growth and preventing violent extremism. Freedom from fear and want and for posterity to enjoy a healthy natural environment all constitute the basic foundation of human and national security (Commission on Human Security, 2003, p. 4). Freedom from want again connotes providing "essential food, water and shelter", whereas freedom from fear embraces the absence of intractable violence and oppression. According to Webersik (2010) all these intersect with Climate change which may lead to fear, want, and public health concerns which endanger the well-being of individuals.

Climate change, also pose existential threat to health and survival. There is a synergy between epidemiology and environmental health of populations. The WHO (1997) has stated that population growth is a key contributor to climate change. Indeed, it has been projected that an increase in carbon dioxide emissions in two decades, that is from 1992 to 2022, will be due mainly to an increase in the population of the number of people on earth. And this will certainly be occurring in developing countries. The WHO (1997) has intimated to world leaders about the implications on public health which is crucial for sustainable development. Costello et al. (2009) have stated that the myriad of effects of climate change on health will impact quite a number of people globally putting the lives and survival of close to a billion people at increased risk. The IPCC has reinforced this by stating that "climate change is projected to increase threats to human health" and this can occur as a result of a wide array of activities. These may include natural disasters, heatwaves, floods, storms and infectious diseases that may affect public health, interruptions in the ecosystem, intractable conflicts due to land and resource scarcity (Pachauri & Reisinger, 2007).

Public health challenges may involve the spread of infectious diseases and an exponential rise in psychosocial challenges mainly as a result of mental stress (WHO, 2014). Insecurity resulting from poor rainfall patterns and varying temperatures may be compounded by unavailability of fertile lands together with land tenure rights (WHO, 2014). This will obviously impact negatively on water available for irrigation, appropriate varieties of plant seeds and appropriate designs from the built environment. Moreover, loss of flora and fauna from biodiversity can compound human security challenges due to scarcity from income from non-timber forest products, herbs as medicine, nutritional supplements when food shortages occur, and seed bank for plant varieties needed (WHO, 2014).

Climate change affects the three most important determinants of health being 'food, air and water.' It is estimated that globally about two billion people on earth do not have access to electricity which invariably affect their daily lives negatively since energy is crucial in the domestic, agriculture and manufacturing industry. Research has also shown that about half of the global population domestic source of energy are traditional biomass, namely charcoal, fire wood, dung and crop waste. These results on exposure to inhouse pollutants which impact on the health of users causing chronic pulmonary disorders and worsening lung infections (Liu et al., 2007), and contributing to childhood pneumonia and lung cancer. Fuel shortages lead to increase in women's load because they are the primary care givers in the domestic setting. Transportation has been deemed to cause growth in greenhouse gas emissions between 1990 and 2020, with its attendant pollution in urban cities which causes about 1.3 million lung cancer related deaths annually (Kovats & Akhtar, 2008).

The Fourth Assessment Report of the IPCC on Climate Change was emphatic that the negative effects of climate change on health occur mostly in less developed economies and on vulnerable populations (McMichael et al., 2008). In sub-Saharan Africa, it is estimated that about 110 million people live in areas that have malaria which is the most cause of outpatients in health facilities. It is estimated that Climate change has the potential to add between 20 and 70 million by the year 2080, with its attendant consequence on public expenditure on health which will exacerbate poverty (McMichael et al., 2008). Africa is susceptible to climate change due mostly to a host of environmental and social challenges. The political economy of most African countries is dependent on agriculture, where about two thirds of the population works and accounts for almost half of incomes from livelihoods in households. Climate scenarios project an increase in mean temperatures by 2100 and the regions that will be heavily affected are southern and northern Africa which are already experiencing drought and a decline in rainfall in the West and East Africa.

Africa has twenty of the countries with the fastest warm temperatures. It is estimated by scientific data that in the past hundred years, temperatures in the African continent have increased over an average of 0.5–2 °C (IPCC, 2019). In addition to this, when thirty countries in the African continent were sampled by about sixty percent (two-thirds) are warming faster than what is happening globally and it is expected to rise in the decades that follow (IPCC, 2019). Cyclone Idai that occurred in 2019 in Southern Africa affected about 2.2 million people in countries such as Zimbabwe, Malawi and Mozambique (World Vision, 2019). Of these, Mozambique suffered the worst consequences with 1.8 million lives being affected, about 600 deaths and 1600 injuries. This also destroyed about $773 million worth of agriculture produce, buildings and other social amenities (UN News, 2019).

In the Sahel region in West Africa, extremists' groups have taken advantage of changes in the climate to unleash terror in communities due to scarcity in land use. Drought in the Southern Africa caused the Victoria Falls to slow which again has heightened discussions on the insecurity posed by climate change. This also has a multiplying effect on livelihood and security in the continent (von Soest, 2020). The Economist (2019) intimated that the violence in Nigeria by Boko Haram is due to reduced livelihoods and so idle hands are being influenced by extremists' groups which have all resulted from the reduction of water inflows into lake Chad. "As the lake receded, people moved towards it, and are then plagued by swarms of tsetse flies. Herds people, farmers and fisherfolk also competed for access to the shrunken supply of water." It is estimated that the world population will increase by 3 billion from 2010 to 2100. 2.5 billion out of this population increase will be in Africa (FAO, 2012), which will invariably bring about pressure on resources such as water, housing and unemployment (Dasgupta, 2018).

Climate change induced drought and increase in population growth will lead to water shortages, which will affect crop yields due to unavailability of water for irrigation leading to low crop yields resulting from 1–2 °C rise in temperatures by 2050 (McMichael et al., 2008). This will add about 12 million Africans to the 200 million already malnourished. Increase in temperature and changes in rainfall patterns is likely to affect tsetse fly growth and increase the spread of trypanosomiasis in livestock. Climate change affecting land use has forced populations to move especially in the Sahel region of pastoralists searching for grazing lands for their animals (McMichael et al., 2008). This has generated tensions in Nigeria, Ghana and Burkina Faso between the Fulani pastoralists and crop farmers. National security is a key function of the governments of countries all over the world. Consequently, the threats to human security due to climate change are implicitly seen as threats to security within the state and territorial sovereignty. Fundamentally, tackling climate change is consequently pivotal to the security of the state (Fig. 1).

The UN Security Council in 2007 had its maiden discussion on the impacts of climate change on peace and security. During this period some states were of the view that, the Security Council was not the appropriate forum to debate such matters, and suggested the most forum convenient was the UN Economic and Social Council (ECOSOC), the UN General Assembly (UNGA) and the UNFCCC (International Institute for Sustainable Development, 2015). But Small Island States supported the debate that was tabled by the United Kingdom, which held the presidency

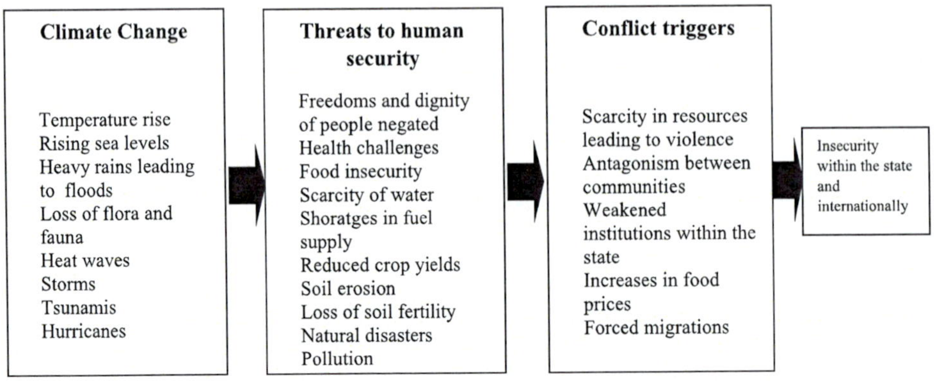

Fig. 1 Climate change and potential threats to human security. *Source* Adapted from Von Soest (2020)

of the Security Council at the time (International Institute for Sustainable Development, 2015). It could be stated categorically now, that the forum was convenient since any disruptions in the environment due to climate change, ultimately affect both national and international security. Also in 2011, the Security Council held another meeting on the effects of climate change on human security. In the said meeting, the Secretary-General of the UN, Ban Ki-moon stated that "climate change not only exacerbates threats to peace and security, it is a threat to peace and security" (United Nations, 2011).

The last assessment report of the IPCC highlighted issues bothering on extremism and human security which was happening for the first time and a sub theme on 'Human Security' provided in the Working Group II's (WGII) contribution on Impacts, Vulnerability and Adaptation. The report brought to the fore that climate change has the potential to exacerbate the displacement of people due to unavailability of resources and warm weather in both the rural and urban centres especially in countries where there is low economic development. This may have the tendency to increase extreme violence inclusive of wars and trigger other drivers of violence such as unemployment and low economic development. Climate change also affects the construction of projects and the integrity of the boundaries of states, and per extension influence national security policies. Specifically, lands being flooded as a result of rise in the sea-level threaten the boundaries of Small Island States and countries closer to oceans. Further, trans boundary impacts can lead to conflicts due to resource scarcity. Even though effective and robust state and governmental organisations may help deal with extreme violence, this may be difficult in states with weaker governance institutions (International Institute for Sustainable Development, 2015).

Malaria is one of the most infectious vector borne diseases in the world which causes about 1–2 million fatalities every year, a greater number of these being infants. Sachs and Malaney (2002) in a study have also confirmed that about 25% of all deaths in children between 0 and 4 years is attributable squarely to malaria. Malaria transmission is contingent on climatic factors and this accounts for the reason why sub-Saharan Africa is susceptible to climate change and occupies the first place in the record books of the diseases of the WHO. The weather in sub-Sahara Africa is conducive for the reproduction of mosquito larva. Beyond the public health implications is the economic costs of sickness, productivity lost due to absence from school and work, decrease in tourism and foreign direct investments (Patz et al., 2007).

According to the WHO (2002) and Ezzati et al. (2004), an estimated number of about 2.4 million children die due to inhalation of pollutants from combustion with wide implications on greenhouse. The Report of the Fourth Assessment in 2007 by the IPCC was emphatic with about ninety percent certainty, that activities of mankind is the main cause of global warming. United States Protection Agency (EPA), has intimated that ozone depletion is the source of a wide range of adverse respiratory conditions such as "airway irritation, coughing, reductions in lung function and aggravation of asthma" in vulnerable individuals (Lautensach & Lautensach, 2020). Climate change induced insecurity threatens livelihoods, compromises culture and identity, exacerbate population movement and diminish the potential of states to provide the necessaries of life to foster human development. The IPCC has identified these four major factors as the main driving factors of human security and changes in climate.

According to Cardwell and Elliott (2013), climate change is designated as the biggest threat to public health globally in the 21st century. There is abundant material evidence to support the claim that climate change threatens public health security due to natural disasters and diseases such as malaria and dengue which are vector-borne (Majra & Gur, 2009). The WHO has stated that 88% of disease burden and natural disasters that stem from climate change afflicts the marginalised and vulnerable in developing countries highlighting another level of inequity (Patz et al., 2007). Article 3 of the UNFCCC stipulates as follows:

> The Parties should protect the climate system for the benefit of the present and future generations of humankind, on the basis of equity and in accordance with their common but differentiated responsibilities in respect to capacities (United Nations, 1992, p. 9).

Climate change through its various effects has led to widespread violent extremism which disturbs the peace and stability of society (Ani et al., 2022). Several coastal areas were experiencing severe issues such as drought, insufficient recharge of the underground water and frequent urban flooding (Kim & Wu, 2021). Climate change also poses danger to the coastline of African countries. The effects of climate change on the peace and security across the borders of countries have been very revealing (United Nations Environmental Programme, 2022) as could be seen from political instability in the Sahel region. Tarif (2022) has opined that there is overwhelming evidence to suggest that the negative impact of climate change has increased the potential threat of violent insurgency in West Africa. The USAID (2017) has posited that West Africa is a key area for insecurity due to climate change. Climate change through its various effects has led to widespread violent extremism which disturbs the peace and stability of society (Ani et al., 2022). Tenure insecurity may also be heightened by the devastating effects of climate change.

4 Climate Change and Sustainable Development

The Brundtland Commission in 1987 defined sustainable development as "development which meets the needs of current generations without compromising the ability of future generations to meet their own needs." The Commission's Report also brought to the fore the need to address developmental and environmental challenges. In 1992, the United Nations Conference on Environment and Development (UNCED), which resulted in the 'Earth Summit' in Rio de Janeiro highlighted the pressing need to adopt sustainable environmental practices or else development will be seriously affected or compromised. And that any achievement in environmental policies will fail in developing countries when a concerted effort is not made to address developmental challenges (Onuoha, 2009). The Kyoto Treaty in 1997 highlighted global climate change concerns and reduction to gas emissions, while the World Summit on Sustainable Development organised by the UN in Johannesburg, South Africa in 2002 also highlighted the need to address sustainable development challenges worldwide (Onuoha, 2009) (Fig. 2).

Natural resources provide man with raw materials that are used for developmental purposes. Sustainability, invariably means that any activity that contributes to needed development should be done in a manner that maintains or increase in the future (Constanza & Daly, 1992). The natural resources are made up of capital which encompasses "biodiversity-aquatic, oceanic, and terrestrial" and other constituents of the physical environment namely soil, air and water (Constanza & Daly, 1992). From the standpoint of climate change and sustainable development, it is important to look at the dependency of biodiversity on elements of the physical environment (Inogwabini, 2019). Byravan and Rajan (2015) have intimated that, "for the tens to hundreds of millions of people living in low-lying areas or on small islands, no physical defense is realistically possible or can be fully protective." And in the view of Huntjens and Nachbar (2015), for people in these areas, climate change adaptive measures is 'about nothing less than the survival of their societies.'

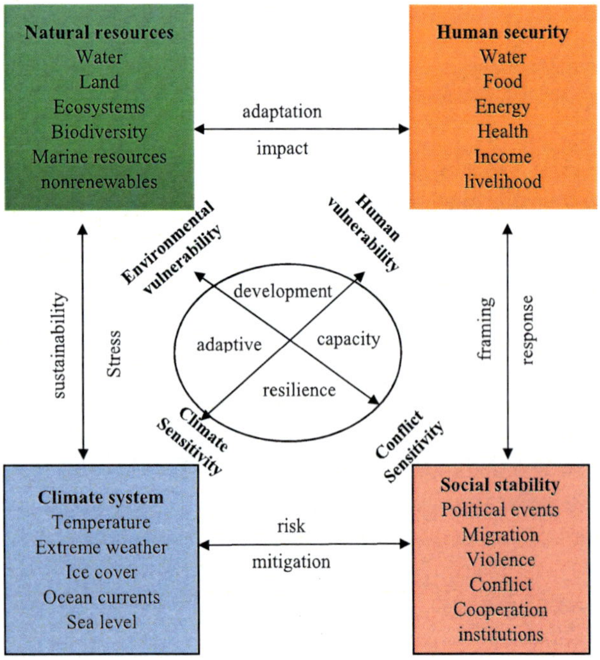

Fig. 2 Interrelationship between the climate system, natural resources and human security. *Source* Jürgen et al. (2012)

Sostaric (2019), has opined that climate variability has effect on all sectors of the economy in all countries. The USAID (2017) has opined that increased temperatures has led to increases in fungal diseases and pests with its attendant consequences on crop yield. Ani et al. (2022) have stressed that the potential effects of climate change as well as its disastrous consequences on mankind spans all the regions of the world. UNEP (2022) has also buttressed that climate change is currently one of the key defining issues of the time with a threat on mankind. Its consequences are felt in all sectors of the economy, and sometimes across borders.

5 Impact of Climate Change on Sustainable Livelihood

There have been growing concerns of mankind on the multiplier effects of climate change on economic, social, environmental, political and international security. Its consequences fall directly on the manufacturing industry, agroforestry, livelihoods, gross domestic product, infrastructure and electoral governance. According to Schneider and Lane (2005), ethically climate change impacts highlight three main inequities. The first is inter country equity, intergenerational equity and lastly, but not the least inter species equity. According to the EM-DAT (International Disaster Database) 346 and 315 natural disasters occurred in 2015 and 2018 respectively. Asia suffered the worse impact of 45%, followed by Latin America then Africa (Koubi, 2019). Even though natural disasters affect negatively both the rich and poor countries, their devastating consequences are felt in developing countries than in the industrial world.

Poverty and climate variability intersect and create hardships for developing economies than in areas with better resources for disaster management (Koubi, 2019). Climate change is affecting the lives of people in communities and is seen as a major threat to the livelihood of people, states and the international community (Ani et al., 2022). The continuous increase in heat waves and global warming has adverse impact on food security. Food security is a poverty issue. It is estimated that, over 1.2 billion people live in communities without adequate water supply and it is expected that by 2025 about 1.8 billion people will experience insecurity in relation to the availability of water. Clean water is critical for human survival (Dasgupta, 2018). Climate-related disasters also affect capital assets, as well as living organisms and destruction of natural as well as the way of life of a people. It also come with indirect losses such as loss of output from capital reduction through reconstruction of infrastructure destroyed which affect a country's Gross Domestic Product (GDP) (Koubi, 2019).

Warm weather conditions have impact on crop production in some geographical regions (Daliakopoulos et al., 2017). This is due mainly to the high volumes of carbon dioxide within the atmosphere which provides grounds for plants to conserve water more efficiently. However, studies by Frieler et al. (2017) have also shown that climate change sometimes affect crop "suitability" in some areas which results in changes in the yield of farm produce. Specifically, high temperature and precipitation patterns has the potential to reduce crop production volumes, with the highest incidence affecting wheat and maize (Frieler et al., 2017), while effects on rice and soybean volumes cannot be determined and can either be positive or negative according to van Oort and Zwart (2018). All of these impacts on poverty, food security, livelihood and public health. The IPCC (2021) has opined that the consequences from changes in the climate reflects in the reduction of the quality of food, low quality of the soil, environmental degradation and constraints in the supply of water.

Temperature, precipitation and extreme weather variability can affect crop yields. The incidence of these will be high in the tropics, where crop yield is close to their thermal limits. It is estimated that crop yield will decrease if warming limit to 1.5 °C compared with 2 °C in 'Sub-Saharan Africa, Southeast Asia and North, Central and South America, inclusive of least developed economies (Schleussner et al., 2018). Climate change will also affect the production of livestock both in numbers and quality (Notenbaert et al., 2017) as well as effects on the livestock industry as a result of feed quality and the spread of pests and diseases (Kipling et al., 2016). Boone et al. (2018) project that worldwide, there will be a reduction in livestock of about 7–10% at about 2 °C of warming, which will result in economic loss between $9.7 and $12.6 billion.

The fishing industry contributes to food security, provides livelihood for a lot of people, and it also provides social, economic as well as nutritional benefits. It is estimated that globally about 3.2 billion people get 20% of their "average per capita intake of animal protein from the sea" (FAO, 2018). Marine fisheries provided 79.3 million tonnes, whereas marine aquaculture provided 28.7 million tonnes in 2016 (FAO, 2018). The fisheries industry (inland and marine) also provided livelihoods for about ten to twelve percent of the world's population according to Barange et al. (2018), with about fourteen percent of fishers and aquaculture workers being women (FAO, 2018). Climate change related natural disasters will obviously impact negatively on people who depend on them for their livelihood and nutritional feed. Employment avenues will also be lost due to climate change (Anyadike, 2009). First and foremost, some sectors of the economy that depend on products from agriculture, the fishing industry and livestock as their raw material supplies will either fold up or cut down on employment to accommodate current happenings (Anyadike, 2009), with its attendant consequences on income reduction.

Muringai et al. (2020) in a study, examined the effects of reduced water supply, increased weather temperature and CO_2 emission on food production. The results of their findings like that of other researchers established an important link between changes in climatic condition and low crop yield. Scientifically, climate change impact food supply in the agricultural production process. Consequently, its effects invariably have impact on all the other segments of food supply. The fishery industry supply chain as noted by Muringai et al. (2020) is also adversely affected by changes in the climate system. Taub et al. (2008) undertook a "meta-analysis" of various scientific experiments on the potential effect of climate change on the production of food and found out that when crops are exposed to excessive concentration of CO_2 the protein content reduces. This then means that, as the concentration of CO_2 in the atmosphere increases the quality of food crops reduces. In another study similar to that of Taub et al. (2008) undertaken by Berhanu and Wolde (2019), the authors found out that the content of mineral elements in the soil is seriously affected by excessive weather conditions like high rainfall, floods and drought. In the view of the authors, heavy rainfall leads to the loss of soil fertility due to leaching of the rich soil minerals whereas floods have the potential to increase the acidic content of the soil. The findings of the study also associated drought with soil toxicity.

In what ways does climate change induces conflict? There exist two potential pathways. In the first instance, climate change poses significant danger to the livelihood of people. This may consequently trigger conflict as a result of clashes that may occur over scarce resources. Undoubtedly, in the second breath climate change contributes significantly to intractable conflict through resource conflict, resulting in reduction in economic output and incomes from agriculture, rising food prices and heightened migration flows (Koubi, 2019). Tarif (2022) in a review on the effects of climate change found out that lack of livelihood opportunities and allied responses to "environmental degradation, changing migration patterns, predatory armed groups and elite behaviour" all contribute significantly to how the impact of climate change exacerbates extreme violence in West Africa.

Krajinovic (2019) has intimated that the Mediterranean has been designated as the "fastest warming" and the "saltiest" which poses dire consequences on marine life as well as human beings. According to the FAO (2021) drought levels to crop destruction with its attendant consequences on food security and sustainable livelihood and income have induced forced migration leading to increased urbanisation. Ogbuchi (2020) has intimated that it is climate change that induced flooding that led to human loss, which affected crops and livestock in addition to displacement in Nigerian cities. In addition, empirical evidence has shown that climate variability has affected agricultural production negatively (Jung & Kunstman, 2007; Kralovec, 2020).

6 Climate Change and Fundamental Human Rights

Climate change affects human rights in several ways with the effects on the right to life, food security, public health being the greatest in various countries (United Nations Environmental Programme, 2015). Rights are freedoms guaranteed the individual without which one cannot grow his or her potential to the fullest. Human rights are universal legal guarantees accorded to individuals and groups against the acts and omissions that tend to interfere with the fundamental freedoms one is entitled to as a human being. The Black's Law Dictionary (2014 p. 858) defines human rights as:

> the freedoms, immunities, and benefits that all human beings demand as a matter of law in the society in which they live according to modern standards, especially at international level and in the constitutionally enshrined liberal democracy.

International human rights law places enormous obligation on states and other public officials to respect, promote, protect and fulfil all human rights (OHCHR, 1993). Human rights are applicable across all borders and are contingent on the inherent dignity of individuals. They are "equal, indivisible, interrelated and interdependent" (OHCHR, 2015). Climate change impacts negatively on the full realisation of human rights. International human rights instruments stipulate that, in times of disasters there should not be a derogation from the obligation of states to protect their citizenry. Disaster management is therefore protected by relevant guiding principles that underlay the rights to "participation and information, transparency, accountability, equity and non-discrimination" (OHCHR, 2015).

Climate change threatens human security through the disruption where people live their source of livelihoods, culture setting and individuality, induces people to move from their original locations without their will, and has the potential to undermine the ability of the government institutions to ensure the provision of the necessary conditions that guarantee the security of citizens (Hadžić, 2022). In Nigeria, disruptions in the weather conditions have affected food production and distribution leading to food scarcity which has led to increases in prices (Oyinloye et al., 2018). The Fifth Assessment Report by the IPCC stated unambiguously that "human influence on the climate system is clear, and recent anthropogenic emissions of green-house gases are the highest in history." It also became evident that climate change activities in contemporary times have had a negative toll on humanity and the ecosystem.

During a discussion by the Human Rights Committee, Flavia Pansieri, the Deputy High Commissioner for Human Rights concluded that "those who have contributed the least to greenhouse gas emissions will be the ones who bear the greatest burden, that is the poorest people, in the poorest countries, their children, and all our children". While the former High Commissioner for Human Rights, Mary Robinson, described climate change as "probably the greatest human rights challenge of the 21st century" (OHCHR, 2015). This then call for efforts to remedy the dire effects through sound environmental governance practices.

7 Sustainable Environmental Governance

There is an urgent need for sustainable environmental management to check climate change. The disastrous consequences of climate change have necessitated international concerns and actions at reducing its impacts in addition to advocating for rules that would limit the actions of man that exacerbate climate change (Ani et al., 2022). First, there must be an integration of climate health risk management across all sectors of the economy in addition to adaptive measures that will contain climate variability. This calls for a resource center where expertise could be shared to either give early warning signs to issues that bother the environment or planners as well as environmental experts who can provide mitigation measures. Furthermore, there should be a concerted effort to have robust environmental governance institutions with the capacity to provide technical advise and to integrate public health and climate related issues into policy, socio-economics and planning. Specific interventions like training these kinds of professionals should be the focus of policy makers (Thomson et al., 2014).

Hadžić (2022), in his view has stressed that if the world wants to minimise global warming below 1.5°–2.0 °C without any associated consequence, then all policies must be geared towards achieving climate neutrality by 2030. The European Commission (2021) has opined that adopting to climate change requires taking diligent measures to negate the changes that may be caused, or alternatively harnessing the potential benefits that may occur with the key object of reducing climate vulnerability. The minimum cost of climate adaptation in 2020 was projected at hundred billion euros and it is expected to rise to two hundred and fifty billion euros per year for the European continent by 2050 (European Trade Union Confederation, 2020).

Additionally, funds should be devoted to research on climate variability together with its related challenges. This will provide impetus for experts to always be alert on areas prone to climate induced disasters and plan ahead to minimize the impact of natural disasters like floods, extreme weather conditions, rise in the sea level as well as hurricanes. If there is data on these, modelling can be done and adequate measures put in place to reduce the after effects of environmental disasters. Policy makers should also devote funds to support integrated health surveillance and climate observation centres to provide information on changes in weather patterns that result from climate change. Moreover, all effort must be made to support the ecosystem by planting a variety of plant species that can withstand heavy storms. Alternative energy sources should be provided to stem the cutting of trees for firewood to provide energy for domestic use. Women and children are usually affected by natural disasters caused by climate change as compared to men. They should therefore be involved in policy making regarding climate change.

Women should be provided with enough education on the burning of fossil fuel which produce a lot of greenhouse gas which ultimately induce climate change. This will in a way reduce climate induced disasters. Dankelman (2008) has expressed the view that women are often neglected in the solution to climate change solutions and the security of all mankind. It is high time they are brought to the table when climate change issues come for discussion since they suffer most in times of disasters. Climate change mitigation and adaptation measures must be put in place by areas that are prone to natural disasters. Sustainable development strategies must be used in the built environment, planning, designs and in the provision of infrastructure and all developmental projects. Sustainable development is an integrative approach that embraces both short- and long-term approaches in decisions that are multi-disciplinary. The development of 2030 Agenda for Sustainable Development at the international level by members of the United Nations as a common policy framework for peace and prosperity for people and planet which incorporates 17 Sustainable Development Goals is a good example (United Nations Department of Economic, 2021). Each member country can then put up a surveillance team to follow the sustainable development goals at the national level (Grant, 2021).

Sustainable development approaches provide avenues to design communities to reduce climate change impacts, which consequently affect public health and the physical development of mankind (US EPA, 2017). Redesigning pedestrian-friendly environments heavily correlate with decreases in negative health effects inclusive of 'obesity, diabetes, hypertension and asthma while dependency on vehicles lead to a 'more sedentary life style' (Centers for Disease Control and Prevention (CDC), 2018; Department of Health & Human Services, 2020). Land and spatial planning policies contributes significantly to desertification which all impact public health (Grant, 2021). Moreover, local land conservation trusts, as well as private conservation institutions tasked with forest and open space restoration can minimise the negative threats posed by climate change (Land Trust Alliance, 2020). Studies by Amadu et al. (2020), Marenya et al. (2020) and Tesfaye et al. (2020) have all intimated that as the changes in climate continue to pose security challenges for countries, one way of resolving this has been the development of 'climate smart agriculture' (CSA). This is composed of soil and water conservation mechanisms through agroforestry and the planting of crops which enables farmers to adapt to climate change.

8 Conclusion

Climate change affect human security and sustainable development in multiple ways. Climate variability has the potential to threaten the social, economic, environmental and political as well as international security of the state. Moreover, climate change also impacts public health. When natural disasters occur, they may disturb the land on which people live forcing them to migrate to other areas which may trigger intractable conflict which invariably lead to loss of lives. It has also been established that climate change has impact on food security, poverty, agriculture and sustainable livelihood. Human rights impacts from climate change have also been documented by several studies. From the foregoing, it could be stated categorically and emphatically that, climate change has a multiplier effect on all facets of human life and it is seen as the biggest threat to mankind in the 21st Century. This calls for concerted effort to deal with the challenges associated with climate induced natural disasters and to also reduce carbon emissions which are the greatest cause of climate variability.

References

Adger, W. N., Pulhin, J. M., Barnett, J., Dabelko, G. D., Hovelsrud, G. K., Levy, M., et al. (2014). *Human security*. Cambridge University Press.

Amadu, F. O., McNamara, P. E., & Miller, D. C. (2020). Yield effects of climate-smart agriculture aid investment in southern Malawi. *Food Policy, 92*, 101869.

Ani, K. J., Anyika, V. O., & Mutambara, E. (2022). The impact of climate change on food and human security in Nigeria. *International Journal of Climate Change Strategies and Management, 14*(2), 148–167.

Anyadike, R. N. C. (2009). Climate change and sustainable development in Nigeria: Conceptual and empirical issues. Enugu forum policy paper. *African Institute for Applied Economics, Nigeria, 10*, 14–18.

Baer, P., Harte, J., Haya, B., Herzog, A. V., Holdren, J., Hultman, N. E., et al. (2000). Equity and greenhouse gas responsibility. *Science, 289*(5488), 2287–2287.

Barange, M., Bahri, T., Beveridge, M. C., Cochrane, K. L., Funge-Smith, S., & Poulain, F. (2018). *Impacts of climate change on fisheries and aquaculture: Synthesis of current knowledge, adaptation and mitigation options*. FAO.

Berhanu, M., & Wolde, A. O. (2019). Review on climate change impacts and its adaptation strategies on food security in sub-Saharan Africa. *Agricultural Socio-Economics Journal, 19*(3), 145–154.

Boone, R. B., Conant, R. T., Sircely, J., Thornton, P. K., & Herrero, M. (2018). Climate change impacts on selected global rangeland ecosystem services. *Global Change Biology, 24*(3), 1382–1393.

Brauch, H. G. (2009). Securitizing global environmental change. In *Facing global environmental change* (pp. 65–102). Springer.

Brundtland Commission. (1987). Report of the World Commission on Environment and Development: Our Common Future. Retrieved June 22, 2022, from: https://sustainabledevelopment.un.org/content/documents/5987our-common-future.pdf

Byravan, S., & Rajan, S. C. (2015). Sea level rise and climate change exiles: A possible solution. *Bulletin of the Atomic Scientists, 71*(2), 21–28.

Cardwell, F. S., & Elliott, S. J. (2013). Making the links: Do we connect climate change with health? A qualitative case study from Canada. *BMC Public Health, 13*(1), 1–12.

Centers for Disease Control and Prevention (CDC). (2018). *Prevalence of overweight, obesity, and severe obesity among adults aged 20 and over: United States, 1960–1962 through 2015–2016*. https://www.cdc.gov/nchs/data/hestat/obesity_adult_15_16/obesity_adult_15_16.htm

Commission on Human Security. (2003). *Human security now: Protecting and empowering people/Commission on Human Security*. United Nations Digital Library. Retrieved March 12, 2022, from https://digitallibrary.un.org/record/503749/files/Humansecuritynow.pdf

Constanza, R., & Daly, H. (1992). Natural and sustainable development. *Revista Conservation Biology, 6*(1), 37–46.

Costello, A., Abbas, M., Allen, A., Ball, S., Bell, S., Bellamy, R., et al. (2009). Managing the health effects of climate change: lancet and University College London Institute for Global Health Commission. *The lancet, 373*(9676), 1693–1733.

Council of the European Union. (2020). Council conclusions on Climate Diplomacy. Retrieved March 17, 2023, from https://data.consilium.europa.eu/doc/document/ST-5033-2020-INIT/en/pdf

Dalby, S. (2013). *Climate change as an issue of human security*. Edward Elgar Publishing.

Daliakopoulos, I. N., Panagea, I. S., Tsanis, I. K., Grillakis, M. G., Koutroulis, A. G., Hessel, R., et al. (2017). Yield response of Mediterranean rangelands under a changing climate. *Land Degradation & Development, 28*(7), 1962–1972.

Dankelman, I. E. M. (2008). Gender, climate change and human security lessons from Bangladesh, Ghana and Senegal.

Department of Health and Human Services. (2020). *Healthy people 2020*. https://www.healthypeople.gov/2020/about/foundation-health-measures/Disparities

Dasgupta, S. (2018). Burden of climate change on malaria mortality. *International Journal of Hygiene and Environmental Health, 22*(5), 782–791.

Dictionary, B. L. (2014). Black's law dictionary. https://dictionary.thelaw.com/truth

Dokos, T., Afifi, T., Bogardi, J., Dankelman, I., Dun, O., Goodman, D. L., et al. (2008). Climate change: Addressing the impact on human security. In *Athens: Hellenic Foundation for European and Foreign Policy and Hellenic Ministry for Foreign Affairs*. Retrieved May 27, 2022, from https://www.eliamep.gr/wp-content/uploads/en/2008/10/climate_change_summary.pdf

European Commission. (2021). Questions and answers: New EU strategy on adaptation to climate change. European Commission. Retrieved June 20, 2021, from https://ec.europa.eu/commission/presscorner/detail/en/qanda_21_664

European Trade Union Confederation. (2020). Guide for trade unions adaptation to climate change and the world of work. Retrieved November 17, 2021, from https://www.etuc.org/sites/default/files/publication/file/2020-09/ETUC-adaptation-climate-guide_HR.pdf

Ezzati, M., Lopez, A. D., Rodgers, A., & Murray, C. J. (2004). Comparative quantification of health risks. In *Global and regional burden of disease attributable to selected major risk factors*. World Health Organization, 1987–1997.

FAO. (2012). *The State of Food and Agriculture 2012: Investment in agriculture for a better future*. Food and Agriculture Organisation.

FAO. (2018). *The state of world fisheries and aquaculture 2018—Meeting the sustainable development goals*. Food and Agriculture Organisation of the United Nations.

FAO. (2021). *The impact of disasters and crises 2021 on agriculture and food security, Food and Agriculture Organization of the United Nations*. Food and Agriculture Organisation. Retrieved November 17, 2021, from http://www.fao.org/3/cb3673en/cb3673en.pdf

Forest, P. A., Heininen, L., Sfraga, M., Virginia, R. A., & Yalowitz, K. S. (2010). Climate Change and Human Security. Retrieved March 18, 2023, from https://www.google.com/url?sa=t&rct=j&q=&esrc=s&source=web&cd=&ved=2ahUKEwjDisKm9eX9AhXhgP0HHYzpDpEQFnoECAwQAQ&url=https%3A%2F%2Ffaculty-directory.dartmouth.edu%2Ffile%2F16146%2Fdownload%3Ftoken%3DlNXcPQvn&usg=AOvVaw1IieLbd1yvZTd2Ml3j60im

Frieler, K., Lange, S., Piontek, F., Reyer, C. P., Schewe, J., Warszawski, L., et al. (2017). Assessing the impacts of 1.5 C global warming–simulation protocol of the inter-sectoral impact model intercomparison project (ISIMIP2b). *Geoscientific Model Development, 10*(12), 4321–4345.

Goodman, D. L., Iltus, S., & Parker, D. (2008). Climate change and children: A human security challenge. *Climate change and children: A human security challenge.*

Grant, I. I. S. (2021). Climate Change and Sustainable Development. Retrieved July 29, 2022, from https://www.extension.purdue.edu/extmedia/ID/ID-524-W.pdf

Hadžić, F. (2022). Impact of climate change in Southeast Europe; adaptation policies, environmental and human security, and normative resolutions. *Marine and Life Sciences, 4*(1), 1–15.

Huntjens, P., & Nachbar, K. (2015). Climate change as a threat multiplier for human disaster and conflict. *The Hague Institute for Global Justice*.

IIED. (2015). *Monitoring and evaluating climate change adaptation and disaster risk reduction in Uganda*. International Institute for Environment and Development.

Inglehart, R. F., & Norris, P. (2012). The four horsemen of the apocalypse: Understanding human security: The 2011 Johan Skytte Prize Lecture. *Scandinavian Political Studies, 35*(1), 71–96.

Inogwabini, B. I. (2019). Climate Change and Sustainable Development. *Encyclopedia of Sustainability in Higher Education*, 215–224.

International Institute for Sustainable Development. (2015). Human Security and Climate Change. Retrieved July 16, 2022, from https://sdg.iisd.org/commentary/policy-briefs/human-security-and-climatechange/#:~:text=It%20concludes%20that%20climate%20change,such%20as%20civil%20war%20and

IPCC. (2001). Climate Change 2001: The Scientific Basis Contribution of Working Group 1 to the Third Assessment Report of the Intergovernmental Panel on Climate Change. Intergovernmental Panel on Climate Change.

IPCC. (2007). *Climate Change 2007: The Physical Science Basis, Contribution of Working Group I to the Fourth Assessment Report of the Intergovernmental Panel on Climate Change*. Cambridge University Press.

IPCC. (2019). Summary for policymakers. Intergovernmental Panel on Climate Change. Retrieved June 20, 2023, from https://www.ipcc.ch/site/assets/uploads/sites/4/2019/12/02_Summary-for-Policymakers_SPM.pdf

IPCC. (2021). Climate Change 2021: The physical science basis. International Panel on Climate Change. Retrieved March 17, 2023, from https://www.ipcc.ch/report/ar6/wg1/downloads/report/IPCC_AR6_WGI_SPM_final.pdf

Jung, G., & Kunstmann, H. (2007). Modelling regional climate change and the impact on surface and sub-surface hydrology in the Volta Basin (West Africa). *IAHS Publication, 313*, 150.

Kim, S & Wu, L. (2021). Economic effects of land conservation spatial characteristics for climate change adaptation. Lincoln Institute of Land Policy. Retrieved March 17, 2023, from https://www.lincolninst.edu/pt-br/publications/working-papers/economic-effects-land-conservation-spatial-characteristics-climate

Kipling, R. P., Bannink, A., Bellocchi, G., Dalgaard, T., Fox, N. J., Hutchings, N. J., et al. (2016). Modeling European ruminant production systems: Facing the challenges of climate change. *Agricultural Systems, 147*, 24–37.

Koubi, V. (2019). Climate change and conflict. *Annual Review of Political Science, 22*, 343–360.

Kovats, S., & Akhtar, R. (2008). Climate, climate change and human health in Asian cities. *Environment and Urbanization, 20*(1), 165–175.

Krajinovic, B. (2019). Al Jazeera, BiH u očekivanju katastrofa: Sve je na papiru za bolja vremena [BiH in anticipation of disasters: Everything is on paper for better times]. Retrieved Nov 17, 2021, from https://balkans.aljazeera.net/teme/2019/9/29/bih-u-ocekivanju-katastrofa-sve-je-na-papiru-za-bolja-vremena

Kralovec, S. (2020). *Food insecurity in Nigeria: An analysis of the impact of climate change, economic development, and conflict on food security*. MA Thesis submitted to the Department of Global Political Studies, Malmo University.

Land Trust Alliance. (2020). Working with landowners. Retrieved June 20, 2023, from https://landtrustalliance.org/resources/learn/topics/stewardship/working-with-landowners

Lautensach, A., & Lautensach, S. (2020). Climate change and human security. In *Human security in world affairs: Problems and opportunities* (2nd edn). Retrieved July 16, 2022, from https://opentextbc.ca/humansecurity/chapter/climate-change/

Liu, J., Dietz, T., Carpenter, S. R., Alberti, M., Folke, C., Moran, E., et al. (2007). Complexity of coupled human and natural systems. *Science, 317*(5844), 1513–1516.

Majra, J. P., & Gur, A. (2009). Climate change and health: Why should India be concerned? *Indian Journal of Occupational and Environmental Medicine, 13*(1), 11–16.

Marenya, P. P., Gebremariam, G., & Jaleta, M. (2020). Sustainable intensification among smallholder maize farmers in Ethiopia: Adoption and impacts under rainfall and unobserved heterogeneity. *Food Policy, 95*, 101941.

Mathez, E., & Smerdon, J. (2018). 8. The climate of the recent past and impacts on human history. In *Climate change* (pp. 229–264). Columbia University Press.

McMichael, A. J., Friel, S., Nyong, A., & Corvalan, C. (2008). Global environmental change and health: Impacts, inequalities, and the health sector. *BMJ, 336*(7637), 191–194.

Muringai, R. T., Naidoo, D., Mafongoya, P., & Lottering, S. (2020). The impacts of climate change on the livelihood and food security of small-scale fishers in Lake Kariba, Zimbabwe. *Journal of Asian and African Studies, 55*(2), 298–313.

Nanthini, S., & Nair, T. (2021). Framing climate change: The need for a human security perspective. Retrieved March 17, 2023, from https://think-asia.org/bitstream/handle/11540/14082/NTS-Insight-IN21-05-Climate-Change-A-Human-Security-Perspective.pdf?sequence=1

Neocleous, M. (2000). Social police and the mechanisms of prevention. *British Journal of Criminology, 40*(4), 710–726.

Norris, P., & Inglehart, R. F. (2012). Muslim integration into Western cultures: Between origins and destinations. *Political Studies, 60*(2), 228–251.

Notenbaert, A., Pfeifer, C., Silvestri, S., & Herrero, M. (2017). Targeting, out-scaling and prioritizing climate-smart interventions in agricultural systems: Lessons from applying a generic framework to the livestock sector in sub-Saharan Africa. *Agricultural Systems, 151*, 153–162.

Ogbuchi, T. C. (2020). Quantitative indicators of production of food crops. *Journal of Tropical Agriculture, 32*(1), 79–88.

OHCHR. (2015). Understanding human rights and climate change. In *Submission of the Office of the High Commissioner for Human Rights to the 21st Conference of the parties to the United Nations Framework Convention on Climate Change* (Vol. 26). United Nations Framework Convention on Climate Change.

OHCHR. (1993). World Conference on Human Rights, Vienna, 1993. United Nations Human Rights Officer of the High Commissioner.

Onuoha, C. M. (2009). Climate change and sustainable development in Nigeria: The mitigating role of green wall Sahara Nigeria programme. *Implications of Climate Change for Economic Growth and Sustainable Development in Nigeria, 33*, 33–41.

Oyinloye, O. D., Akinola, O. O., Akande, Y. O., Akinyele, A. A., & Mosimabale, M. M. (2018). Food insecurity in Africa. *Journal of Humanities and Social Science, 23*(9), 68–75.

Pachauri, R. K., & Reisinger, A. (2007). *IPCC fourth assessment report*. IPCC.

Patz, J. A., Gibbs, H. K., Foley, J. A., Rogers, J. V., & Smith, K. R. (2007). Climate change and global health: Quantifying a growing ethical crisis. *EcoHealth, 4*(4), 397–405.

Ramin, B. M., & McMichael, A. J. (2009). Climate change and health in sub-Saharan Africa: A case-based perspective. *EcoHealth, 6*(1), 52–57.

Reid, P., & Cogel, C. (2006). Living and responding to multiple stressors in South Africa Glimpses from KwaZulu-Natal. *Global Environmental Change, 16*(2), 195–206.

Rothschild, E. (1995). What is security? *Daedalus, 124*(3), 53–98.

Sachs, J., & Malaney, P. (2002). The economic and social burden of malaria. *Nature, 415*(6872), 680–685.

Scheffran, J., Brzoska, M., Kominek, J., Link, P. M., & Schilling, J. (2012). Climate change and violent conflict science *336*(6083), 869–871. https://doi.org/10.1126/science.1221339.

Schleussner, C. F., Deryng, D., D'haen, S., Hare, W., Lissner, T., Ly, M., et al. (2018). 1.5 C hotspots: Climate hazards, vulnerabilities, and impacts. *Annual Review of Environment and Resources, 43*, 135–163.

Schneider, S., & Lane, J. (2005). Integrated assessment modeling of global climate change: Much has been learned—Still a long and bumpy road ahead. *Integrated Assessment, 5*(1), 41–75.

SOAS University of London. (2018). Climate Change and Development. Retrieved July 31, 2022, from https://www.soas.ac.uk/cedep/online-programmes/msc-climate-change-anddevelopment/file123037.pdf

Sostaric, T. (2019). Al Jazeera, BiH u očekivanju katastrofa: Sve je na papiru za bolja vremena [B&H in anticipation of disasters: Everything is on paper for better times]. Retrieved November 17, 2021, from https://balkans.aljazeera.net/teme/2019/9/29/bih-u-ocekivanju-katastrofa-sve-je-na-papiru-za-bolja-vremena

Stern, N. (2007). *The economics of climate change: The Stern review*. Cambridge University Press.

Tarif, K. (2022). *Climate change and violent conflict in West Africa: Assessing the evidence*. SIPRI. https://doi.org/10.55163/VHIY5372

Taub, D. R., Miller, B., & Allen, H. (2008). Effects of elevated CO2 on the protein concentration of food crops: A meta-analysis. *Global Change Biology, 14*(3), 565–575.

Tesfaye, W., Blalock, G., & Tirivayi, N. (2020). Climate-smart innovations and rural poverty in ethiopia: Exploring impacts and pathways. *American Journal of Agricultural Economics, 103*(3), 878–899.

The Economist. (2019). How Climate Change Can Fuel Wars. Retrieved March 24, 2022, from www.economist.com/international/2019/05/23/how-climate-change-can-fuel-wars

The United Nations Conference on Environment and Development (UNCED). (1992). United Nations Conference on Environment and Development. Retrieved July 25, 2022, from https://www.un.org/en/conferences/environment/rio1992

Thomson, M. C., Mason, S., Platzer, B., Mihretie, A., Omumbo, J., Mantilla, G., et al. (2014). Climate and health in Africa. *Earth Perspectives, 1*(1), 1–12.

Uitto, J. I., Puri, J., & Van den Berg, R. D. (2017). *Evaluating climate change action for sustainable development*. Springer Nature.

UN News. (2019). Tropical Cyclone Idai affects 1.5 million across Mozambique and Malawi, as UN ramps up response. Retrieved June 20, 2023, from https://news.un.org/en/story/2019/03/1034741

UNDP. (1994). *Human Development Report 1994*. United Nations Development Programme. Retrieved July 29, 2022, from https://hdr.undp.org/system/files/documents//hdr1994encompletenostatspdf.pdf

UNDP. (2007). *Human Development Report 2007/2008 Fighting Climate Change: Human Solidarity in a Divided World*. United Nations Development Programme.

United Nations Environmental Programme. (2015). Climate Change and Human Rights. Retrieved July 31, 2022, from https://wedocs.unep.org/bitstream/handle/20.500.11822/9530/-Climate_Change_and_Human_Rightshuman-rights-climate-change.pdf.pdf?sequence=2&amp%3BisAllowed

UNDP. (2020). *Human Development Report 2020*. United Nations Development Programme. Retrieved February 21, 2022, from https://hdr.undp.org/content/human-development-report-2020

UNEP. (2022). *Emissions Gap Report 2022*. United Nations Environmental Programme.

United Nations Environmental Programme. (2022). *Climate Change 2022: Impacts, Adaptation and Vulnerability, the Working Group II contribution to the Sixth Assessment Report*. United Nations Environmental Programme.

United Nations. (1992). United Nations Framework Convention on Climate Change. United Nations.

United Nations. (2011). *Warning of climate change's threat to global security, Ban urges concerted action*. United Nations News.

United Nations. (2017). *United Nations Human Security Handbook*. United Nations. Retrieved April 21, 2022, from https://www.un.org/humansecurity/wp-content/uploads/2017/10/h2.pdf

United Nations Department of Economic. (2021). World social report 2021: reconsidering rural development. United Nations.

United State Agency for International Development (USAID). (2017). Monitoring, and evaluation support activity (measure-B&H) expanding USAID's network of development, Final Report. Retrieved March 17, 2022, from http://measurebih.com/uimages/THE202010-201320USAID20BiH20DGP20EVALUATION20FINAL20REPORT2C20edited2C2012072017.pdf

US EPA. (2017). *Smart growth fixes for climate adaptation and resilience*. United States Environmental Protection Agency. Retrieved June 21, 2023, from https://www.epa.gov/sites/default/files/2017-01/documents/smart_growth_fixes_climate_adaptation_resilience.pdf

Van Oort, P. A., & Zwart, S. J. (2018). Impacts of climate change on rice production in Africa and causes of simulated yield changes. *Global Change Biology, 24*(3), 1029–1045.

von Soest, C. (2020). A heated debate: Climate change and conflict in Africa. Retrieved March 17, 2023, from https://www.giga-hamburg.de/en/publications/giga-focus/a-heated-debate-climate-change-and-conflict-in-africa

Webersik, C. (2010). *Climate change and security: A gathering storm of global challenges: A gathering storm of global challenges*. Abc-Clio.

WHO. (2014). Achieving Universal Health Coverage and Health Security in Africa: The Africa We Want to See. WHO Africa Forum.

WHO. (1997). *Health and environment in sustainable development: Five years after the earth summit health and environment in sustainable development*. World Health Organisation. Retrieved August 03, 2022, from http://apps.who.int/iris/bitstream/10665/63708/1/WHO_EHG_97.12_eng.pdf

World Vision. (2019). 2019 Cyclone Idai: Facts, FAQs, and how to help. Retrieved June 20, 2023, from https://www.worldvision.org/disaster-relief-news-stories/2019-cyclone-idai-facts#:~:text=Catastrophic%20flooding%20from%20the%20storms,rains%20that%20led%20to%20flooding

Yasmeen, S. (2021). Climate change and human security: In context of global economy, culture and conflict. *Journal of the Research Society of Pakistan, 58*(3), 203–216.

Role of Value Orientation and Belief in Shaping Indian Pre-service Teachers' Personal Norms to Address Climate Change

Divya Chandrasenan, Sajilaraj Selvaraj and Reshma John Kuleenan

Abstract

A thorough examination of all available data reveals that a significant portion of the observed increase in global average warming over the past 50 years or more cannot be explained by natural processes alone and must instead be attributed to human activity. Since the beginning of the industrial period, human activities have had a global warming influence on temperature. It also contributed significantly to climate change by reflecting or absorbing solar energy. These elements are intricately tied to human conduct, which is governed by people's values and beliefs. The value-belief-norm theory and the norm activation model define personal norms as self-expectations of pro-environmental behaviour that are affected by the way values and beliefs assign responsibility and explain consequences. Individuals tend to have a moral commitment to engage in ecologically responsible behaviour when accountability and consequence consciousness criteria are met. Developing environmentally friendly personal norms is an important first step in combating global warming. Personal norms are influenced by one's value orientation and beliefs. The results of a survey given to pre-service teachers in Kerala, India, were examined to see whether there was any relationship between value orientation and belief; value orientation and personal norms; and belief and personal norms with regard to climate change and were shown to have strong positive relations.

D. Chandrasenan (✉) · S. Selvaraj
Department of Education, University of Kerala, Thiruvananthapuram, India
e-mail: divyacsenan@keralauniversity.ac.in

R. J. Kuleenan
Department of Environmental Sciences, All Saints' College, University of Kerala, Thiruvananthapuram, India

Keywords

Value orientation · Beliefs · Personal norms · Climate change · Pre-service teacher

1 Introduction

Climate change is no longer "just one of many environmental and regulatory problems," and scientists are no longer still trying to figure it out. The UN Secretary General says it is the "single biggest problem" that environmental regulators have to deal with. Climate change will affect more than just the economy. It will affect health, safety, food supply, and national security, among other things. Even though there are a lot of things that affect the climate, scientists have found that most of the warming that has happened in the last 50 years is now mostly due to what people have done. Climate change caused by humans is one of the biggest problems the world is confronted with right now, and it needs big changes in how individuals and groups interact with the natural world. A lot of scientific studies back up the idea that humans are really causing big changes in the climate. Some of these studies, called "detection and attribution" studies, may be the most important (IDAG, 2005). A recent analysis of 88,125 climate-related studies found that more than 99.9% of peer-reviewed scientific journals believe that people are the main cause of climate change (Ramanujan, 2021).

Understanding the social dynamics that motivate individuals to care about and act on climate change remains crucial. To gain widespread support, we must prevent climate change from becoming an existential threat. A lot of research has already been done on how human values affect how people seem to think about and react to climate change (Corner et al., 2014; Dietz et al., 2007; Poortinga et al., 2019). People's actions and perceptions about climate change are likely to be greatly affected by how much they know about how climate change events hurt

B.-J. He et al. (eds.), *Resilient Horizons: Building Sustainable Environments for Climate Adaptation and Health*, Advances in Science, Technology & Innovation, https://doi.org/10.1007/978-3-031-46109-5_15

the environment and the amount they expect to be hurt by these events (Marshall et al., 2019). Several studies show that different responses to climate change are influenced by the society factors or value orientations (Bouman et al., 2018; Leombruni, 2015; Price et al., 2014; van der Linden, 2015).

2 Motivation

In recent years, Kerala has experienced an increase in temperature, an unpredictable monsoon, and a water shortage. In latest years, however, crises have taken the shape of life-threatening natural disasters. The repercussions of climate change have been exacerbated by continuous human activities. In recent years, the state has endured an onslaught of torrential rains, floods, landslides, and droughts. Rainfall data for the IMD stations of Kerala from 1871 to 2008 (140 years) shows a lowering tendency in annual and southwest monsoon rainfall over the previous 60 years and a rising trend in post-monsoon rainfall, indicating probable changes in rainfall patterns. A study of the seasonal precipitation pattern in Kerala over the course of fifty years (1954–2003) revealed that floods and water scarcity are markers of climate change caused by seasonal extremes in rainfall (Indrani Pal & Abir Al Tabbaa, 2008).

Climate change threatens food security, agriculture, health, and marine resources in the Indian state of Kerala. Persistent rain wreaked havoc on Kerala's agriculture business. People's emission of carbon into the atmosphere since the Industrial Revolution are one of the causes of climate change. Climate change has already had an impact on some areas and sectors in Kerala. As time passes, it becomes obvious that these kinds of local consequences are a powerful catalyst for changing people's perspectives of the global extent of climate change and the risks associated with it (Hansen & Cramer, 2015; Lee et al., 2015; Myers et al., 2012). Individuals who are less likely to believe in climate change, engage in pro-environmental behaviours, or experience ecological grieving are assumed to have distinct psycho-cultural orientations, failing to recognize or care about changes in environmental conditions or their own personal effect (Marshall et al., 2019).

The current study attempts to determine whether a person's value orientation and beliefs have an effect on his or her personal norms. We tested these hypotheses in the Indian state of Kerala, where climate change is already having a significant impact on extreme natural catastrophes such as floods and landslides.

3 Value Orientation, Beliefs, and Personal Norms on Climate Change

Human behaviour research has attempted to demonstrate that acts are ultimately the result of a person's core values. Studies on this topic have proven that a person's perspectives are composed of a succession of interconnected events involving values, beliefs, and conventions (Ball-Rokeach et al., 1984; Homer & Kahle, 1988). The VBN theory (Stern & Dietz, 1994; Stern et al., 1999) has really been empirically examined extensively and focuses on people's values, beliefs, and personal standards as behavioural determinants (Ruepert et al., 2016; Yeboah & Kaplowitz, 2016). As noted by Hatch and Cunliffe (2006), people's values shape the community's social principles, goals, and standards. Significantly, biospheric, altruistic, and egoistic value orientations are related to pro-environmental conduct (De Groot & Steg, 2008). Biospheric value orientation emphasizes concern for all biosphere components, including living and nonliving organisms (Schultz et al., 2005). Individuals with a biospheric value orientation consider the broader ecological benefits of their actions (De Groot & Steg, 2008). A person with altruistic value orientation is concerned with the beneficial consequences their activities have on others. Before engaging in a behaviour, the socially altruistic examine whether it has a positive impact on humanity. A person with an egoistic value orientation emphasizes the benefits or consequences of an action for himself or herself (Stern & Dietz, 1994). Egocentric individuals care about the environment to suit their own desires and interests. A person with an egoistic value orientation is focused with achieving personal success.

Belief is an essential parameter that is taken into account in the VBN hypothesis and is shown to influence the development of environmentally conscious behaviours. Belief in the VBN hypothesis is comprised of the individual's worldview, awareness of the repercussions of the harmful environmental event, and feelings of responsibility for causing or halting the phenomena. The connection that exists between people and the natural world gives rise to the idea that human activity can throw off the natural order, that there are constraints on human expansion, and that humans have the power to exercise authority over the natural world (Dunlap et al., 2000). As a result, environmentally acceptable behaviour is based on the idea that humans and nature have a reciprocal relationship with one another. After a person's core values and beliefs, the next variable in the VBN theory to be considered is their personal norms. Personal Norms has a direct impact on the behaviour that will actually be carried out as a result of its influence (Stern, 2005). Personal Norms played an important role in determining consumer buying behavior (Minton & Rose, 1997), people's willingness to pay more for an ecologically friendly product (Thøgersen, 2002), and the relevant decision (Minton & Rose, 1997; Thgersen, 2002). Personal Norms also played an integral role in the process of making the relevant decision (Hunecke et al., 2001; Minton & Rose, 1997; Nordlund & Garvill, 2003; Thøgersen, 2002).

Thus, it seems reasonable to assume that value orientation, beliefs, and personal norms would be positively related each other, specifically about climate change which led to the hypothesis of the study conducted in the state of Kerala, India among the teacher education students who are trying to address the climate change curriculum in the classrooms. Hence the hypothesis of the current study is stated as 'The predictive efficiency of Value Orientation and Belief on climate change of Teacher education students to predict Personal Norms towards climate change is significant'.

Table 2 Factor loading for the value orientation scale

Question numbers	Component		
	Factor 1	Factor 2	Factor 3
Q1		0.746	
Q2		0.645	
Q3		−0.649	
Q4		0.618	
Q5		0.410	
Q6			0.732
Q7			0.767
Q8		0.424	
Q9	0.679		
Q10	0.562		
Q11	0.653		
Q12	0.680		
Q13	0.722		
Q14			0.662
Q15	0.656		
Q16	0.353		

4 Materials and Methods

This survey-based study aims to provide a descriptive correlation between two variables. The Value Orientation measure was created specifically for the study, while the Personal Norms and Beliefs scales were adapted. The Value Orientation scale was created using value scales from prior research (De Groot & Steg, 2008; Dietz et al., 2005). Respondents were asked to rate the importance of each value on a 5-point scale that range from "value goes against the principle" (1) to "value is important" (3) to "value is very important" (4) to "value is the most important" (5). The scale was implemented among the adequate sample group. Using SPSS16, the acquired data was subjected to a principal component factor analysis. Via Kaiser-Meyer-Olkin (KMO) statistics, a sampling adequacy test was conducted (Table 1). Since KMO values exceed 0.6, it is acceptable, according to Kaiser and Rice (1974). The Kaiser-Meyer-Olkin measure of sampling adequacy with a value of 0.78 was therefore acceptable. Barlett's Test of Sphericity significant (2 = 706.642, df = 120, p 0.000) demonstrates that the values are significant and therefore acceptable, indicating that non-zero correlations existed at the significance level of 0.000; therefore, it provided a sufficient basis for proceeding with the factor analysis.

Table 1 KMO value and bartlett's test of sphericity

Kaiser-Meyer-Olkin measure of sampling adequacy		0.784
Bartlett's test of sphericity	Approx. Chi-Square	706.642
	Df	120
	Sig.	0.000

These outcomes indicated that the data were adequate to proceed with the EFA (Anderson, 2010; Mvududu & Sink, 2013). The factors in each component were identified using Exploratory Factor Analysis. It is obvious from Table 2 that the value orientation is composed of three components. The distribution of the three components was as follows: Factor 1 (Biospheric value), Factor 2 (Altruistic value), and Factor 3. (Egoistic value).

Principal component analysis was used to minimise the number of components determined through exploratory factor analysis. The outcome of integrating eigenvalue and scree plot (Fig. 1) suggested that the measurements represent three criteria for evaluating value orientation on climate change. Seven items were used to measure biospheric value, six items were used to measure altruistic value, and three items were used to measure egoistic value. Cronbach's alpha () = 0.82 represents an adequate reliability level, indicating a high degree of reliability (Cresswell, 2005). Given the factor loadings and reliability findings, all sixteen items were kept. All items are capable of evaluating the value orientation in relation to climate change.

Responses to questions about respondents' views towards climate change were ascertained using the "New Environmental Paradigm" (NEP) scale and the "New Ecological Paradigm" (NEP) scale (Dunlap & Van Liere, 1978; Dunlap et al., 2000). To assess each respondent's level of belief and personal norms, ten questions from the

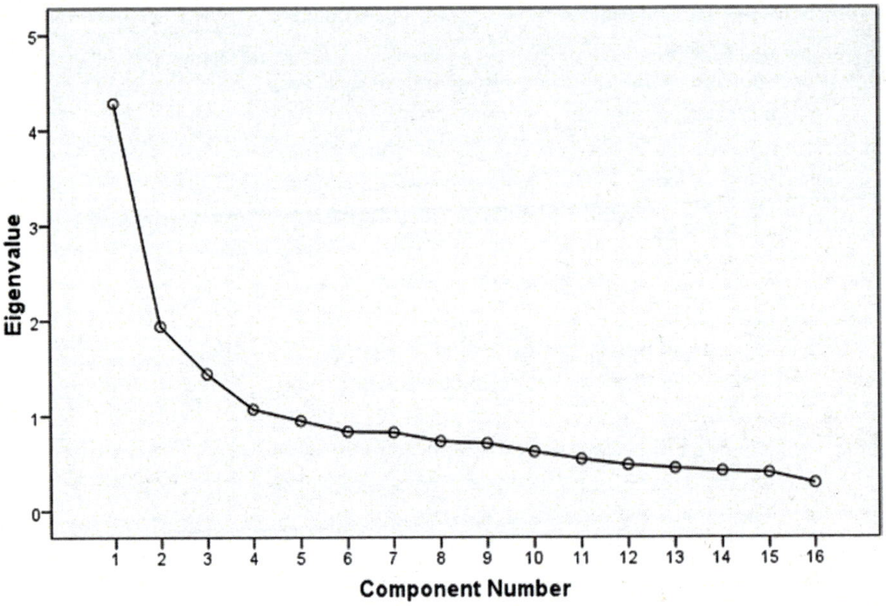

Fig. 1 Scree plot diagram

study by Ignell et al. (2019) were adapted. Cronbach's alpha was ()=0.82 for the Belief scale and ()=0.77 for the personal norms scale.

4.1 Survey Implementation and Sample Description

The Faculty of Education (N=178) of the University of Kerala, India, was sampled by random to collect the data. 200 initial surveys were sent out, and 178 were returned. The survey included three scales: Value orientation, Beliefs, and Personal norms. The questionnaires were distributed to teacher education students during a class session. The data collection took approximately two weeks. The individuals who completed the questionnaire had thirty minutes to do so and were informed that their responses would remain anonymous and confidential. The data were examined for absences and anomalies (Hair et al., 2010).

The researchers discovered that the majority of respondents exhibited the following characteristics based on our preliminary examination of the questionnaire data and the fundamental make-up of the study sample. In comparison to men, women made up a substantial majority of survey replies (75.28%) (24.32%). 82.6% of college students are enrolled in Bachelor of Education programmes, with the remainder enrolling in Master of Education programmes.

5 Data Analysis

The obtained data were analyzed using mean, SD, Karl Pearson's product moment correlation, multiple correlations, regression and multiple regressions. The data have been analyzed by SPSS Package and interpretation of data is given below.

5.1 Result

It was found that the mean score and standard deviation of value orientation on climate change among students in higher education were correspondingly 60.82 and 6.59. According to the normative standards of the scale's qualitative categories, it may be deduced that they have an average level of value orientation. The sample's mean scores for beliefs and personal norms were found to be 41.29 and 40.21 respectively, which implies that higher education students have a lower-than-average degree of belief and personal norms regarding climate change.

The combined influence of the independent variables value orientation and beliefs on the dependent variable, personal norms, was assessed using correlation and regression analysis. Table 3 shows the correlation coefficients between the variables, Value orientation, beliefs, and personal norms concerning climate change.

Table 3 Correlation coefficients among value orientation, beliefs and personal norms towards climate change

		Value orientation	Beliefs	Personal norms
Value orientation	Pearson correlation	1	0.340**	0.426**
Beliefs	Pearson correlation	0.340**	1	0.462**
Personal norms	Pearson correlation	0.426**	0.462**	1

** The correlation is significant at the 0.01 level

To determine whether the value orientation of teacher students in higher education is predicted by their views and personal norms, regression analysis was done and the same is shown in Table 4. According to the study of multiple regression, the dependent or predicted variable—personal norms—can be explained by two independent variables, value orientation and belief, when integrated in the multiple equation in the least squares sense, by 21.0% ($R=0.546$, $R2=0.21$).

5.2 Discussions

There were significant positive associations between value orientation and belief, value orientation and personal norms, and values and belief and personal norms. The findings revealed that value orientations and beliefs have a positive effect on the formation of personal norms about climate change. According to the statistics, Value Orientation and Climate Change Belief of Teacher Education Students are important predictors of Personal Climate Change Norms. The study presents a framework for developing climate change curricula and training teacher education students to teach about climate change in the classroom.

According to the findings of a marine conservation research study conducted by Engel et al. (2020), activities aiming to motivate people to involve in sustainable behaviours should be based on relational values such as care and concern for the ocean. According to De Groot and Steg (2007), the differentiation between the three value orientations is valid and beneficial for studying environmentally related behaviour. Karpudewan and Mohd Ali Khan (2017) investigates the relationship between values, beliefs, personal norms, and climate change behaviours before modelling climate conservation behaviours. There were substantial positive correlations between values and belief, values and personal norms, values and behaviour, belief and personal norms, belief and behaviours, and personal norms and behaviours, according to the data. The current study complements the previously mentioned research studies by emphasizing the need of strengthening value orientation, belief, and personal norms of teacher education students towards tackling climate change through innovative climate change education.

6 Conclusion

Improvements in environmental value orientations, beliefs, and personal norms in regards to climate change will undoubtedly lead to a deeper comprehension of the connections between humans and their natural surroundings, as well as the means to mitigate the effects of climate change brought about by human activity. Belief in both natural and human-caused climate change, rather than just one or the other, is something to keep in mind when conducting future studies. It appears from these findings that environmental managers may be able to harness people's environmental value orientations to influence climate change reactions in the direction of environmental stewardship and sustainability.

Table 4 Regression analysis

Model		Unstandardized coefficients		Standardized coefficients	t	Sig.
		B	Std. error	Beta		
1	(Constant)	11.012	3.511		3.137	0.002
	Value orientation	0.253	0.056	0.303	4.511	0.000
	Beliefs	0.329	0.061	0.363	5.401	0.000

The model is determined to be statistically significant ($F=11{,}012$, $p=0.01$, $R2=0.21$). The study indicated that Value Orientation (= 0.303, p 0.01) and Beliefs (= 0.363, p 0.01) are significant predictors of Personal Norms. Using the following regression equation, it is possible to predict the scores of pre-service teachers' personal norms regarding climate change

Y(PN) = 11.012 + 0.303(VO) + 0.363(B)

Where, Y (PN) = Predicted Value of Personal Norms, VO = Value Orientation and B = Belief

References

Anderson, C. (2010). Presenting and evaluating qualitative research. *American Journal of Pharmaceutical Education, 74,* 141. https://doi.org/10.5688/aj7408141

Ball-Rokeach, S. J., Rokeach, M., & Grube, J. W. (1984). *The great American values test: Influencing behaviour and belief through television.* Free Press.

Bouman, T., Steg, L., & Zawadzki, S. J. (2020). The value of what others value: When perceived biospheric group values influence individuals' pro-environmental engagement. *Journal of Environmental Psychology, 71,* 101470.https://doi.org/10.1016/j.jenvp.2020.101470

Corner et al. (2014). Public engagement with climate change: the role of human values. *WIREs Climate Change 2014, 5,* 411–422. https://doi.org/10.1002/wcc.269

Creswell, J. W. (2005). *Educational research: Planning, conducting and evaluating quantitative and qualitative research* (2nd ed.). Merrill/Prentice Hall.

De Groot, J. I. M., & Steg, L. (2008). Value orientations to explain environmental attitudes and beliefs: How to measure egoistic, altruistic and biospheric value orientations. *Environment and Behavior, 40*(3), 330–354.

Dietz, T., Fitzgerald, A., & Shwom, R. (2005). Environmental values. *Annual Review of Environment and Resources, 30*(1), 335–372. https://doi.org/10.1146/annurev.energy.30.050504.144444//

Dietz, T., Fitzgerald, A., & Shwom, R. (2007). Environmental values. *Annual Review of Environment and Resources, 30*(1), 335–372. https://doi.org/10.1146/annurev.energy.30.050504.144444//

Dunlap, R. E., & Van Liere, K. D. (1978). The "New Environmental Paradigm." *The Journal of Environmental Education, 9*(4), 10–19. https://doi.org/10.1080/00958964.1978.10801875

Dunlap, R. E., Van Liere, K. D., Mertig, A. G., & Jones, R. E. (2000). New trends in measuring environmental attitudes: Measuring endorsement of the new ecological paradigm: A revised NEP scale. *Journal of Social Issues*.

Hair, J. F., Black, W. C., Babin, B. J., & Anderson, R. E. (2010). *Multivariate data analysis* (7th ed.). Prentice Hall.

Hansen, G., & Cramer, W. (2015). Global distribution of observed climate change impacts. *Nature Climate Change, 5(3),* 182–185. https://doi.org/10.1038/nclimate2529

Hatch, M. J., & Cunliffe, A. L. (2006). Organization theory: Modern, symbolic, and postmodern perspectives (2nd ed.). New York: Oxford University Press.

Homer, P., & Kahle, L. R. (1988). A structural equation test of the value-attitude-behavior hierarchy. *Journal of Personality and Social Psychology, 54*(4), 638–646.

Hunecke, M., Blöbaum, A., Matthies, E., & Höger, R. (2001). Responsibility and environment: Ecological norm orientation and external factors in the domain of travel mode choice behavior. *Environment and Behavior, 33*(6), 830–852. https://doi.org/10.1177/00139160121973269

Ignell, C., Davies, P., & Lundholm, C. (2018). A longitudinal study of upper secondary school students' values and beliefs regarding policy responses to climate change. *Environmental Education Research, 25*(5), 615–632. https://doi.org/10.1080/13504622.2018.1523369

Kaiser, H. F., & Rice, J. (1974). Little Jiffy, Mark Iv. *Educational and Psychological Measurement, 34*(1), 111–117.https://doi.org/10.1177/001316447403400115

IDAG (2005) Detecting and attributing external influences on the climate system: a review of recent advances. *Journal of Climate 18,* 1291–1314.

Lee, T. M., Markowitz, E. M., Howe, P. D., Ko, C. Y., & Leiserowitz, A. A. (2015). Predictors of public climate change awareness and risk perception around the world. *Nature Climate Change, 5*(11), 1014–1020. https://doi.org/10.1038/nclimate2728

Leombruni, L. V. (2015). How you talk about climate change matters: A communication network perspective on epistemic skepticism and belief strength. *Global Environmental Change, 35,* 148–161.https://doi.org/10.1016/j.gloenvcha.2015.08.006

Marshall, N. A., Thiault, L., Beeden, A., Beeden, R., Benham, C., Curnock, M. I., Diedrich, A., Gurney, G. G., Jones, L., Marshall, P. A., Nakamura, N., & Pert, P. (2019). Our Environmental Value Orientations Influence How We Respond to Climate Change. *Frontiers in Psychology, 10.* https://doi.org/10.3389/fpsyg.2019.00938

Minton, A. P., & Rose, R. L. (1997). The effects of environmental concern on environmentally friendly consumer behavior: An exploratory study. *Journal of Business Research, 40*(1), 37–48. https://doi.org/10.1016/S0148-2963(96)00209-3

Mvududu, N. H., & Sink, C. A. (2013). Factor analysis in counseling research and practice. *Counseling Outcome Research and Evaluation, 4*(2), 75–98. https://doi.org/10.1177/2150137813494766

Myers, T. A., Nisbet, M. C., Maibach, E. W., & Leiserowitz, A. A. (2012). A public health frame arouses hopeful emotions about climate change. *Climatic Change, 113*(3–4), 1105–1112. https://doi.org/10.1007/s10584-012-0513-6

Nordlund, A. M., & Garvill, J. (2003). Effects of values, problem awareness, and personal norm on willingness to reduce personal car use. *Journal of Environmental Psychology, 23*(4), 339–347.

Pal, I., & Al-Tabbaa, A. (2010). Assessing seasonal precipitation trends in India using parametric and nonparametric statistical techniques. *Theoretical and Applied Climatology, 103(1–2),* 1–11. https://doi.org/10.1007/s00704-010-0277-8

Poortinga, W., Whitmarsh, L., Steg, L., Böhm, G., & Fisher, S. (2019, March). Climate change perceptions and their individual-level determinants: A cross-European analysis. *Global Environmental Change, 55,* 25–35.https://doi.org/10.1016/j.gloenvcha.2019.01.007

Price, J. C., Walker, I. A., & Boschetti, F. (2014). Measuring cultural values and beliefs about environment to identify their role in climate change responses. *Journal of Environmental Psychology, 37,* 8–20. https://doi.org/10.1016/j.jenvp.2013.10.001

Ramanujan, K. (2021). More than 99.9% of studies agree: Humans caused climate change.

Ruepert, A., Keizer, K., Steg, L., Maricchiolo, F., Carrus, G., Dumitru, A., Garcia Mira, R., Stancu, A., & Moza, D. (2016). Environmental considerations in the organizational context: A pathway to pro-environmental behavior at work. *Energy Research & Social Science, 17,* 59–70.

Schultz, P. W., Gouveia, V. V., Cameron, L. D., Tankha, G., Schmuck, P., & Franěk, M. (2005). Values and their relationship to environmental concern and conservation behavior. *Journal of Cross-Cultural Psychology, 36*(4), 457–475. https://doi.org/10.1177/0022022105275962

Stern, P. C., & Dietz, T. (1994). The value basis of environmental concern. *Journal of Social Issues, 50*(3), 65–84. https://doi.org/10.1111/j.1540-4560.1994.tb02420.x//

Stern, P. C. (2005, January 1). Understanding individuals' environmentally significant behavior. 35 Envtl. L. Rep. News & Analysis 10785

Stern, P. C., Dietz, T., Abel, T., Guagnano, G. A., & Kalof, L. (1999). A value-belief-norm theory of support for social movements: The case of environmentalism. *Research in Human Ecology, 6*(2), 81–97.

Thøgersen, J. (2002). Direct experience and the strength of the personal norm-behavior relationship. *Psychology & Marketing, 19*(10).

van der Linden, S. (2015). The social-psychological determinants of climate change risk perceptions: Towards a comprehensive model. *Journal of Environmental Psychology, 41,* 112–124. https://doi.org/10.1016/j.jenvp.2014.11.012

Yeboah, F. K., & Kaplowitz, M. D. (2016). Explaining Energy Conservation and Environmental Citizenship Behaviors Using the Value-Belief-Norm Framework. *Human Ecology Review, 22*(2), 137–160. https://www.jstor.org/stable/24875161